Cambridge Studies in Biological and Evolutionary Anthropology

Series editors

HUMAN ECOLOGY
C. G. Nicholas Mascie-Taylor, University of Cambridge
Michael A. Little, State University of New York, Binghamton
GENETICS
Kenneth M. Weiss, Pennsylvania State University
HUMAN EVOLUTION
Robert A. Foley, University of Cambridge
Nina G. Jablonski, California Academy of Science
PRIMATOLOGY
Karen B. Strier, University of Wisconsin, Madison

University of Chester

This book is to be returned on or before the last date stamped below. Overdue charges will be incurred by the late return of books.

Macaque Societies

Animal and human societies are multifaceted. In order to understand how they have evolved, it is necessary to investigate each of the constituent facets including individual abilities and personalities, life-history traits, mating systems, demographic dynamics, gene flow, social relationships, ecology, and phylogeny. By exploring the nature and evolution of macaque social organization, this book develops our knowledge of the rise of societies and their transformation during the course of evolution. Macaques are the most comprehensively studied of all monkey groups, and the 20 known species feature a broad diversity in their social relationships making them a particularly good group for exploring the evolution of societies. This book will be of primary interest to those studying animal behavior and primatology, but will also be useful to those involved in the study of human societies.

BERNARD THIERRY is Research Director at the Centre National de la Recherche Scientifique in Strasbourg, France. He has studied the social behavior of non-human primates for the past 25 years, and is particularly interested in how internal constraints channel the evolutionary changes of social organization.

MEWA SINGH is Professor of Psychology at the University of Mysore, India. His main research focus is on the evolution of sociality, and he is especially interested in bridging the gap between conservation biology and behavioral biology.

WERNER KAUMANNS is Curator of Primates and Head of the Primatology Working Group at Cologne Zoo, Germany. His special interest is also in conservation biology, and he has been involved in research on liontailed macaques with special reference to the effects of habitat fragmentation.

Macaque Societies

A Model for the Study of Social Organization

Edited by

Bernard Thierry
*Centre National de la Recherche Scientifique and
Université Louis Pasteur, Strasbourg*

Mewa Singh
University of Mysore

Werner Kaumanns
Zoologischer Garten Köln

CAMBRIDGE
UNIVERSITY PRESS

PUBLISHED BY THE PRESS SYNDICATE OF THE UNIVERSITY OF CAMBRIDGE
The Pitt Building, Trumpington Street, Cambridge, United Kingdom

CAMBRIDGE UNIVERSITY PRESS
The Edinburgh Building, Cambridge, CB2 2RU, UK
40 West 20th Street, New York, NY 10011–4211, USA
477 Williamstown Road, Port Melbourne, VIC 3207, Australia
Ruiz de Alarcón 13, 28014 Madrid, Spain
Dock House, The Waterfront, Cape Town 8001, South Africa

http://www.cambridge.org

First published 2004

Printed in the United Kingdom at the University Press, Cambridge

Typeface Times 10/12.5 pt.   *System* LATEX 2$_\varepsilon$ [TB]

A catalogue record for this book is available from the British Library

Library of Congress Cataloguing in Publication data
Macaque societies: a model for the study of social organization / edited by Bernard
Thierry, Mewa Singh, and Werner Kaummanns.
    p. cm. – (Cambridge studies in biological and evolutionary anthropology)
Includes bibliographical references (p.).
ISBN 0 521 81847 8
1. Macaques – Behavior.   2. Animal societies.   I. Thierry, Bernard.   II. Mewa
Singh.   III. Kaumanns, Werner.   IV. Series.

QL737.P93M29   2004
599.8′64156 – dc22   2003069749

ISBN 0 521 81847 8 hardback

Contents

The color plates are situated between pages 5 and 6.

Contributors

Christophe Abegg
Centre d'Ecologie, Physiologie et
    Ethologie, UPR 9010, Centre
    National de la Recherche
    Scientifique, Strasbourg, France, &
    Abteilung Reproduktionsbiologie,
    Deutsches Primatenzentrum,
    Göttingen, Germany

Filippo Aureli
School of Biological and Earth
    Sciences, Liverpool John Moores
    University, Liverpool, UK

Fred B. Bercovitch
Center for Reproduction of Endangered
    Species, Zoological Society of San
    Diego, San Diego, California, USA

Carol M. Berman
Department of Anthropology, State
    University of New York, Buffalo,
    New York, USA

Marina Butovskaya
Center of Evolutionary Anthropology,
    Institute of Ethnology and
    Anthropology, Russian Academy of
    Sciences, Moscow, Russia

Josep Call
Department of Developmental and
    Comparative Psychology, Max
Planck Institute for Evolutionary
    Anthropology, Leipzig,
    Germany

John P. Capitanio
Department of Psychology and
    California National Primate
    Research Center, University of
    California, Davis, California,
    USA

Bernard Chapais
Département d'Anthropologie,
    Université de Montréal, Montréal,
    Canada

Christophe Chauvin
Centre d'Ecologie, Physiologie et
    Ethologie, UPR 9010, Centre
    National de la Recherche
    Scientifique, Strasbourg,
    France

Matthew A. Cooper
Center for Behavioral Neuroscience,
    Department of Psychology, Georgia
    State University, Atlanta, USA

Frans B. M. de Waal
Living Links, Yerkes National Primate
    Research Center and Department of
    Psychology, Emory University,
    Atlanta, Georgia, USA

Wolfgang Dittus
Primate Biology Program, Institute of
Fundamental Studies, Kandy, Sri
Lanka, and Department of
Conservation Biology, National
Zoological Park, Smithsonian
Institution, Washington, District of
Columbia, USA

Ardith A. Eudey
Primate Specialist Group, Species
Survival Commission, International
Union for Conservation of Nature,
Upland, California, USA

Jessica C. Flack
Living Links, Yerkes National Primate
Research Center and Department of
Psychology, Emory University,
Atlanta, Georgia, and Santa Fe
Institute, Santa Fe, New Mexico,
USA

Hélène Gachot-Neveu
Centre d'Ecologie, Physiologie et
Ethologie, UPR 9010, Centre
National de la Recherche
Scientifique, Strasbourg, France

Maurice Godelier
Ecole des Hautes Etudes en Sciences
Sociales, Paris, France

Nancy C. Harvey
Center for Reproduction of Endangered
Species, Zoological Society of San
Diego, San Diego, California, USA

Charlotte K. Hemelrijk
Theoretical Biology, Centre For
Ecology and Evolutionary Studies,

University of Groeningen, Haren, the
Netherlands

David A. Hill
School of Biological Sciences,
University of Sussex, Falmer,
Brighton, UK

Werner Kaumanns
Zoologischer Garten Köln, Köln,
Germany

Dario Maestripieri
Laboratory of Comparative
Development, Institute of Mind and
Biology, University of Chicago,
Chicago, Illinois, USA

William A. Mason
California National Primate Research
Center, University of California,
Davis, California, USA

Nelly Ménard
Ethologie, Evolution et Ecologie, UMR
6552, Centre National de la
Recherche Scientifique, Université
de Rennes I, Paimpont, France

Yasuyuki Muroyama
Field Research Center, Primate
Research Institute, Kyoto University,
Inuyama, Aichi, Japan

Kyoko Okamoto
Faculty of Humanities, Tokai-Gakuen
University, Tenpaku, Nagoya, Japan

Andreas Paul
Institut für Zoologie und
Anthropologie, Universität

Göttingen, Göttingen,
Germany

Signe Preuschoft
Cooperation and Communication Study
    Group, Home of Primates Europe,
    Safaripark Gänserndorf,
    Gänserndorf, Austria

Gabriele Schino
Istituto di Scienze e Tecnologie della
    Cognizione, Consiglio Nazionale
    delle Ricerche, Rome, Italy

Mewa Singh
Biopsychology Laboratory, University
    of Mysore, Mysore, Karnataka,
    India

Anindya Sinha
National Institute of Advanced Studies,
    Indian Institute of Science,
    Bangalore, and Centre for Ecological
    Research and Conservation, Nature

Conservation Foundation, Mysore,
Karnataka, India

Joseph Soltis
Laboratory of Comparative Ethology,
    National Institute of Child Health
    and Human Development, National
    Institutes of Health, Department of
    Health and Human Services,
    Poolesville, Maryland, USA

Bernard Thierry
Centre d'Ecologie, Physiologie et
    Ethologie, UPR 9010, Centre
    National de la Recherche
    Scientifique, & Centre de
    Primatologie, Université Louis
    Pasteur, Strasbourg, France

Paul L. Vasey
Department of Psychology and
    Neuroscience, University of
    Lethbridge, Lethbridge, Alberta,
    Canada

Acknowledgments

This volume arises from a conference on *Macaque Societies and Evolution* sponsored by the Wenner–Gren Foundation for Anthropological Research and held at the University of Mysore in March 2001. The conference provided a perfect place for free-thinking and stimulating discussion owing to the continuous support of the Foundation and the keen assistance of the staff of the Department of Psychology of the University of Mysore. During the conference, it became apparent that the time was ripe to synthesize different lines of investigation on macaque social organization and to start considering how their various dimensions could be integrated.

Many people have participated in bringing this book to fruition. The editors would like to express their heartfelt gratitude to the contributors of the volume. They met the challenge to produce innovative syntheses within the scope of a tight editorial schedule. They generously assisted as reviewers of other contributors. All have agreed to join the editors in donating their share of royalties to a primate conservation fund.

We are indebted to Roland Seitre and Christophe Abegg for offering photographs free of charge. We are grateful to James Anglin and Chittampalli Ravichandra for high-quality language advice. We owe additional thanks to Tracey Sanderson at the Cambridge University Press for unremitting commitment during the editorial process. Last but not least, we would like to acknowledge Odile Petit, Pierre Uhlrich, Philippe Ropartz and Yvon Le Maho for unfailing support.

Acknowledgments by contributors

Chapter 2. Thanks to S. Gosling, M. Prather, R. Robins and W. Mason for helpful comments, to C. Brennan, K. Floyd and E. Tarara for technical assistance, and to the National Institutes of Health for their support (MH49033 and RR00169).

Chapter 3. Thanks to J. Capitanio, C. Schaffner and the three editors for useful discussion and comments.

xiii

Chapter 4. Thanks to W. Mason, M. Singh, J. Soltis and B. Thierry for comments on the manuscript.

Chapter 5. The research on toque macaques was supported by a series of grants from the US National Science Foundation, the National Geographic Society, the Harry Frank Guggenheim Foundation, the Smithsonian Institution Scholarly Studies Fund and the Institute for Field Studies (Earthwatch). The Smithsonian National Zoological Park provided logistic and administrative support. Thanks to the government of Sri Lanka for permission to carry out the research and in particular the directors of the Institute of Fundamental Studies, A. Kovoor and K. Tennakone. Thanks to S. Gunatilake and K. Liyanage for long-term assistance with field research, N. Basnayake, U. Chandra for assistance in data collection, V. Dittus for help in project management, and W. Kaumanns, N. Ménard, M. Singh, B. Thierry and A. Watson for critical reading of the manuscript.

Chapter 6. Thanks to B. Thierry, A. Paul, J. de Ruiter and G. A. Hoelzer for helpful comments on an earlier draft, and to the Bettencourt–Schueller Foundation for financial support.

Box 6. Thanks to W. Kaumanns, B. Thierry, and P. Vasey for comments, and M. Singh and his collaborators, whose efforts made the Mysore conference both pleasant and successful.

Chapter 7. Thanks to M. Singh, B. Thierry, and W. Kaumanns for inviting me to the conference from which this book grew. The work on Japanese macaques was supported by the National Science Foundation and the Japan Society for the Promotion of Science, and I wrote the chapter while receiving an intramural training award from the National Institutes of Health.

Chapter 8. Thanks to B. Thierry, B. Chapais, M. Butovskaya, A. Paul, and W. Kaumanns for helpful comments. The first author acknowledges the Wenner–Gren Foundation for support during the project.

Chapter 9. Thanks to the editors for inviting me to contribute to this volume, and to M. Cooper, L. Isbell, W. Kaumanns, D. Maestripieri, J. Prud'homme, S. Teijeiro, and B. Thierry for helpful comments.

Chapter 10. Thanks to the editors for inviting us to contribute to the volume, and to B. Thierry, D. Custance, F. de Waal, M. Drapier, W. Kaumanns, and K. Watanabe for helpful discussion and comments.

Chapter 11. Thanks to the editors for inviting me to participate, and to D. Hill, B. Thierry, M. Singh, P. Deleporte, and C. Chauvin for constructive comments.

Chapter 12. Thanks to C. Marengo, C. Abegg and O. Petit for providing suggestions and data, to C. Hemelrijk, F. Aureli, and R. Noë for helpful comments, and to A. Iwaniuk and S. Pellis for carrying out phylogenetic analyses.

Chapter 13. Thanks to the editors for inviting me to contribute to the volume, to R. Pfeifer and R. Martin for continuous support, and to the A. H. Schultz Foundation for financial support.

Introduction

1   *Why macaque societies?*

BERNARD THIERRY, MEWA SINGH,
AND WERNER KAUMANNS

Societies are built through the interactions of the individuals who compose them. In a sense, they look like transparent organisms, which allow us to observe their mechanisms (Kummer, 1984). Significant parts of them, however, remain hidden. We have only indirect access to the motives and strategies pursued by individuals, and the structures and functions produced by the social dynamics go far beyond the behaviors of individuals.

To analyze societies, we commonly distinguish between social networks and demographic structures. Social networks represent the sets of interactions and relationships that link individuals. Demographic structures refer to the size and age-sex composition of social units. Attempts to find out which come first – the social networks or the demographic structures – have given headaches to numerous students. We end up with the same chicken-or-egg problem when we attempt to uncover causal relationships between these structures and other entities like mating and rearing systems.

The main issue is, what is a social organization? In this book we will use the terms social organization and society as synonymous. We define them as sets of conspecifics that are distributed and behave in a structured manner. This is admittedly a minimal definition, but societies are the very object of our studies and we are still learning about their nature. Some go as far as to ask whether societies exist at all (Rowell, 1993). The reason for such a question is that the organization is partly in the eye of the observer (Ashby, 1962; Strum & Latour, 1987). It is the observer who recognizes structures. We do not perceive the social organization per se, we only see what may be called *sociodemographic forms* (Thierry, 1994a). Human observers may disagree with each other about the forms they identify according to their theoretical assumptions and the analytic tools they use. But there is more to it than that. The actors in a society may themselves differ in their views on social organization. What is significant in the physical world of animals depends on their perceptual and cognitive abilities,

Macaque Societies: A Model for the Study of Social Organization, ed. B. Thierry, M. Singh and W. Kaumanns. Published by Cambridge University Press. © Cambridge University Press 2004.

and the same holds true for their social world (von Uexkühl, 1956; Gibson, 1979; Cheney & Seyfarth, 1989). For instance, whether or not a monkey can read the social relationships that link its group-mates will affect the number of dimensions of the society in which it lives (see Dasser, 1988a). There is still more than the observers' and actors' perspectives. By investigating how societies are built, we aim to gain knowledge of how they arise, how they work and how they change through the course of evolution. Evolution primarily shapes societies by acting upon individuals. But how do selective processes see a social organization? We cannot just presume that any sociodemographic forms we perceive are direct targets of selective processes.

The biological study of behavior is usually approached from the four questions envisioned by Tinbergen (1963): immediate factors, ontogenetic development, adaptive function and evolutionary history. Any behavior should be explained in terms of the above questions. These questions are often grouped following the classical distinction between proximate and ultimate causes (Mayr, 1961). Proximate causation includes immediate and developmental causes that take place once the program encoded in DNA is actualized in the individual. In ultimate causation, evolutionary processes determine the genesis of the program itself. Heuristic though this dichotomous scheme may be, it nevertheless has its limits (Dewsbury, 1992). Even leaving aside the difficulties inherent in the concept of genetic program (Oyama, 1985), a considerable problem remains, which is to discern causes and consequences. This is especially true for complex systems like social organizations, which are made up of individuals linked by information flow. Numerous feedback loops make it difficult to distinguish between features relevant to selection and those which are merely their side effects, and between the direct and indirect advantages or disadvantages they may entail. We need thorough knowledge concerning the proximate mechanisms of societies in order to get back to their ultimate causes (Lott, 1991).

To understand the determinism of multifaceted organizations, we have to consider in turn how every facet relates to the core of societies: individual abilities and personalities, life-history traits, mating systems, demographic dynamics, gene flow, social relationships, intergenerational transmission of behaviors, self-organization processes, ecological factors and phylogenetic correlates. A huge amount of knowledge has accumulated on many of these issues in humans and their societies, but *Homo sapiens* is the only extant representative of its genus, and it is further characterized by wide intercultural variations. This precludes any comparative enterprise that would aim to trace social diversity back to its evolutionary foundations. Among animals, knowledge is basically represented by fragmented information, some facets of social organizations are documented in one species or genus, and other facets in another species or genus. In this context, the genus *Macaca* appears as an outstanding exception.

This is the best-studied group of monkeys. There is no insuperable gap in our knowledge regarding their biology and societies. Macaques also feature broad behavioral diversity. They provide us with a Rosetta stone, a unique model allowing researchers to contrast all aspects of primate societies. By model we mean as well the reference animals, the comparative framework attached to their study, and the example to be extended to other species.

A brief look at the history of work on macaques shows that this research has brought a number of prime scientific discoveries. The early studies of rhesus macaques yielded a wealth of information regarding physiology and reproduction (Bourne, 1975). We still use the rhesus name as a label for the Rh blood group. The finding that infant rhesus macaques may prefer warmth to nurture, and that an early separation from mother induces irreversible psychological damage, drastically changed our understanding of feelings in nonhuman primates (Harlow & Harlow, 1965). Japanese macaques were the first wild animals that were individually identified and followed for their whole lifetime by observers. This allowed Japanese primatologists to discover that newly acquired behaviors might be socially transmitted among macaques (Kawamura, 1959; Kawai, 1965). The observations changed our views regarding innovation and change in primate organization. These studies also made us aware of the pivotal role of kin relationships in macaques (Kawai, 1958; Kawamura, 1958). In groups of rhesus and Japanese macaques, females form matrilines, i.e., subgroups of relatives who help one another in contests. As a result of this, the dominance status of individuals depends on the support of their allies, and strict rules of rank inheritance determine the social status of females. Combined with the conclusions reached from the study conducted on baboons and chimpanzees, this, for a time, led to the belief that most primate societies are governed by dominance and nepotism (Strier, 1994). Though there were early hints that some macaques might be 'nicer' than others (Rosenblum *et al.*, 1964), two decades passed before interspecific contrasts in the conciliatory dispositions of macaques were directly addressed (Thierry, 1986a, 2000; de Waal & Luttrell, 1989). During the same period, several socioecological models were proposed to explain differences in the social relationships and mating systems of primates by environmental variations (Wrangham, 1980; Caldecott, 1986a; van Schaik, 1989; Sterck *et al.*, 1997).

Macaques are mainly frugivorous, semi-terrestrial primates. They inhabit a wide range of habitats, from equatorial to temperate ecosystems, and from evergreen primary forests to grasslands, mangrove swamps, semi-deserts or areas settled by humans (Fooden, 1982a; Richard *et al.*, 1989). They differ both in their morphology and behavior. As shown in the Plate 1, their morphological diversity looks like variations on the same theme. We may expect similar homologous variations with regard to their behavioral diversity. Macaque species differ in

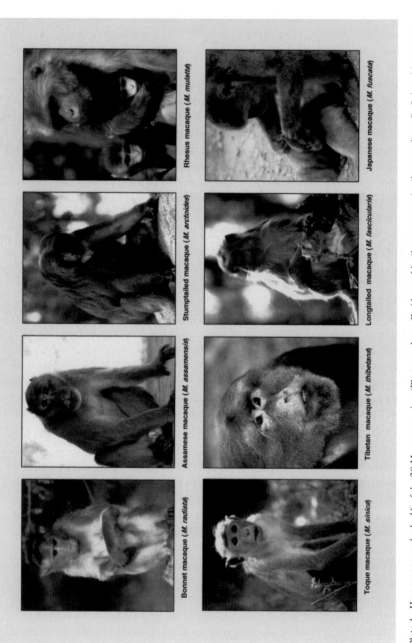

Plate 1. *How macaques look: 16 of the 20 Macaca species*. (Photographers: C. Abegg. *M. silenus, pagensis siberu, radiata*; R. Seitre. *M. maurus, hecki, nemestrina, sinica, assamensis, arctoides*; B. Thierry, *M. sylvanus, nigra, tonkeana, thibetana, fascicularis, mulatta, fuscata*.)

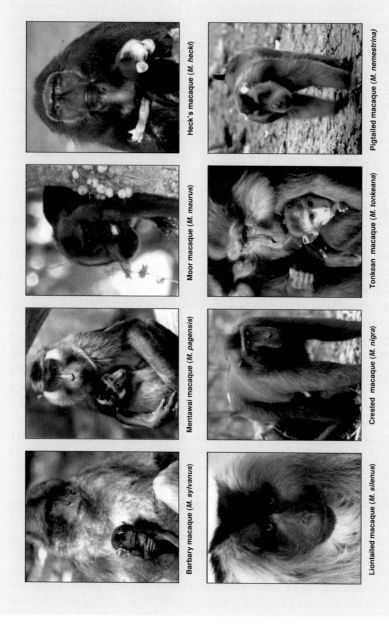

Heck's macaque (*M. hecki*)

Pigtailed macaque (*M. nemestrina*)

Moor macaque (*M. maurus*)

Tonkean macaque (*M. tonkeana*)

Mentawai macaque (*M. pagensis*)

Crested macaque (*M. nigra*)

Barbary macaque (*M. sylvanus*)

Liontailed macaque (*M. silenus*)

Plate 1. (*cont.*)

their styles of affiliation, aggression, dominance, nepotism, maternal behavior and socialization. However, they share the same basic patterns of organization. They form multimale, multifemale groups, that is, groups which permanently contain both adult males and females with offspring. Neighboring groups have overlapping home ranges. The adult sex ratio is biased toward females. Females constitute kin-bonded subgroups within their natal group while most males transfer between groups at maturation.

We are now in a position not only to test the ecological fit of social behaviors, but also to assess their reproductive consequences and phylogenetic correlates. Steroid assays allow recognition of the females' reproductive state from fecal and urine samples collected in the field (Heistermann *et al.*, 1995). DNA analyses enable us to identify males' paternity and measure genetic diversity (Melnick & Hoelzer, 1996). The advent of computer simulations allows us to explore how structures may arise from self-organization processes (Camazine *et al.*, 2001). Phylogenetic analyses make it possible to test the influence of the evolutionary past upon social organization patterns (Brooks & McLennan, 1991; Harvey & Pagel, 1991). We have enough knowledge of macaque systematics and phylogeny (Fooden, 1976; Hoelzer & Melnick, 1996), to provide a firm foundation for examining macaque societies in an historical perspective (Matsumura, 1999; Thierry *et al.*, 2000).

The genus *Macaca* (*Mammalia*: *Cercopithecidae*) is one of the most successful primate radiations. It has the widest geographical range of primates after *Homo*. We presently recognize 20 species of macaques, which are distributed in South and East Asia, with the exception of the Barbary macaque in North Africa (Table 1.1 & Fig. 1.1). The number of species may differ according to taxonomic decisions. If the two subspecies of pigtailed macaques (*Macaca nemestrina nemestrina* and *M. n. leonina*) and the two subspecies of Mentawai macaques (*M. pagensis pagensis* and *M. p. siberu*) are ranked as full species, that would increase the number of macaque species by two (see Groves, 2001). Also, if we recognize six or eight species of macaques on Sulawesi island (Groves, 2001; Froehlich & Supriatna, 1996) instead of seven (Fooden, 1969), that would change the number of species by one. A comprehensive review of the various classification schemes for macaques proposed in the last century may be found in Fa (1989).

Macaques represent a monophyletic group (Delson, 1980; Morales & Melnick, 1998). They are placed within the tribe Papionini, which also includes baboons, geladas, mangabeys, drills and mandrills. The fossil record indicates that macaques diverged from other Papionini in northern Africa in the late Miocene 8–7 millions years ago (Delson, 1980). They invaded Eurasia about 5.5 million years ago probably via the Near East. They then split into several phyletic lineages that have been identified from morphological (Fooden, 1976;

Table 1.1. *Species, phyletic lineages and geographic distribution of the genus* Macaca *Lacépède, 1799*

Species	Distribution
***silenus–sylvanus* lineage**	
Barbary macaque (*M. sylvanus*)	Algeria, Morocco
Liontailed macaque (*M. silenus*)	Southwest India
Crested macaque (*M. nigra*)	North Sulawesi
Gorontalo macaque (*M. nigrescens*)	North Sulawesi
Heck's macaque (*M. hecki*)	North Sulawesi
Tonkean macaque (*M. tonkeana*)	Central Sulawesi
Moor macaque (*M. maurus*)	Southwest Sulawesi
Booted macaque (*M. ochreata*)	Southeast Sulawesi
Muna-Butung macaque (*M. brunnescens*)	Southeast Sulawesi
Mentawai macaque (*M. pagensis*)	Mentawai
Pigtailed macaque (*M. nemestrina*)	Indochinese peninsula, Sumatra, Borneo
***sinica–arctoides* lineage**	
Toque macaque (*M. sinica*)	Sri Lanka
Bonnet macaque (*M. radiata*)	South and West India
Assamese macaque (*M. assamensis*)	Continental Southeast Asia
Tibetan macaque (*M. thibetana*)	East and Central China
Stumptailed macaque (*M. arctoides*)	South China, Indochinese peninsula
***fascicularis* lineage**	
Longtailed macaque (*M. fascicularis*)	Indochinese peninsula, Indonesia, Philippines
Rhesus macaque (*M. mulatta*)	Continental South and East Asia
Japanese macaque (*M. fuscata*)	Japan
Taiwanese macaque (*M. cyclopis*)	Taiwan

(From Fooden, 1976, 1982; Delson, 1980; Groves, 2001.)

Delson, 1980) and molecular evidence (Hoelzer & Melnick, 1996). Three main species groups are distinguished among extant macaques (Fooden, 1976): the *silenus* group, which includes the liontailed, the Sulawesi, the Mentawai and the pigtailed macaques; the *sinica* group, which includes the toque, the bonnet, the Assamese and the Tibetan macaques; and the *fascicularis* group, which includes the longtailed, the rhesus, the Japanese and the Taiwanese macaques. The taxonomic position of two further species is still debated. The Barbary macaque is likely the most ancient taxon of the genus, and is in fact its last African representative. It is alternatively classified as being either the only member of its own species group (Delson, 1980) or one belonging to the *silenus–sylvanus* group (Fooden, 1976). Similarly, the stumptailed macaque is either ascribed to its own species group (Fooden, 1976) or included in the *sinica–arctoides* group (Delson, 1980) (see Deinard & Smith, 2001). We may retain that there are three broad macaque lineages (Table 1.1). Modern macaques most

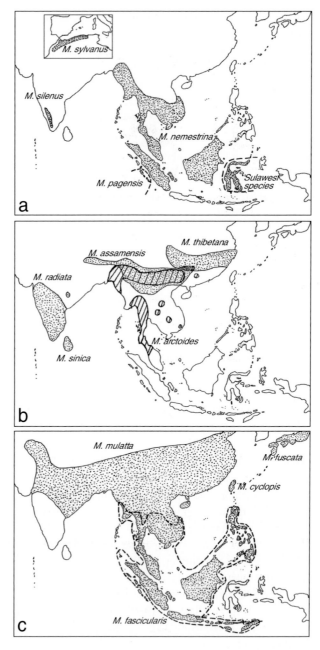

Fig. 1.1. Present geographical distribution of macaques (from Fooden, 1982a).
(a) *silenus–sylvanus* lineage, (b) *sinica–arctoides* lineage, (c) *fascicularis* lineage.

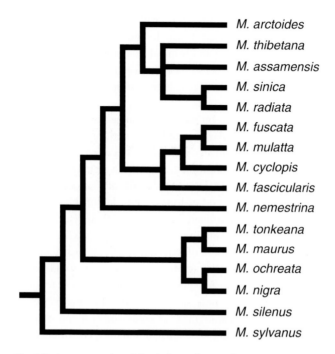

Fig. 1.2. A reconstruction of the phylogenetic tree of macaques (16 of the 20 species), based on a meta-analysis of previously published data, both morphological and molecular (from Purvis, 1995).

likely arose from speciation events that occurred within the last 2 million years (see Fa, 1989). The phylogenetic tree constructed by Purvis (1995) provides a tentative branching order for macaque species (Fig. 1.2).

From the present geographic distribution, it may be inferred that macaques dispersed in three successive waves in Asia (Fooden, 1976). The *silenus–sylvanus* lineage originated from the first macaque radiation. It has the most fragmented geographical distribution (Fig. 1.1a). Its 11 species are situated far away from each other and often present a relict distribution, indicating an early dispersal. The Barbary macaque lives in semi-deciduous montane forests of North Africa. The liontailed macaque is found in the evergreen hill-country forests of South India. The pigtailed macaque is present in the Indochinese peninsula and fringing continental islands of Indonesia that are connected by shallow-waters. Other species inhabit the far-reaching evergreen forests of Sulawesi and Mentawai deep-water oceanic islands. The *sinica–arctoides* lineage is thought to be the second to have dispersed. This lineage has a moderately fragmented distribution in peninsular India and subtropical Southeast Asia (Fig. 1.1b). Bonnet, Tibetan, Assamese and stumptailed macaques are continental. Only the toque macaque is present on a fringing island, Sri Lanka. The *fascicularis* lineage is likely the third to have dispersed. It has the most broadly continuous distribution from equatorial Indonesia and tropical Indochina (longtailed macaque) to as far as subtropical and temperate Asia between China and Pakistan (rhesus macaque), in Taiwan (Taiwanese macaque) and in Japan (Japanese macaque) (Fig. 1.1c).

The replacement of older lineages by younger ones was for some time explained by the occurrence of interspecific competition (Fooden, 1976; Delson, 1980). The most recent colonizers were supposed to have been better adapted to the new environmental conditions. It has been pointed out, however, that the climatic changes of the Pleistocene significantly affected the distribution of primates (Eudey, 1980; Brandon-Jones, 1996; Jablonski, 1998). The aridity associated with glaciations eliminated species from their distribution ranges and split up areas where relict populations were able to survive. A reappraisal of the circumstances of the deployment of macaques in Southeast Asia indicates that the disappearance of the first lineage from large areas was caused by glacially induced deforestation (Abegg & Thierry, 2002). When forests recovered during wetter and warmer periods, some species could not disperse because of geographical obstacles and oceanic gaps, whereas others found migration routes that allowed them to colonize a wide range of habitats. Historical contingencies therefore have played a major role in the deployment and present distribution of macaque species.

We have a wealth of knowledge regarding the evolutionary history and behavior of macaques. It is today possible to consider how the processes and events of evolution shape their social organizations. This volume examines the various facets of macaque societies with the aim of unraveling their architecture and determinism. Because of the diversity of perspectives, chapters include boxes dealing with specific topics. The volume is divided into five parts: (1) individual attributes; (2) demography and reproductive systems; (3) social relationships and networks; (4) external and internal constraints; and (5) an outside viewpoint.

Part I *Individual attributes*

Introduction

Individuals are the building blocks of which societies are made. A determinis-
tic daemon who possessed perfect knowledge about the general characters of
animals, their physical attributes, their physiological and reproductive patterns,
their temperament, their psychological states, motives, cognitive and communicative abilities, would be able to predict which interactions and feedbacks
would occur between each of the animals, and what kind of societies they
would produce when gathered together. Such an ontology is unattainable by us,
if only because we have no means of testing an individual's social disposition
outside the social situation itself (Thierry, 1994a). Notwithstanding this caveat,
a better acquaintance with individual attributes may help us to understand which
social organization is or is not possible for given animals. Individual personalities and emotions have neurobiological and physiological bases. Establishing
attachment bonds and relationships brings about needs, desires and the ability
to carry out trade-offs and social tactics (Chapters 2 and 3). Life-history traits
further funnel the social organization. Traits such as lifespan, age at first birth
and interbirth interval, for instance, influence the size of families and the age–
sex composition of populations (Altmann & Altmann, 1979; Dunbar, 1988).
Female macaques may give birth every one or two years during a 10 to 20-year
reproductive life period, thus simultaneously having adult daughters who have
their own offspring. Such overlaps between generations allow the constitution
of powerful matrilines the members of which can help each other out in various situations. Life-history traits are key adaptations in the implementation of
demographic and social dynamics (Chapter 4).

2   *Personality factors between and within species*

JOHN P. CAPITANIO

Introduction

What is the basis of the differences in aggressiveness, tendency to affiliate, and maternal care within the genus *Macaca* (Fig. 2.1) (Thierry, Singh & Kaumanns, Chapter 1)? This is a deceptively simple question, because at its heart are some fundamental issues regarding what behavior is, how individuals are organized psychologically, what natural selection acts on, and how characteristics of societies emerge from the interactions of individuals. It is my thesis that "personality" is one construct (though certainly not the only one: see Call, Box 2, and Aureli & Schino, Chapter 3) that represents an important source of differences between members of different macaque species. This construct is not only useful in providing a *description* of species differences in social organization, but may actually be an *explanation* of those differences. That is, in the process of evolution, natural selection may have acted on response dispositions rather than on specific responses (behaviors). The result is that members of different species have different "modal tendencies" to respond (Mendoza & Mason, 1989), and these differences contribute to the variations seen on the basic macaque "pattern" of social organization.

I will try to develop some of these ideas in this chapter. The focus will be less on empirical issues and more on conceptual ones, inasmuch as our knowledge of personality in any nonhuman primate species is rather limited. An important starting place is the literature on human personality, which has addressed some of the questions that bear on my thesis: psychological organization of individuals, how personality relates to behavior, and what the biological underpinnings of personality may be. Moreover, the human personality literature can suggest important issues for consideration in the study of nonhuman primate personality, such as personality's multidimensional nature, and the extent to

Macaque Societies: A Model for the Study of Social Organization, ed. B. Thierry, M. Singh and W. Kaumanns. Published by Cambridge University Press. © Cambridge University Press 2004.

a

b

Fig. 2.1. The first comparative study of characteristic patterns of interindividual distance in macaques. Pigtailed macaques (a) demonstrate significantly less passive contact than do bonnet macaques (b). (Reprinted from Rosenblum *et al.* (1964) by permission of the Academic Press, an imprint of Elsevier Science.)

which personality is trait-like and/or motive-like. Following this overview, I will discuss issues pertaining specifically to nonhuman primate personality and its relationship to social behavior.

Personality in humans

Gordon Allport, a prominent early personality theorist, gave 50 definitions of the term "personality" in his seminal 1937 book, but finally settled on the following:

"the dynamic organization within the individual of those psychophysical systems that determine his unique adjustments to his environment" (Allport, 1937: 48). There are three important aspects of this definition that deserve comment: the distinction between personality and behavior, the involvement of "psychophysical systems," and the idea that personality reflects "dynamic organization."

First, the definition implies that personality and behavior ("adjustments to the environment") are somewhat different phenomena. Personality is expressed behaviorally and is inferred from behavior, but personality is not behavior. Allport (1937) suggests that "Personality *is* something and *does* something. It is not synonymous with behavior or activity . . . It is what lies *behind* specific acts and *within* the individual" (p. 48, emphasis as in the original). In this view, behavior is more of a process than a trait, mediating the needs and wishes of the organism with the opportunities that the environment affords. Personality encompasses the goals that the individual is trying to attain, as well as the manner in which he tries to attain them (Winter *et al.*, 1998; more on these two aspects of personality below). As such, personality is a higher-level phenomenon than behavior, and characteristics of the individual's personality interact with the characteristics of the situation to allow the animal to express a variety of behaviors to achieve some adaptive outcome. Thus, an extraverted person may prefer going to parties rather than staying home to read a novel, may prefer working in groups to working alone, and may speak more in social situations, compared to an introverted person. These are very different behaviors, yet they are *about* the same thing – a positive orientation toward others.

This notion of hierarchical organization is central to the idea of personality and deserves further comment. I have presented a simple hierarchy, with behavior at a lower level and personality characteristics at a higher level. Many human personality theorists conceptualize personality as more hierarchically complex than that. For example, Eysenck (1990) conceptualizes personality organization as involving four levels: (1) personality *dimensions* (e.g., extraversion), which reflect intercorrelations of (2) *traits* (e.g., sociability), which reflect intercorrelations of (3) *habitual acts* (e.g., going to parties), which reflect repeated occurrences of (4) *single acts* (e.g., going to a specific party). Similarly, the Five Factor Model (FFM) of personality (described below) distinguishes between five major dimensions or *domains*, described as "multifaceted collections of specific cognitive, affective, and behavioral tendencies" and *facets*, which "designate the lower level traits corresponding to these groupings" (Costa & McCrae, 1995: 23). In the FFM, extraversion is a domain that comprises the six facets of warmth, gregariousness, assertiveness, activity, excitement seeking, and positive emotions. No equivalently complex conceptualization of nonhuman primate personality has been proposed.

It is important to reiterate that personality is manifested in a context – if the individual is marooned on a desert island, her or his extraversion may never be manifested. The inability to see extraverted behavior cannot necessarily be taken to mean that the person is low on this dimension, or that the dimension does not exist. This point has methodological implications, in that a large sample of behavior must be obtained to identify personality dimensions.

A second point regarding Allport's definition is that personality involves "psychophysical systems." This implies that there is some physical basis to personality, although Allport (1937: 48) reminds us that personality is "neither exclusively mental nor exclusively neural." There is a substantial and growing literature describing the physical basis of personality (e.g., Zuckerman, 1991; Pickering & Gray, 1999). One prominent psychobiological theory of personality, for example, postulates three dimensions, each of which is based upon the three major monoamine neurotransmitter systems in the brain: Cloninger (1986) describes "novelty-seeking" as involving frequent exploration, and is associated with dopaminergic activity; "harm avoidance" as reflecting a tendency to respond strongly to aversive stimuli, and is associated with serotonergic activity; and "reward dependence" as a tendency to respond strongly to reward, and is associated with noradrenergic activity. Moreover, a variety of behavior–genetic studies have revealed that individual differences in personality traits, such as the personality dimensions identified in the FFM (and which are often referred to as the "Big Five") show heritability coefficients in the 0.4 to 0.5 range (McGue & Bouchard, 1998). Recently, evidence has been presented that the facet-level traits of the Big Five are also heritable (Jang *et al.*, 1998). Evidence of the physical basis of personality is important for the argument that natural selection has acted on personality characteristics.

The final point to be made about Allport's definition is that personality reflects a "dynamic organization" within the individual. This is an underappreciated aspect of personality (Pervin, 1994a). As I will discuss below, one important conceptualization of personality focuses on traits (which can be defined broadly as "dispositions to respond"), and factor analysis is the methodology that is most commonly used to identify traits. This procedure, however, emphasizes relationships between variables (behaviors) *across* individuals, whereas "personality," at the level of the individual, reflects a more or less coherent, unified, and integrated set of dispositions *within* the individual (Epstein, 1994; John & Robins, 1994; Pervin, 1994b). The organization of these dispositions (i.e., personality) is dynamic in the sense that they can interact in producing behavior. "Extraversion" may be manifested differently by two people, though both may be equally extraverted. The difference in behavioral expression may be a reflection of their differences in another dimension, such as agreeableness. That is, extraversion and agreeableness may interact, such that behavioral expressions may be a reflection of the *combination* of traits. This aspect of personality was discussed by Allport more than 60 years ago when he wrote "The adaptive act of the moment is only partially a function of the one trait. It is determined as well by many other traits . . . and by all manner of specific attitudes, by mood, and by momentary conditions" (1937: 330). The relevance of this discussion to the task at hand is that it emphasizes the complex relationship between personality and behavior – the same behavior, in two different individuals, may be manifestations of different aspects of personality, and the same aspect of personality may manifest itself differently in different individuals.

Conceptualization of personality

I have already indicated one major way that personality has been conceptualized, and that is as a set of traits. Traits reflect dispositions to respond, and the general orientation in trait research is on consistencies in behavior. Each trait is conceptualized as continuously (and probably normally) distributed in a population, every individual can be located along the distribution of each trait and traits themselves cluster into higher-order categories referred to, variously, as factors, dimensions or domains. In human personality research, many different schemes have been proposed, ranging from Eysenck's (1990) three factors of neuroticism (emotional stability vs. instability), psychoticism (kind and considerate vs. aggressive and antisocial), and extraversion (internally vs. externally oriented) to Cattell's (1966) 16 factor theory. A strong current opinion is that the structure of personality is best described by five factors (i.e., the Big Five): extraversion, agreeableness, conscientiousness, neuroticism, and openness to experience, each of which comprises more specific facets (Costa & McCrae, 1995). Despite very different orientations, there is substantial overlap (both conceptually and empirically) in the factors that have been identified by various theorists; for example, in John and Srivastava's (1999) comparison of the major personality theories, Eysenck's extraversion and neuroticism factors are considered comparable to those similarly-named in the Big Five taxonomy, and Eysenck's psychoticism dimension reflects both agreeableness and conscientiousness.

The second historical approach to personality has emphasized motives. Motives refer to "the disposition to be concerned with and to strive for a certain class of incentives or goals" (Emmons, 1989: 32). While some motives are fundamental in terms of survival – i.e., hunger, sex – personality psychologists (e.g., Emmons, 1997) consider the motives of achievement, affiliation, and power (sometimes referred to as "social motives" or the "Big Three") as fundamental aspects of human psychological functioning (see also Murray, 1938).

Whereas traits typically emphasize the consistencies in behavior, motives have been considered responsible for the variability in behavior (Winter *et al.*, 1998) – behavior that is goal-directed should show considerable variation depending on the circumstances.

Recently, a rapprochement has been attempted between these two venerable (and sometimes conflicting) schools of thought in personality psychology. Winter and collaborators (1998) have suggested that the two conceptualizations are *both* critical aspects of personality, each addressing somewhat different (and independent) aspects of human psychological functioning. These authors suggest that "motives involve wishes, desires or goals (often implicit or nonconscious), whereas traits channel or direct the ways in which motives are expressed in particular actions" (Winter *et al.*, 1998: 231). Thus, a broad view of personality includes those things that are of importance to us and that we will work toward, as well as the manner in which we try to achieve our goals. As Winter and collaborators (1998) suggest: motives drive behavior, traits channel behavior.

One point must be emphasized concerning these historical conceptualizations – whether one was a subscriber to the trait approach or to the motive approach to personality, it was obvious that personality comprised multiple components. As described above, the number of factors proposed in the more well-known trait theories ranged from 3 to 16, although the prevailing view is that there are five factors, each of which comprises many lower-level traits. Similarly, a variety of motives have been suggested, ranging from a large list by Murray (1938) to a smaller set of more "physiological" (e.g., hunger, thirst, sex) and more "psychological" (e.g., affiliation, power, achievement) motives. The importance of this point cannot be overemphasized for two reasons. First, a multidimensional view of personality can create enormous complexity. As I suggested earlier, the expression of a particular trait may depend on the combination of the other traits that an individual possesses. For example, if we subscribe to the idea that there are five basic personality factors, then proximity-seeking, as a behavior displayed in particular contexts, may be an expression not only of the degree to which one is extraverted, but also the degree to which one is conscientious, agreeable, neurotic, and open to experience. That is because "proximity-seeking" occurs in contexts in which activation/expression of the other traits may also be relevant. This is the idea that personality reflects dynamic organization, and even slight differences between individuals may result in larger differences in behavioral output (akin to a systems-theoretic concept of "sensitive dependence on initial conditions", Barton, 1994). The second reason for emphasizing this point is that much of the research in nonhuman primate personality has focused on single dimensions of personality. This is somewhat limiting to the task at hand.

Personality and temperament

Before leaving the discussion of research on human personality, it is important to discuss briefly the concept of temperament. Generally speaking, stable behavioral dispositions are usually referred to in the literature on nonhumans as "temperament" and rarely as "personality." In contrast, research concerning humans typically restricts the use of the term "temperament" to the infancy period, and refers to such dispositions in older children and adults as "personality." What is the relationship between these two constructs?

Temperament usually refers to "relatively stable, primarily biologically based individual differences in reactivity and self-regulation"(Rothbart in Goldsmith *et al.*, 1987: 510). A substantial literature exists in psychology that explores the relationship between these constructs (see discussion in Goldsmith *et al.*, 1987). Allport (1937: 53) describes the view, still common today, that temperament forms the basis of personality:

> Temperament . . . might be said to designate a certain class of raw material from which personality is fashioned. Strictly speaking there is no temperament apart from personality, nor any personality devoid of temperament. It is merely convenient to employ the term in speaking of dispositions that are almost unchanged from infancy throughout life (dispositions saturated with a constant emotional quality, with a peculiar pattern of mood, alertness, intensity, or tonus). The more anchored a disposition is in native constitutional soil the more likely it is to be spoken of as temperament.

Thus, temperament is often seen as forming the basis of personality, relatively unchanging from infancy, more "biological" than personality, and focused on broad patterns of emotional responsiveness. Relationships between personality and temperament in humans have been demonstrated by Kagan, who has identified a pattern of high reactivity, characterized by vigorous motor activity, fretting, and crying in response to novel situations (Kagan *et al.*, 1998). This pattern of reactivity shows some stability over time, and has been related to patterns of social interaction at later ages: e.g., Kagan and collaborators (1998) have reported that "high-reactive" 4-month-olds were more likely to show an "inhibited" social style at 4.5 years of age. Other data from Kagan's laboratory have demonstrated a link between "inhibited" temperament at 2 years of age and social anxiety in childhood and adolescence (Kagan & Snidman, 1999; Schwartz *et al.*, 1999; Biederman *et al.*, 2001).

Kagan's results demonstrate a clear link between temperament and characteristics more usually associated with personality (e.g., emotional stability, extraversion). Although researchers in both human and nonhuman primates have generally considered temperament as "biological," "hereditary," and "unchanging," and personality as much more influenced by experience and susceptible to change, data suggest these distinctions are not strong. On the one hand, personality dimensions have been found to be stable over time, to demonstrate reasonably high heritabilities, and to have a neurochemical basis (see discussion above), characteristics more usually associated with temperament. On the other hand, nonhuman primate data have shown that both prenatal and postnatal experience affects variables that are commonly associated with temperament. For example, offspring of rhesus macaque females exposed to unpredictable noise during mid- to late-gestation – an environmental manipulation – show a more "reactive" temperament: greater hypothalamic–pituitary–adrenal reactivity (Clarke *et al.*, 1994b), more disturbance behavior and less exploration in a novel environment at 6 months of age (Schneider, 1992), and more "anxious" social behavior (clinging), and less contact at 18 months of age (Clarke & Schneider, 1993). In terms of postnatal experience, rearing with animate compared to inanimate companions results in differences in cardiovascular and cortisol responsiveness to novelty (Mason, 1978a; Mason *et al.*, 1991), common indicators of temperament. Together, these and other data suggest that "temperament" may be no more fundamental or "biological" than is personality.

Personality in nonhuman primates

The foregoing has suggested that personality is not the same as behavior, it has a particulate basis (and so can be acted on through natural selection), has both trait-like and motivational components, reflects dynamic organization of multiple dimensions that can produce complex behavioral outcomes, and may not be fundamentally different from most peoples' views of a related construct, temperament. In this section, I will review issues pertaining specifically to the study of personality in nonhuman primates. Where possible, I will try to touch bases with the issues discussed above, but it should quickly become clear that there are large differences in the approaches taken by primatologists and human personality psychologists, and that the knowledge base that has arisen is also very different. I will then present data suggesting how personality's contribution to social behavior can be studied intraspecifically, in a way that might shed light on interspecific differences.

Approaches to studying personality in nonhuman primates

In general, two distinct approaches have been taken to studying personality in nonhuman primates. The first approach can be considered a "behavioral" approach (Gosling, 2001, refers to this as a "coding" approach). Here, the focus is on recording specific behaviors displayed by individuals from the same (or different) species, usually (though not always) in specific and highly controlled situations. A good recent example of this approach is the research of Fairbanks and colleagues (2001) in vervets involving an "intruder challenge" test. An unfamiliar animal was placed in a small holding cage outside the living enclosure of a group of conspecifics. Data were collected on latencies of the residents to approach the intruder, as well as behaviors displayed by the residents during the 30-minute trials. Analyses revealed a single dimension labeled "social impulsivity" characterized by short latencies to approach, touching, and displaying to the intruder. In general, the assumption of the behavioral approach is that variation in the specific behavioral outcome measure is taken to reflect variation in an underlying psychological construct.

The second approach taken has involved "subjective assessments" by humans who are very knowledgeable of the animals (referred to as a "rating" approach by Gosling, 2001). This approach was described and elaborated by Stevenson-Hinde and colleagues in the late 1970s (e.g., Stevenson-Hinde & Zunz, 1978), but actually dates back to the 1930s, when Crawford took a strikingly similar approach with chimpanzees (Crawford, 1938). This approach involves asking humans who have spent considerable time watching the animals to rate each animal on several adjectives that might describe it. Ratings are typically performed using a scale ranging from one to seven, where the anchor points reflect something like "not at all characteristic of the animal" to "very characteristic of the animal," respectively. Some or all of the adjective ratings will then be subjected to a statistical procedure (factor analysis or principal components analysis) that reduces the data set to a small number of dimensions that describe the interrelations among the ratings. Similar procedures have been used in human studies (e.g., John *et al.*, 1994), particularly with young children who may be unable to answer questionnaires.

There are three important points to be made about these different approaches. First, while it is clear that both approaches rely on behavior, the behaviors are "about" different things. In the "behavioral" approach, the behavior that is observed is typically focused on a particular problem of the experimenter's choosing – e.g., dealing with an intruder, with being restrained, with being in an unfamiliar place. That is, the animals' goals are set by the experimental paradigm, and the behavior being displayed by the animals presumably reflects stylistic differences in how to "solve" the "problem." (Nevertheless, we must be careful in our assumption that all animals in standardized situations such as these are perceiving the problem in the same way, and that the same goals, such as escape, are being aroused equally in all animals.) In the "subjective assessment" approach, on the other hand, the behaviors that are seen by the human observers are usually not focused on specific, experimenter-defined problems.

Often, the humans' knowledge of the animals is the result of observing the animals in their social groups where the animals can freely engage in behaviors of their choosing. Thus, the "problems" to be solved in free-social situations are defined more by the animals themselves. By becoming familiar with the individuals, the observers learn something about not only *stylistic consistencies* in behavior – how the individuals behave across many different situations – but also what goals may be important to them, such as being in proximity to other animals. Viewed in this way, it is perhaps not surprising that this procedure captures something about both motives and traits. For example, one trait that has been identified repeatedly through use of the rating methodology has been labeled "sociability." Animals high on this dimension have been rated highly on the adjective "sociable" (among other adjectives), typically defined as "seeks companionship of others" (e.g., Stevenson-Hinde & Zunz, 1978). One could argue that this particular adjective may have more in common with the human *motive* of "affiliation", in the sense of it reflecting more of a goal, than with the *trait* of extraversion, in the sense of reflecting simply a positive orientation toward others.

Second, both approaches can be used in either a descriptive or explanatory fashion. Often, these two uses are not made clear. When a statistical procedure produces a result that shows the interrelations of variables, we usually have a descriptive result. From this, we might infer explanation (e.g., that a trait, so-identified, causes the variables), but this is somewhat circular. The value of traits as explanatory constructs would be enhanced if they were employed predictively, that is, if traits, identified in one situation and with one type of measure, were shown to predict a theoretically related outcome in another situation (and possibly with another type of measure). Few primate studies have demonstrated more than the descriptive use of these constructs, although there are some exceptions – see Bolig *et al.*, 1992; Stevenson-Hinde *et al.*, 1980b; Capitanio, 1999; and a long-standing research program by Mason and collaborators (e.g., Anzenberger *et al.*, 1986) comparing squirrel and titi monkeys.

Finally, if one goal in nonhuman personality research is direct species comparisons, then the two approaches will likely not be equally valuable. If, for example, we want to determine whether bonnet macaques are more affiliative than pigtailed macaques (Fig. 2.1), the behavioral approach might prove to be the better one. Members of different species can be observed under similar conditions (e.g., given the same "problem" to solve) and frequencies and durations of behavior can be obtained. These data can be compared directly (although, again, care must be taken in assuming that the situations are being perceived and responded to in the same fashion by members of the two species). Such an approach was taken by Clarke and colleagues who demonstrated, across a series of identical situations, that longtailed macaques were more behaviorally responsive than were rhesus or bonnet macaques; and whereas rhesus tended to respond with hostility, longtailed macaques tended to respond fearfully. Rhesus also were the most active in these situations, while bonnets were the most passive (e.g., Clarke & Mason, 1988; Clarke *et al.*, 1988; see also Clarke & Boinski, 1995). In contrast to such a behavioral approach, the subjective assessment procedure is more relativistic – the rater is instructed to indicate how character-istic a particular adjective is for each animal, but we have no idea whether the same metric is used when different individuals from different laboratories rate members of different species, even if the same adjective list is used. One would expect, for example, that a rating of "7" (reflecting "highly characteristic") for an adjective such as "protective" might well be given to some female bonnet macaques and to some female pigtailed macaques. Data obtained using a behavioral approach, however, have shown that bonnet and pigtailed macaques are quite different in terms of maternal protectiveness (e.g., Kaufman & Rosenblum, 1969). While the subjective assessment procedure may have less value in direct cross-species comparisons, it is useful in uncovering the *structure* of personality within a species. This is because it involves becoming familiar with the animals, a process that is likely to reveal something about their motives and traits. Consequently, rating-derived information can be useful in identify-ing dimensions; this information might then be used to construct situations in which more directly comparable behavioral measures can be obtained for direct species comparisons.

Personality dimensions in nonhuman primates

Because the behavioral approach tends to focus on collecting data under a spe-cific type of situation (typically of the experimenter's choosing), it is usually the case that a single personality/temperament dimension is identified, reflecting the manner in which the animals "solve" the particular problem. (Here I use the terms "dimensions" and "traits" interchangeably.) In contrast, the subjective assessment approach typically identifies multiple personality dimensions, as is the case in human research. Together, these procedures have suggested a hand-ful of traits that characterize personality in Old World monkeys. These have been variously labeled as affiliation, aggressiveness, anxiety, confidence, emo-tionality, equability, excitability, fearfulness, hostility, impulsivity, irritability, reactivity, sociability, social impulsivity, and sociality (Chamove *et al.*, 1972; Stevenson-Hinde *et al.*, 1980a; Clarke & Mason, 1988; Higley & Suomi, 1989; Schneider *et al.*, 1991; Bolig *et al.*, 1992; Figueredo *et al.*, 1995; Mehlman *et al.*, 1995; Higley & Linnoila, 1997; Capitanio, 1999; Fairbanks, 2001). Presumably, many of these descriptors refer to similar or identical underlying dimensions (in fact, labeling of personality dimensions in humans is a concern as well, see John & Srivastava, 1999). For example, affiliation, sociability, and sociality may all reflect a tendency to affiliate, whereas anxiety, emotionality, excitability, and reactivity may all reflect a tendency toward rapid, emotional responsiveness in situations.

The foregoing has suggested that the major dimensions of personality in macaques may include impulsivity, excitability/reactivity, sociability, and confidence/aggressiveness, and there are probably additional dimensions (e.g., Gosling, 2001, suggests other dimensions might include curiosity/exploration and activity). The relationships between the dimensions identified in the various primate studies remain unknown, however. Recall that human personality theory has resulted in complex hierarchical models of personality, in which relatively narrow traits are subsumed by broader dimensions. To my knowledge, no one has approached personality in any nonhuman species from this perspective. Identifying and understanding the hierarchical organization of personality in nonhuman species remain important tasks.

The biological basis of nonhuman primate personality

Earlier, I described research suggesting that human personality dimensions had a neurochemical and genetic basis. Several lines of evidence suggest that personality characteristics in nonhuman primates also demonstrate a biological basis.

First, the aforementioned behavioral differences between rhesus, bonnet, and longtailed macaques, reported by Clarke and colleagues, have physiological correlates. The more behaviorally responsive longtailed macaques showed the highest and most sustained heart rate of the three species as well as the greatest corticosteroid response, while rhesus macaques demonstrated the lowest levels for both measures (Clarke *et al.*, 1988, 1994a).

Second, older (e.g., Kling, 1972), as well as more recent (Emery *et al.*, 2001) research has suggested that particular brain areas, especially the amygdala, are important in socioemotional behavior. For example, adult male rhesus macaques with lesions of the amygdala show higher frequencies of approach and positive social behaviors directed toward unfamiliar stimulus animals. Amygdala-lesioned animals are also rated as less tense, more confident, and more affiliative in their interactions, compared to control animals (Emery *et al.*, 2001), characteristics that suggest a role for the amygdala in sociability.

Third, a variety of studies have examined the role of central monoamine neurotransmitters (and particularly serotonin) in personality. Serotonin function is typically assessed in one of two ways, either by measuring concentrations of its major metabolite 5-hydroxyindoleacetic acid (5-HIAA) in cerebrospinal fluid (lower concentrations of which presumably reflect reduced serotonin neurotransmitter activity), or by pharmacological challenge (e.g., with fenfluramine), which changes plasma concentrations of an easily measured hormone, such as prolactin, whose release is associated with serotonin neurotransmission. Serotonin activity has been associated, in both the human and nonhuman literature with behavioral inhibition (e.g., Cloninger, 1986), and in nonhuman primates, reduced serotonin activity has been associated with social impulsivity (*Chlorocebus aethiops*: Fairbanks *et al.*, 2001); impulsive aggression (*M. mulatta*: Higley *et al.*, 1996a); contact aggression and social withdrawal (*M. fascicularis*: Botchin *et al.*, 1993); and lower durations of grooming and proximity (*M. mulatta*: Mehlman *et al.*, 1995). Concentrations of other monoamine neurotransmitters (dopamine and norepinephrine) can be assessed in a similar fashion, but have been less well-studied (see, however, Kraemer *et al.*, 1989). One exception has come from our laboratory, where we found the dimension "sociability" to be related to both serotonergic and dopaminergic activity, and "irritability/impulsivity" to be related to noradrenergic and serotonergic activity (unpublished observations). It is also worth noting that longitudinal studies of monoamine neurotransmitter metabolite concentrations have revealed substantial temporal stability; that is, individual differences in concentrations of these metabolites have been found to be remarkably consistent in rhesus macaques (Higley *et al.*, 1992; Kraemer *et al.*, 1989).

Fourth, neuropeptides have been implicated in emotional behavior in humans and rodents, and data are beginning to show similar relationships in nonhuman primates. For example Rosenblum, who first described differences in affiliation between bonnet and pigtailed macaques (Fig. 2.1), has recently demonstrated that the more affiliative and less strongly hierarchical bonnet macaques have significantly higher cerebrospinal fluid (CSF) concentrations of the peptide oxytocin, and significantly lower CSF concentrations of the peptide corticotropin-releasing factor than do the less affiliative and more strongly hierarchical pigtailed macaques (Rosenblum *et al.*, 2002).

Fifth, relationships are beginning to be reported between genetic data, personality characteristics, and monoamine function, though only in rhesus macaques at this time. Significant heritabilities have been found for concentrations of monoamine neurotransmitter metabolites that have been associated with personality characteristics, particularly for 5-HIAA (Higley *et al.*, 1993). Moreover, an attempt has begun to examine specific allelic variation relating to monoamine neurotransmitter systems. Efforts have focused on polymorphisms in the promoter region of the serotonin transporter gene, which produces a protein that is involved in the process of reuptake following synaptic release of serotonin. In humans, the short allele appears to be associated with reduced transcriptional efficiency of the promoter protein, and, presumably, impaired serotonin activity (Heils *et al.*, 1996). Human research has found associations between the presence of a short allele and psychopathology, and recent studies in rhesus macaques have demonstrated that this allele is associated with earlier natal dispersal (Trefilov *et al.*, 2000), an outcome that previous research had shown to be associated with lower 5-HIAA concentrations (Mehlman *et al.*, 1995).

Finally, an important study has examined the relationship between neurochemistry and aggression in a comparative framework. Westergaard and collaborators (1999) found that, compared to female pigtailed macaques, female rhesus displayed higher frequencies of extreme aggression, and also showed lower concentrations of CSF 5-HIAA. These data confirmed, in a cross-species design, the previously-known intra-specific association between intense aggression and low 5-HIAA concentrations.

While these data suggest that personality has genetic and neural components, it is important to emphasize the experiential influences on physiological and behavioral aspects of personality. Personality, like many other genetically-influenced traits, is not static and unchangeable. It is now recognized, for example, that social restriction early in life can have long-lasting influences on social behavior and, presumably, personality (Caine *et al.*, 1983; Capitanio, 1986). Such experiences can also result in lasting changes in neurochemical systems that have been associated with personality characteristics (e.g., Kraemer *et al.*, 1989; Lewis *et al.*, 1990; Higley *et al.*, 1991; Martin *et al.*, 1991; Coplan *et al.*, 2001). Even more interesting, Bennett and collaborators (2002) have reported that the relationship between serotonin transporter genotype and CSF concentrations of 5-HIAA was dependent on early experience: mother-raised macaques that were heterozygous for the short allele in the promoter region of the transporter gene did not have different 5-HIAA concentrations compared to mother-raised monkeys that were homozygous for the long allele. In contrast, heterozygous peer-reared monkeys had significantly lower 5-HIAA than did peer-reared monkeys homozygous for the long allele. These data show the considerable plasticity that exists in biobehavioral systems, and illustrate the complex relationship between genotype and phenotype. Typically, the social environment will "support" the production of the species-typical dispositions. At other times, however, environmental conditions may favor a phenotype that is not at the species-typical mode, and this phenotype may arise due to an altered developmental trajectory associated with these changed conditions. A consequence of this could be alterations in gene frequencies after many generations, as biobehavioral systems "settle" into a new equilibrium. Such plasticity thus reflects the variation upon which natural selection can work, and could be a means whereby speciation occurs. West-Eberhard (1989) discusses these issues more extensively.

Personality dimensions and social behavior

If we agree that personality is a multidimensional phenomenon (while still being unsure about what the dimensions are, their hierarchical organization, and the extent to which they are trait-like, motive-like, or some combination of both), how can personality differences result in different social organizations across species? To be sure, all species of macaques arose from the same common ancestor, and so probably possess the same basic "equipment;" that is, the personality dimensions themselves probably do not differ among the various macaque species. Rather, the species probably differ in their modal 'location' along each personality dimension. For example, although there would be substantial overlap in the distributions of bonnet and pigtailed macaques along a dimension of "sociability" (e.g., Rosenblum *et al.*, 1964), it is likely that bonnets as a species have a modal value that is higher than that of pigtailed macaques. Complicating the picture even further, it is likely that within a species, the modes for each dimension differ for adult males and adult females – adult female rhesus macaques might be considered more "sociable" than adult males. Thus, in terms of species comparisons, differences in societies might reflect differences in the "strength" of various dispositions by adult males and adult females.

It is, of course, impossible to know how natural selection has acted on dispositions to produce inter-specific differences in societies. Examination of existing species, particularly in naturalistic settings, must be done cautiously to avoid circularity (e.g., bonnet macaques are more affiliative because they are more sociable, which we measure by recording the frequency of affiliative acts). The causal role of personality in social organization can be studied experimentally, however, which may shed light on the complexity of the problem, particularly regarding the potentially synergistic, non-additive effects of different personality characteristics. As part of a larger project examining the effect of social factors in simian immunodeficiency virus disease (Capitanio *et al.*, 1998, 1999), we have collected some pertinent data.

Personality assessments were performed on 45 adult male rhesus macaques living with their natal groups in half-acre enclosures at the California National Primate Research Center, and four personality dimensions were found: sociability (reflecting a tendency to affiliate); equability (which seems to be the opposite of impulsivity); confidence (reflecting self-assurance and aggressiveness); and excitability (reflecting emotional reactivity). Following the personality assessments, animals were relocated to individual housing indoors, and underwent a series of social group formations. In all cases, social groups were formed without regard to personality characteristics (see details in Capitanio, 1999). In one situation (when animals ranged from 6.4 to 9.5 years of age) we placed 15 previously unfamiliar animals together in three-member groups for 100 minutes per day for more than a year. Despite the fact that animals were

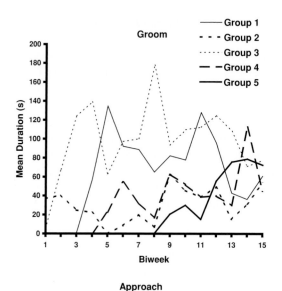

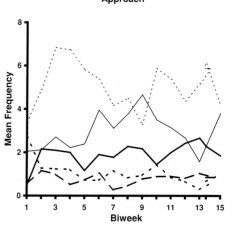

Fig. 2.2. Duration of groom and frequency of approach for five three-member groups of adult male rhesus macaques observed across 15 biweekly periods (means).

approximately the same age, had the same background, and were observed under identical conditions by the same behavioral observers, significant group differences were found between the five three-member social groups for five of the seven behaviors analyzed. Figure 2.2 shows group means for two behaviors, approach and duration of groom. Group 3 showed significantly higher frequencies of approach and greater durations of groom, compared to Group 5, which showed significantly higher frequencies of threat (not shown). Group 5

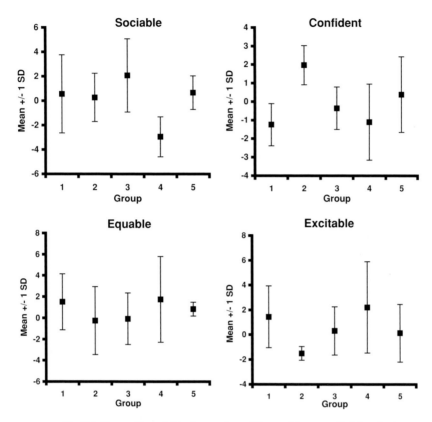

Fig. 2.3. Personality variables for members of five social groups of adult male rhesus macaques (means and standard deviations).

monkeys also demonstrated more *sustained* conflictual behavior: threats were recorded in no more than 8 of the 15 biweeks for monkeys in Groups 1 to 4, whereas threats were recorded for 13 of the 15 biweeks for Group 5 animals.

How do these behavioral data relate to the personality profiles of individuals? Recall that groups were formed randomly with respect to personality characteristics – group differences in both mean values for the personality measures, as well as the standard deviations, were not statistically significant (Fig. 2.3). Using group ($n = 5$) as the unit of analysis, Spearman rank-order correlation coefficients were computed for each biweek between mean frequency or duration of behavior on the one hand, and mean or standard deviation of personality measures on the other hand. Results showed that groups that had a higher mean value for sociability also had higher frequencies of approach – correlation coefficients between these two measures were greater than 0.7 for 14 of the 15 biweekly periods. The particular "mix" of personalities was also related to behavior: groups whose members were more variable on the dimension of sociability (i.e. had a larger standard deviation, indicating some members were highly sociable and others less sociable) had higher durations of grooming – 8 of 15 correlation coefficients were greater than 0.7. Finally, sustained conflict – defined as a greater number of biweekly periods in which threats were recorded – was associated with greater variation in confidence and lower variation in equability.

These data, of course, are limited in a number of ways: the number of groups was small as was the number of animals per group; only adult males were studied; the animals were not together 24 hours per day; and there were few resources to compete over. Nevertheless, I believe several aspects of these results are relevant to the issue of personality and macaque social evolution. First, the amount of social behavior was related to average "level" of personality: groups whose members were more sociable showed higher frequencies of approach. This suggests that species differences in personality might be causal to species differences in the affiliative aspects of social organization. That is, were a selective advantage to arise for an increased tendency for group-members to affiliate, the society that such individuals produce could be different in measurable ways from the societies of previous generations. Such an increased tendency, of course, then sets up additional selective pressures that can either reinforce or counter the directional push for that dimension, as well as, perhaps, affecting the mean "level" of other existing dimensions. For example, depending on the feeding ecology of a species, an increased tendency to affiliate might be countered by increased aggression (and perhaps conciliatory tendencies) owing to greater feeding competition from animals that now are in greater proximity to each other. Hemelrijk (2000a, Chapter 13) discusses unexpected changes in social behavior resulting from different assumptions in her individual-based models of social dominance.

Second, these data suggest that significant differences in behavior can be generated by nonsignificant differences in personality. While this idea may seem counter-intuitive, it is important to reiterate that personality is not the same as behavior. Since personality "drives and channels" behavior, slight changes/differences in the higher-level phenomenon could have large effects on the lower-level one (Feibleman, 1954). In systems theory, this phenomenon is referred to as "sensitive dependence on initial conditions" (Barton, 1994). Even if one species' modal tendency to affiliate is only slightly more than another's, potentially large differences in behavior could result.

Third, these data point to the importance of variability among individuals in personality. Duration of groom at the group level, for example, was associated with variation in sociability – more grooming was seen in groups in which members were more different from each other on this dimension. While such a result may have implications for group selectionist models of social evolution (Wilson & Sober, 1994), at the individual level the data remind us that complementarity is an important component of social relationships (Hinde, 1976a). Societal characteristics are a reflection of the relationships that are formed between individuals, and those relationships reflect a "meshing" of needs and goals of group-members. Unlike in this experiment, animals in the wild are active participants in choosing with whom they associate. Animals that choose unwisely may suffer reduced fitness. Perhaps animals that are uniformly high in sociability end up competing with each other for opportunities to affiliate, with the result being more conflict than affiliation. As species evolve, a "dynamic stability" will have to be met between the motives and styles of the individuals on the one hand, and the "problems" that exist in the environment – abiotic, biotic, and conspecific.

Conclusion

I have argued that personality may have been an important target of selection in the evolution of macaque societies. This belief stems from the view that individuals are organized entities, with behavior mediating the needs and desires of the individual on the one hand and the opportunities and constraints of the environment on the other. It is probably rare that natural selection acts on specific behaviors (e.g., running away from a predator); rather, it probably acts on more general dispositions, thresholds, and "tendencies to respond" (e.g., being more reactive, which may be manifested in a shorter latency to run away following detection of the predator), which I refer to broadly as "personality." Research on humans has reinforced the idea that there exists a small number of superordinate dimensions that "drive and channel" behavior. An individual can be thought of as inhabiting a multidimensional space defined by his "location" along each personality dimension. Personality is dynamic in the sense that certain dimensions are more likely to be influential at some times than at other times. At the level of the individual, the total "mix" of all personality characteristics influences the individual's behavioral expressions. At the level of the group, the particular mix of characteristics represented by the group members can have an important influence on the types of behavior one sees in the group and, presumably, on the organization of the group.

I believe it is possible to extend these ideas from the individual within a group and the group within a species to the level of species within a genus. Although there are individual differences within a species, it is likely that each species has a modal value on each of a small number of dimensions. It is still unclear what the dimensions are, how many dimensions there are, their hierarchical organization, or even if we should consider that adult males and females might have different modal values. I suspect, however, that affiliation, aggression, impulsivity and reactivity are among the dimensions. As macaques exploited novel areas during evolutionary time, natural selection could have favored animals of particular personality types. For example, if animals were moving into an area of greater predation, individuals that were more reactive and/or more affiliative may have had a selective advantage. This could lead an evolving population to a different modal point on these dimensions than was characteristic of the progenitor animals. Eventually, a new species could come to occupy its own place in a multidimensional space defined by this small number of dimensions. The result could be the variation we see today in social organization of the 20 species of macaques.

Much is still unknown. In addition to understanding the dimensions and their relationships within a species, direct comparative work is essential for understanding just what the species differences are that have been superimposed on the general "macaque" pattern of social organization. For this task, attention must be paid to getting a good sample of behavior. The data collected in our laboratory (described above) can serve as a useful model for understanding how personality factors affect social organization, but they also provide a warning that intraspecifically, groups can differ from each other in important ways; thus, multiple groups for the different species must be examined. Finally, methodological issues are important. Personality is a psychological system, and like other systems, one learns the most about how a system works by challenging it. This often involves experimental work under controlled laboratory conditions. But such work can also be conducted, to some extent, under field conditions. The key procedure, in my opinion, is to present the animals with standardized situations – social and nonsocial – to examine their responses both to the situations and to each other while in those situations. For example, use of paradigms developed in captive situations – intruder paradigms (Fairbanks), responsiveness to human observers (Clarke) – as well as utilization of data commonly collected in the field – presentation of predator models (e.g., Coss & Ramakrishnan, 2000) or alarm calls, induction of competition (as in dominance testing), assessment of time budgets – could, if used in standardized fashion by different observers, in different settings, and with different species, be of enormous use in identifying species differences in modal tendencies.

To summarize and conclude, I have argued that a focus on personality factors as influential in the evolution of social organization forces us to examine how and why behavior is expressed by individuals, and what natural selection might have acted upon to produce the variation one sees within this (or any) genus. I believe that personality factors are likely candidates for consideration.

Box 2 Social intelligence 33

Personality appears to be multidimensional, has a neurobiological basis and is related to behavior in complex ways. It seems a safe assumption that social organization of groups within a species reflects the interaction of individuals that differ along a small number of personality dimensions. Both intraspecific and interspecific approaches to the study of personality will be useful in our understanding of social evolution in macaques.

Box 2   Social intelligence

Josep Call

The Machiavellian intelligence hypothesis states that animals have evolved cognitive mechanisms that permit them to compete and cooperate with their conspecifics for access to limited resources (Byrne & Whiten, 1988; Byrne, 1997). Those mechanisms are particularly important for animals, like macaques, who live in complex groups with repeated opportunity for inter-action. One important component of social problem solving is knowledge about the social environment. Research on social knowledge in macaques has focused on two complementary aspects: behavior and psychological states of others, and social relationships.

Knowledge about behavior

The behavior of others is a rich source of information that can help indi-viduals to orient toward particular relevant stimuli and to assess the goals of others. Of particular importance for social interactions is the ability to assess what others can and cannot see, and even take the perspective of others. Recent studies have shown that macaques can follow the gaze of conspecifics (Emery *et al.*, 1997; Tomasello *et al.*, 1998) and humans (Ferrari *et al.*, 2000) to distant locations. Tonkean macaques can even use visual and odor cues such as food remains on a conspecific mouth to travel to one of two loca-tions matching the conspecific's cues (Drapier *et al.*, 2002). There is also neurobiological evidence that shows that single neurons respond strongly to certain eye orientations such as direct stare (Perrett & Mistlin, 1990). This sensitivity to gaze and head direction contrasts with the failure of macaques to find hidden food when its location is indicated by means of head and gaze direction (Anderson *et al.*, 1996).

Kummer and collaborators (1996) found that longtailed macaques could be trained to reduce their drinking behavior from a nipple when the experi-menter was facing them as opposed to having his/her back turned. However, in a transfer phase the macaques failed to select one of two nipples that was behind a barrier during trials in which the experimenter faced them throughout. Thus, macaques did not actively hide to perform the forbidden activity. Yet, in rhesus and Japanese macaques, Cheney and Seyfarth (1990) found that adult females behaved differently toward infants of a dominant mother depending on whether the mother was visible. Similarly, Ducoing and Thierry (2003) found that subordinate Tonkean macaques stopped approaching a hidden fruit (whose location they knew) when they were being monitored by a dominant animal (who did not know the precise location of the food). However, this differential behavior could be interpreted as an inhibitory response to the dominant's presence rather than as a case of knowledge attribution. Cheney and Seyfarth (1990) also found that macaque mothers did not alter their behavior depending on whether their offspring were either ignorant or knowledgeable about the presence of food or a predator. That is, macaque mothers did not inform their ignorant offspring about the presence of food or a predator. Similarly, Povinelli and collaborators (1991) found that rhesus macaques failed to select the cup indicated by an experimenter that had witnessed the baiting as opposed to an experimenter that had not witnessed the baiting. Povinelli and collaborators (1992) also found that rhesus macaques that had learned to cooperate to obtain food each doing a different task (one monkey indicated the location of food and the other one operated a handle so that both could get a reward), were disrupted when their roles were reversed. Taken together the current evidence indicates that although macaques are sensitive to social cues, there is no evidence suggesting that they possess the more abstract concept of what others can or cannot see or other psychological states.

Knowledge about social relationships

The pattern of interactions among individuals has been used to infer the type of social relationships that individuals perceive within their groups. Agonistic interactions, in particular, offer a unique opportunity to investigate how individuals distribute their help among friends and foes when these are attacked by third parties (Fig. 2.4) (Das, 2000; Watts *et al.*, 2000). There is ample evidence showing that macaques take kin relations (or familiarity based relations) into consideration in post-conflict situations. Some studies show that victims of aggression redirect aggression against their opponent's kin (Judge, 1982; Aureli & van Schaik, 1991; Aureli *et al.*, 1992). Aureli and collaborators (1992) found an even more complex situation in Japanese macaques in which the kin of victims retaliated against the aggressor's kin. Similarly, Silk (1992) found that male bonnet macaques joined coalitions that targeted individuals with whom they had had a previous conflict, something

Box 2 Social intelligence 35

Fig. 2.4. Two adult male stumptailed macaques engaged in a hold-bottom behavior during the course of an agonistic interaction with a third party (Yerkes National Primate Research Center, USA). Triadic interactions can be used to infer some of the social knowledge that individuals have about their groups. (Photograph by J. Call.)

the author termed revenge. There is also some evidence of an increase in affiliation following conflicts toward the opponent's kin in several macaque species. Macaque aggressors of various species increased their contacts with the kin of their victims (Judge, 1991; Das *et al.*, 1997; Call *et al.*, 2002). Likewise, stumptailed macaque victims also increase their contacts with the aggressor's kin (Call *et al.*, 2002).

Outside agonistic contexts, Sinha (1998) found that pairs of bonnet macaques engaged in grooming interactions assess the relative rank of approaching third parties. In most cases, it was the subordinate animal of the pair who left upon a dominant's approach. Moreover, those cases in which the dominant animal in the pair left, thus not adhering to this rule, could be explained by the degree of affiliation between the subordinate of the pair and the approaching dominant. Sinha (1998) argued that this showed that macaques assessed both the relative dominance rank and the level of friendship among individuals. Dasser (1988a,b) examined the ability of longtailed macaques to recognize mother–infant relations using a slide discrimination procedure. Two longtailed macaques were shown pictures of certain mother–infant pairs belonging to their own social group. Once they were trained in those slides they were presented with new slides of mother–infant pairs who also belonged to the group but which subjects had not seen during the first phase of the experiment. These slides were compared with slides depicting non-mother–infant pairs chosen so that they would resemble the physical appearance of mothers and infants (i.e., adult and immature animals). Results indicated that both subjects selected the slides depicting a mother–infant pair or sibling relationships.

Although these studies suggest that monkeys may recognize triadic relationships among group members, alternative explanations have not been ruled out yet. One possibility is that macaques may simply remember past interactions ("who did what to whom") and use that information to predict future interactions. This explanation would not require positing any knowledge about triadic relationships, but simply a memory for past interactions. One serious limitation of this mechanism is that macaques may be incapable of predicting the outcome of novel interactions such as when the "balance of power" in the group is altered by removing some key individuals. This latter manipulation is precisely what Chapais (1992a, Chapter 9) did in a series of experiments to assess rank acquisition and maintenance in Japanese macaques. He found that high-ranking juveniles whose kin had been removed from the group lost their rank because they challenged subordinate animals who still had their supporters with them. Had the dominant juveniles avoided conflicts they may have been able to conserve their high rank. Thus, this suggests that juvenile Japanese macaques may not be able to flexibly adapt to changes in the social environment. This is not to say that macaques cannot adapt to other novel social changes. In fact, some studies have shown that even macaques reared with non-animate surrogate mothers can improve the complexity of their initially deficient social interactions with conspecifics after repeated exposure to them (Anderson & Mason, 1974; Capitanio, 1985). Whether this increased complexity also translates into a qualitatively different level of social intelligence is still an open question.

In summary, macaques follow gaze and can learn to cooperate but there is no evidence that they take the perspective of others, their knowledge states, or reverse roles. Individuals behave as if they know the social relations of third parties, in particular affiliation (or kinship) and dominance relations. Yet, the precise mechanism responsible for this behavior is unknown, although there is some evidence suggesting that macaques may not foresee the consequences of their actions. Thus, the evidence available suggests that macaque social intelligence is based on observable behavior (not psychological states) and past interactions with conspecifics. Note, however, that this tentative conclusion is based on a less than ideal sample. Most studies have

Box 2 Social intelligence 37

been conducted with macaques with non-egalitarian societies. It is conceivable that macaques whose social organization is more fluid (Thierry, 1986a, Chapter 12) may also show a different set of cognitive abilities. Finally, it is important that future studies investigate the precise influence of variables such as personality (see Capitanio, this Chapter) or rearing history (Anderson & Mason, 1974) on the development of social intelligence in macaques.

3 The role of emotions in social relationships

FILIPPO AURELI AND GABRIELE SCHINO

Introduction

Differences in the quality of social relationships play a major role in the variation of social organization among gregarious animals. Understanding the differences between social relationships can be achieved by carefully studying the social interactions that group members exchange (Hinde, 1979). The same individual has different social relationships with two group members when it interacts in a different way with each of them. On a broader scale, groups or species differ in their social organization when relationships between group members are overall different, i.e., when individuals interact with the average group member or distribute interactions among other group members in different ways.

In the case of macaques, the variation in dominance style (*sensu* de Waal, 1989a; Flack & de Waal, Chapter 8) among these closely related species is reflected at the interaction level by interspecific differences in the degree of tolerance toward the average group member (Thierry, 1985a, 2000; de Waal & Luttrell, 1989). Similarly, interspecific differences in the degree of kin bias in social interaction have been reported (Thierry, 1990a; Aureli *et al.*, 1997; Demaria & Thierry, 2001) and are likely to underlie much variation in macaque social organization. Understanding the mechanisms that bring about such differences would be an important contribution to the full comprehension of the relation between the social organization and the individual inputs and, in a broader sense, the relation between society and individual minds. In this chapter we attempt to contribute to such an understanding by exploring the role of emotions in explaining the variation in social interactions. Harlow's pioneering work has clearly shown the importance of the emotional bases of social exchanges starting from the mother–infant relationship (Harlow, 1958). The

Macaque Societies: A Model for the Study of Social Organization, ed. B. Thierry, M. Singh and W. Kaumanns. Published by Cambridge University Press. © Cambridge University Press 2004.

differences in tolerance and kin bias across macaque species could be related to possible variations in other aspects such as the degree of impulsivity, anxiety, and the need for emotional support from others. Systematic research on interspecific variations in these aspects has not been carried out yet, although initial work has been promising (Clarke & Boinski, 1995).

We begin the chapter by reviewing the different positions on studying emotions in animals and define how we use the concept of emotion in the remaining part of the chapter. We then briefly present physiological and behavioral indicators of emotions that are particularly useful in animal studies. The indicators are used to review evidence of emotions in macaques. Particular attention is given to emotions as causes and consequences of social life and to their role in relationship assessment and mediation of social interactions. Finally, we propose a framework for future research in which we emphasize that the comparative study of emotional responses could provide insights into the selective pressures at the basis of the origin of social diversity across macaque species.

Animal emotions

During the last two decades research on emotions has grown enormously but the majority of research has focused on humans (Cacioppo & Gardner, 1999; Aureli & Whiten, 2003). Terms with emotional connotations are still rarely to be found in studies of animal behavior. Although the recent development of the discipline of *affective neuroscience* (Davidson & Sutton, 1995; Panksepp, 1998) has produced a boost in research using animal models, relatively few studies have systematically investigated behavioral phenomena related to emotions in nonhuman animals.

The paucity of research on animal emotions is possibly mainly due to the fact that emotions have been traditionally described in terms of subjective experiences. For example, it is common to find in introductory psychology textbooks emotion being defined as an evaluative response of the organism involving physiological arousal, expressive behavior, and conscious experience. Most students of animal behavior have thus preferred to avoid terms related to emotions, seeing them as overly anthropomorphic. They have attempted instead to relate animal behavioral sequences to contextual contingencies and motivating factors (van Hooff & Aureli, 1994). The prevalent view has been that subjective experience, unlike behavior, cannot be observed, thus only behavior, and not emotion, can be scientifically studied.

There have been voices of dissent as to whether the impossibility of determining if nonhuman animals have subjective experiences should prevent the investigation of animal emotion (see Panksepp, 1989). The dissent has increased in recent years and an alternative view has been put forward. The fact that scientists may never be able to investigate the subjective experience of other animals should not prevent the study of animal emotions (Aureli & Whiten, 2003). After all, we have difficulty in knowing whether our own subjective emotional experience is similar to that of other human beings, yet scientific research on human emotions has nonetheless progressed greatly. We certainly need to acknowledge that the degree of subjective feeling associated with emotion may vary, depending on the extent to which a species possesses the capacity of conscious awareness (LeDoux, 1995). In any case, subjective feeling, though important, is only one aspect of emotion (Öhman, 1993; Davidson & Sutton, 1995; LeDoux, 1995, 1996; see Damasio, 2000).

Following MacLean's (1952) evolutionary approach to brain anatomy, emotions cannot be assumed to be uniquely human traits because most of the brain structures involved in emotions are essentially the same in all mammals and perhaps in all vertebrates (LeDoux, 1996; Panksepp, 1998). MacLean's conceptual framework has been strongly supported by neurophysiological evidence (Panksepp, 1989, 1998; Brothers, 1990; Davidson & Sutton, 1995; LeDoux, 1996; Rolls, 1999). We can therefore use the term "emotion" without implying the conscious feeling that humans associate with it – other animals may experience it as well, but we may never be able to demonstrate this.

There is a growing consensus among scientists on the general function of emotions in terms of adaptive response to environmental demands, preparing the individual to cope with them and increasing survival (e.g., Ekman, 1984; Frijda, 1986; Lazarus, 1991; Öhman, 1993; LeDoux, 1996; Damasio, 2000). At the most basic level, emotions are part of the homeostatic regulation that assures the survival of the organism. This basic function is probably achieved by relying on a relatively simple evaluative system that differentiates between hostile and hospitable stimuli and allows rapid orienting responses of approach or withdrawal (Cacioppo & Gardner, 1999; see Davidson, 2000). This characteristic can explain why emotions are inseparable from the concepts of reward or punishment, of pleasure or pain, of advantage or disadvantage (Damasio, 2000) and why brain systems critical for reward–punishment evaluation are involved in emotional experience (Rolls, 1999). This conceptualization may explain why the dichotomous classification of "positive" and "negative" emotions, although potentially arbitrary, appears intuitively solid: it is actually based on very basic adaptive responses. It follows that the concept of positive and negative emotions, so pervasive in human psychology, can be confidently applied to the study of animal emotions.

A critical functional issue is the specific role of emotions in motivating organisms to act (Rolls, 1990; LeDoux, 1996). Emotions can be viewed as interfacing between sensory inputs and motor outputs in a way that allows *flexibility* in the response, in contrast to (genetically programmed or learned) fixed action patterns as a reaction to a stimulus (Gray, 1975; Panksepp, 1989; Rolls, 1995). The essential function of emotions can therefore be summarized as gearing a particular type of motivational control to the perception of critical circumstances (Aureli & Whiten, 2003). This leads the individual to take a particular motivational stance, which severely constrains its further behavior, i.e., it constrains decision-making for some time appropriate for its referent (see Johnson-Laird & Oatley, 1992; Damasio, 1994). Although decision-making is constrained, this is only for an appropriate and temporary period. Overall flexibility of response is not hampered, but is achieved by changing the motivational stance elicited by different circumstances.

Emotions as intervening variables

Intervening variables are constructs used to explain complex webs of causation. For example, many different aspects of drinking behavior (e.g., effort to obtain drink, amount drunk) can be caused by many different factors (e.g., time since last drink, salt load). The causal linkage between a multiplicity of such factors and the many aspects they may influence can be explained most economically by hypothesizing a central intervening variable, in this case "thirst" (Miller, 1959). Although we may not be able to observe an animal's state of thirst directly, we can infer it on the basis of observable phenomena and having done so we can well predict how the animal will behave in a variety of contexts. We can do so without having to decide whether thirst *feels* the same in ourselves and the animal of interest (Aureli & Whiten, 2003). This kind of approach to identifying phenomena that are not directly observable is normal and common scientific practice. A similar approach has been used for "states of mind," such as thought and belief, of other individuals (Whiten, 1996; Call, 2001). The same logic can apply to the recognition of emotional states in animals (Hinde, 1972; Aureli & Whiten, 2003). The view of emotion as an intervening variable is highly compatible with the concept of flexible responses derived from emotional interfacing between sensory inputs and motor outputs, involving a process of canalization of motivational stance and decision-making (see above).

The understanding of the relationship between neurophysiological processes and the nature of emotions can benefit from viewing emotions as intervening variables (Fig. 3.1). On the one hand, neurophysiological phenomena may be incorporated into the web of causal links that justifies the attribution of a particular state of emotion, which has the explanatory power of an intervening variable. On the other hand, direct inferences from neurophysiological variables become more powerful when an inter-correlated array of them is identified, rather than a single one (Cacioppo *et al.*, 1993), and this indeed reflects the logic behind the concept of an intervening variable (Aureli & Whiten, 2003).

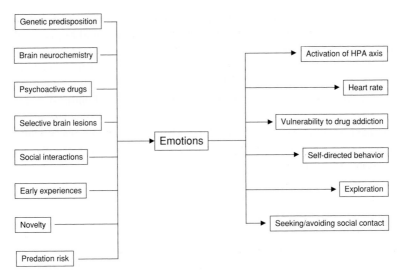

Fig. 3.1. Emotions as intervening variables relating causal factors to behavioral and physiological responses. (After Hinde, 1972; Aureli & Smucny, 2000.)

The integration of neurophysiological mechanisms with the perspective of emotions as intervening variables is also indicated from another theoretical position. Economical representations of events (such as intervening variables), although elegant, may not reflect reality if too simplistic. Biological solutions do not emerge *ex-novo* as the most parsimonious possibilities, but develop from pre-existing structures and therefore are constrained by their evolutionary past. Biologically relevant explanations of emotions as intervening variables should therefore aim to incorporate information about the underlying neurophysiological processes (Aureli & Whiten, 2003; see Zupanc & Lamprecht, 2000, for motivation as an intervening variable). In the remaining part of the chapter, we use the concept of emotion as an intervening variable involving neurophysiological processes but without implying subjective feeling.

Indicators of emotions

In this section, we review research on emotional states and emotional expressions in human and nonhuman primates to illustrate various indicators of emotions. Given the recent growth of the field, the review cannot claim to be exhaustive. We especially focus on research that highlights indicators that have been used in studies on macaques. Here, however, we deliberately cover examples mainly from primates other than macaques because studies on macaques are reviewed in detail in the next section. We start with the indicators of brain processes involved in emotions, and then we review indicators of peripheral physiological activity, followed by evidence of behavioral indicators.

Indicators of brain processes

Various methods have been used to explore the role of structures of the central nervous system and different neurotransmitters in the regulation of emotion. Traditionally these methods have been invasive (e.g., electrical stimulation, chemical manipulation, cerebrospinal fluid sampling, and lesions of specific brain areas). Through the use of these methods, much has been learned, for example, on the role of the amygdala in regulation of fear and other emotions (LeDoux, 1996; Aggleton & Young, 2000; Davis & Whalen, 2001; Amaral, 2002). More recently, noninvasive methods have been used, such as positron emission tomography (PET) and functional magnetic resonance imaging (fMRI). These techniques allow scientists to relate particular neuroanatomical areas and brain activity with emotional responses and emotional memory (Dolan & Morris, 2000; Reiman *et al.*, 2000; Davis & Whalen, 2001). Activity of specific brain areas and its possible asymmetric nature have been successfully investigated with electroencephalography (EEG) and other techniques (e.g., brain temperature). These findings suggest a lateralization of brain processing of emotions in human and nonhuman primates (Davidson, 1995; Canli *et al.*, 1998; Parr & Hopkins, 2000).

The plasticity underlying the flexibility of emotional states is achieved by the actions of neuromodulators. An important example of neuromodulators is provided by brain opioids. These are substances that produce pleasant sensations and that are usually released during "positive" emotions associated with rewarding situations (Dum & Herz, 1987; Panksepp *et al.*, 1997). For example, in talapoin monkeys the receipt of grooming increases the concentration of endogenous brain opioids (Keverne *et al.*, 1989).

Physiological peripheral indicators

Common techniques for the monitoring of peripheral physiological parameters include the measurement of functions such as heart rate, blood pressure, hand temperature, skin resistance, muscle tension, and hormonal levels (reviewed in Cacioppo & Tassinary, 1990; Bauer, 1998). Some studies use only one of these measures, whereas others monitor several physiological functions simultaneously. An example of a multi-measure study is the classical work by Ekman and collaborators (1983) on human subjects while eliciting six emotions. Physiological changes depending on the quality of the elicited emotion have also been reported for nonhuman primates. For example, a study monitoring cardiac activity of infant chimpanzees differentiated their emotional responses to two conspecific vocalizations (Berntson *et al.*, 1989).

An increase in heart rate has been shown to be associated with acute experience of anxiety in humans (Öhman, 1993; Berntson *et al.*, 1998) and is also produced by treatment of rhesus macaques with anxiety-eliciting drugs (Ninan *et al.*, 1982). Biotelemetry has been used to monitor physiological changes in free-moving nonhuman primates. In one such study, dominant male hamadryas baboons showed an anticipatory increase in heart rate in the seconds before starting an aggressive interaction (Smith *et al.*, 2000).

Various physiological events are characteristic of the "stress response" (von Holst, 1998) which results in the rapid mobilization of energy: heart rate, blood pressure, and breathing rate all increase to transport nutrients and oxygen where needed. The response is adaptive as it enables individuals to take immediate action in threatening situations. Measures of variation in hormones and neurotransmitters associated with the stress response (e.g., cortisol, norepinephrine) have provided biological support for individual differences in emotional profiles of children and olive baboons (Kagan *et al.*, 1988; Sapolsky, 2000).

Behavioral indicators

Since Darwin's (1872) original emphasis, *facial expressions* have been among the emotional indicators primarily studied in humans (Ekman, 1993). There is evidence that facial expressions are also a means to convey emotions in nonhuman primates (Parr, 2001; Preuschoft, Box 3). Darwin (1872) also pointed to vocal signals as primary carriers of emotional valence. There is substantial evidence that emotional experience produces changes in respiration, phonation, and articulation, which affect acoustic features, and that the acoustic features of *vocal emotional expression* are similar in human and nonhuman primates (Jürgens, 1998). Detailed acoustic analyses of human speech and primate vocalizations have revealed that vocal parameters not only index the degree of emotional intensity, but also differentiate emotional valence (Bachorowski & Owren, 1995; Fichtel *et al.*, 2001).

A less conventional set of behavioral indicators of emotional experience has been used recently in nonhuman primates. *Self-directed behaviors*, such as self-scratching and self-grooming, obviously have a hygienic function, but they have also been considered as displacement activities, i.e., activities which are apparently irrelevant to an individual's ongoing behavior and which reflect motivational ambivalence or frustration (Tinbergen, 1952; McFarland, 1966). Recent ethological studies on human and nonhuman primates have documented an increased frequency of these behaviors in situations of uncertainty, social tension, or impending danger (reviewed in Maestripieri *et al.*, 1992a; Troisi, 2002). Furthermore, pharmacological manipulations have demonstrated that self-directed behavior is strongly associated with anxiety in various primate species (Ninan *et al.*, 1982; Schino *et al.*, 1991, 1996; Maestripieri *et al.*, 1992b; Cilia & Piper, 1997; Barros *et al.*, 2000).

Emotions in macaques

Neurobiological bases

The neurobiological bases of emotional responses in macaques have been the subject of extensive investigations. "Positive" emotions associated with social contact appear to be mediated by brain opioids. Infant rhesus macaques separated from their mothers decrease their distress vocalizations ("coo" calls) when treated with low, non-sedative doses of morphine, while distress vocalizations increase in frequency after treatment with naloxone, an opioid antagonist (Kalin & Shelton, 1989). Interestingly, opioids appear to modulate both bonds between mothers and offspring and bonds between group mates: administration of naloxone to group-living juvenile longtailed macaques caused a generalized increase in the requests for affiliation directed to both the subject's mother and to other groupmates (Schino & Troisi, 1992; see also Martel *et al.*, 1995).

"Negative" emotions such as fear of a staring human (probably perceived as a "predator") appear to be modulated by the gamma aminobutyric acid (GABA)-benzodiazepine receptors. Diazepam decreased fearful responses to the human "predator" in isolated infant rhesus macaques (Kalin & Shelton, 1989). The amygdala seems to be the key brain structure involved in the mediation of acute fear and anxiety: in rhesus macaques, selective lesions of the amygdala caused a blunting of acute fear in response to the presentation of stimuli such as a snake or a threatening unfamiliar individual (Emery *et al.*, 2001; Kalin *et al.*, 2001).

There is suggestive evidence supporting the lateralization of emotions in rhesus macaques (Hauser, 1993). In the same species, fearful temperament is associated with asymmetric frontal brain activity (as assessed by electroencephalography) and to elevated concentrations of corticotropin-releasing hormone (CRH) in the cerebrospinal fluid (Kalin *et al.*, 1998, 2000). The genetic bases of interindividual differences in emotionality and brain monoamine functioning are just beginning to be investigated in both macaques (Champoux *et al.*, 1999; Bennett *et al.*, 2002) and baboons (Kaplan *et al.*, 2001; Rogers *et al.*,

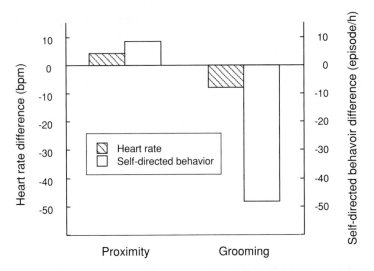

Fig. 3.2. Effects of social interactions on behavioral and physiological measures of emotions. Proximity to higher-ranking animals (within 0.5 or 1 m depending on the study) causes an increase, whereas receiving social grooming causes a decrease in both self-directed behavior and heart rate. Both measures are expressed as differences from control observations. The figure combines data from three separate studies on longtailed and rhesus macaques. Original results and methodological details can be found in Schino *et al.* (1988), Pavani *et al.* (1991) and Aureli *et al.* (1999). Note that self-directed behavior refers to scratching only in the study of the effect of proximity, and to the sum of scratching, self-grooming, body shaking, and yawning in the study of the effect of social grooming.

2001). Significant interactions between genetic predisposition and early rearing experiences seem to characterize the development of emotionality in macaques (see Suomi, 1999; Capitanio, Chapter 2).

Emotions as consequences of social interactions and relationships

Social interactions have been shown to cause short-term emotional reactions. Proximity to dominant individuals, insofar as it is related to an increased uncertainty and risk of aggression, is associated in longtailed macaques with increases in self-directed behavior such as scratching and self-grooming (Troisi & Schino, 1987; Pavani *et al.*, 1991; see also Castles *et al.*, 1999, for similar data on olive baboons). Similarly, in a biotelemetry study of emotional responses of rhesus macaques during social interactions, Aureli and collaborators (1999) monitored the heart rate of free-moving individuals, recording their response to being approached by different group members, while controlling for movement. The approach of a dominant individual caused an increase in heart rate (Fig. 3.2), whereas the approach of a subordinate or kin did not cause consistent heart rate changes. The increases in self-directed behavior and heart rate as a consequence of potentially threatening interactions are mirrored by the decreases observed during and following friendly interactions, possibly associated with positive emotions. Receiving social grooming reduces both self-directed behavior and heart rate (Fig. 3.2) (Schino *et al.*, 1988; Boccia *et al.*, 1989; Aureli *et al.*, 1999). Similarly, heart rate and self-directed behavior increase during the post-conflict period and are reduced following reconciliation, i.e., the friendly reunion of former opponents that may occur after conflict (Fig. 3.3) (Aureli & van Schaik, 1991; Smucny *et al.*, 1997; Aureli & Smucny, 2000).

Other evidence of emotional responses to social interactions is provided by studies investigating dominance effects. Different changes in heart rate are found between dominant and subordinate longtailed macaques in response to varying proximity with group members during the initial phases of group forma-tion (Manuck *et al.*, 1986). Furthermore, intermediate-ranking rhesus macaques scratch themselves more than others at feeding time probably because they expe-rience higher uncertainty about how to behave than do dominant individuals, who easily monopolize the food, and subordinate individuals, who wait for access to the food (Diezinger & Anderson, 1986).

Evidence that facial displays are perceived as expressions of emotional states is provided by an elegant (although problematic for animal welfare) series of experiments involving pairs of rhesus macaques in the role of "informer" and "responder." Miller (1971) reported that the view of a silent bared-teeth display exhibited by the informer was sufficient for the responder to press a lever and avoid an electric shock. The informer responded with this facial display when seeing a warning signal indicating the upcoming electric shock, but the warning signal was not visible to the responder who had to base its response only on the fear conveyed by the display of the partner.

Over a longer time span, social relationships can modulate the emotional responsiveness of macaques to stressful events. Schino and Troisi (2001; see also DeVinney *et al.*, 2001) described how yearling Japanese macaques often respond to the birth of a sibling (and to the associated abrupt reduction in maternal availability) with depression and emotional withdrawal, as evidenced by greatly reduced social play. The likelihood of a depressive response in the yearling macaques was predicted by the quality of the mother–yearling rela-tionship *before* the sibling birth: yearlings having a warmer relationship with their mother were less likely to become depressed following the birth of a sibling.

Social relationships can also induce changes in brain functioning and neuro-chemistry, possibly influencing emotional responsiveness. For example, social dominance has been shown to alter brain dopaminergic functionality as assessed by PET. After becoming dominant, longtailed macaques showed an increase in

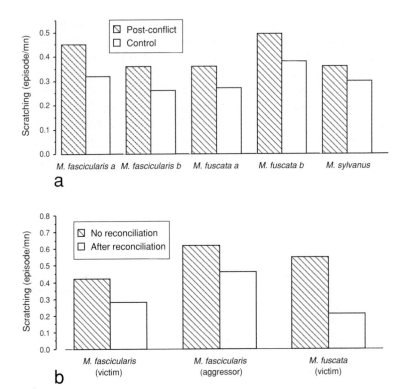

Fig. 3.3. Effects of aggression and reconciliation on rates of scratching shown by macaques (rates of the two periods are significantly different in all comparisons). (a) Scratching displayed by victims of aggression during post-conflict and control periods in five populations of macaques. Sources of data: *Macaca fascicularis* a from Aureli & van Schaik, 1991; *M. fascicularis* b, from Aureli, 1992; *M. fuscata* a, from F. Aureli, H. Veenema & C. van Panthaleon van Eck (unpub. data); *M. fuscata* b, from Rosati, 1996; *M. sylvanus* from Aureli, 1997. (b) Scratching displayed by former opponents during post-conflict periods with no reconciliation and during post-conflict periods following reconciliation in three populations of macaques. Sources of data: *M. fascicularis* (victim) from Aureli & van Schaik, 1991; *M. fascicularis* (aggressor) from Das *et al.*, 1998; *M. fuscata* (victim) from Kutsukake & Castles, 2001.

indices of brain dopaminergic activity and a consequent decrease in the reinforcing properties of cocaine (Morgan *et al.*, 2002). The experience of social dominance and the associated higher degree of perceived control over environmental and social events may modify emotional responses mediated by brain dopaminergic activity.

Finally, quality of early mother–infant relationships has repeatedly been shown to have long-term effects on juvenile and adult emotionality in a variety of

Fig. 3.4. A Japanese macaque infant approaches a playmate while its mother watches and scratches herself (Rome Zoological Park, Italy). (Photograph by G. Schino.)

macaque and other primate species (Hinde & Spencer-Booth, 1970; Andrews & Rosenblum, 1994; Schino *et al.*, 2001, for data on macaques; see Fairbanks & McGuire, 1988, 1993; Dettling *et al.*, 1998, for data on other primates). Both correlational and experimental studies concur in documenting an association between maternal style experienced as infants and later emotionality as juveniles/adults. Interestingly, behavior under baseline, non-stressful conditions is not affected by early experiences, while they potently affect responses to mildly stressful situations such as interactions with unfamiliar conspecifics or objects, response to aggression and to the proximity of dominant individuals. Emotional responses investigated in these studies ranged from self-directed behavior to urinary cortisol, to latency to approach unfamiliar objects or locations.

Emotions as causes of social interactions and relationships

Anxiety has long been suggested as a proximate factor underlying interindividual differences in macaque maternal style (e.g., Hooley, 1983). Empirical support for this hypothesis has been provided by the significant correlations that have been demonstrated to exist between behavioral indicators of anxiety, such as scratching and visual monitoring (Fig. 3.4), and protective maternal styles (Troisi *et al.*, 1991; Maestripieri, 1993).

A more direct test of the role of anxiety in modulating macaque social behavior derives from ethopharmacological studies that investigated the effects of psychoactive drugs on animals living in social groups. Maestripieri and collaborators (1992b) found that administration of an anxiogenic drug (beta-CCE) to infant rhesus macaques increased their contact seeking behavior toward

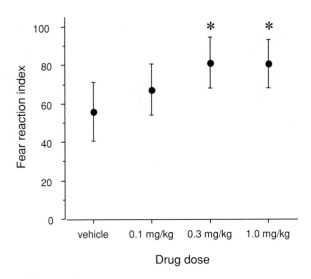

Fig. 3.5. Effect of various doses of an anxiogenic drug (FG 7142, a benzodiazepine inverse agonist) on the fear reaction shown by mature male longtailed macaques to the approach of the dominant male. The asterisks indicate significant differences from the vehicle control. The fear reaction index is calculated as the percentage of received approaches to which the subject responds with bared-teeth or an immediate leave. (After Schino *et al.*, 1996.)

their mothers, while anxiolytic drug treatment (with midazolam) facilitated playful interactions with group mates, especially with older individuals. An anxiolytic drug (diazepam) increased the rate of friendly exchanges during reunion of familiar juvenile rhesus macaques, whereas an anxiogenic drug (d-amphetamine) had the opposite effect and increased submissive behavior (Thierry *et al.*, 1984). Schino and collaborators (1996; see also Bellarosa *et al.*, 1980; Knobbout *et al.*, 1996) observed increased fear reactions to the approach of a dominant individual as a consequence of anxiogenic treatment (with FG 7142) in mature male longtailed macaques (Fig. 3.5).

Interindividual differences in emotionality (and here we move very close to issues relevant to personality/temperament: see Capitanio, Chapter 2) as assessed by basal plasma cortisol were shown to be linked to differences in the styles of dominance among high-ranking wild olive baboons (Sapolsky & Ray, 1989; Sapolsky, 2000). No comparable data are available for macaques, but consistent interindividual differences in cortisol concentrations and heart rates have been reported for adolescent rhesus macaques (Rasmussen & Suomi, 1989). Similarly, among infant rhesus macaques consistent interindividual differences in reactivity to mildly stressful stimuli have been described

(Suomi, 1999). Highly-reactive infants share behavioral and physiological characteristics with "shy" children (Kagan *et al.*, 1988).

Furthermore, interindividual differences in impulsivity seem to underlie much variation in primate social behavior and even life history. In male vervet monkeys impulsivity, as measured by means of an intruder challenge test, showed large interindividual variation and peaked at age 4 years, the typical age of emigration in this species. High-ranking males showed moderate levels of impulsivity while low-ranking males were either extremely or very slightly impulsive (Fairbanks, 2001). In rhesus macaques, heart rate was lower in males who had recently emigrated than in males of comparable age who were still in the natal group (Rasmussen *et al.*, 1988).

Primate impulsivity has also been linked to the functionality of the serotonergic system. Low serotonergic activity is correlated with severe aggression, excessive mortality and early emigration in rhesus and longtailed macaques (Botchin *et al.*, 1993; Mehlman *et al.*, 1994, 1995; Kaplan *et al.*, 1995; Higley *et al.*, 1996a,b,c; Capitanio, Chapter 2) and with high impulsive scores in male vervets during the intruder challenge test (Fairbanks, 2001). Pharmacological manipulation provided experimental confirmation of some of these correlational results (Fairbanks, 2001). Recently, it has even been suggested that the interspecific differences in the tendency to emigrate observed in male hamadryas and olive baboons may be linked to differences in serotonergic activity (Kaplan *et al.*, 1999). No comparative data of this kind for interspecific differences are available for macaques.

The mediating role of emotions

From the examples presented in the previous two sections a clear picture emerges in which emotions can be viewed both as causes and consequences of social interactions. In this respect, emotions may play an important part in linking social interactions and therefore may have a mediating role in social exchanges between individuals. Such a mediating role is implicit in the "intervening variable" concept illustrated in Fig. 3.1.

There has been growing attention to the mediating role of emotions in the human literature (Frijda, 1986; Panksepp, 1989; Rolls, 1995) and in research on other animals (Crook, 1989; Lott, 1991; Pryce, 1996; Owren & Rendall, 1997; Aureli & Whiten, 2003). The emotional experience (without implying subjective feeling) of an individual is certainly affected by the frequency and quality of previous interactions with group members (see above). Emotional states may express a crucial integration of the information contained in the various interactions between two partners and may change over time depending on the interactions exchanged. The emotional experience can then be functionally equivalent to the processes of bookkeeping of the various interactions with a partner, computation of their relative frequencies, and conversion of their quality and associated information into a common currency, all needed for relationship assessment (Aureli & Schaffner, 2002). The resulting emotional experience is partner-dependent. Thus, emotional differences can be at the core of the observed variation in social interactions reflecting the variation in relationship quality across partners.

An example of the mediating role of emotions in macaques can be extracted from the study of conflict resolution. Aggression causes an increase in post-conflict anxiety in both the victim and the aggressor as indexed by self-directed behavior and heart rate (Aureli *et al.*, 1989; Aureli & van Schaik, 1991; Aureli, 1992; Smucny *et al.*, 1997; Das *et al.*, 1998). The post-conflict anxiety appears to be due both to the risk of renewed aggression incurred by the victim (Aureli & van Schaik, 1991; Aureli, 1992) and to the disturbance of the (valuable) social relationship between victim and aggressor (Aureli, 1997; Katsukake & Castles, 2001). Anxiety seems to be a proximate factor facilitating reconciliation which, in turn, reduces the risk of renewed aggression (Aureli *et al.*, 1989; Aureli & van Schaik, 1991; Cords, 1992; Katsukake & Castles, 2001), restores the social relationship (Cords, 1992) and thus causes a reduction in anxiety (Aureli *et al.*, 1989; Aureli & van Schaik, 1991; Smucny *et al.*, 1997; Das *et al.*, 1998; Katsukake & Castles, 2001). Post-conflict anxiety and reconciliation appear therefore to be part of a homeostatic mechanism, which regulates and stabilizes macaque social relationships threatened by aggression. In particular, post-conflict anxiety may mediate the occurrence of reconciliation by motivating opponents to exchange friendly behavior in the aftermath of aggressive interactions (Aureli & Smucny, 2000).

Since reconciliation appears to repair the relationship between former opponents disturbed by the previous conflict, it makes sense that reconciliation occurs more often between individuals with more valuable relationships (de Waal & Aureli, 1997; Aureli *et al*, 2002). The same individual therefore engages in friendly post-conflict reunions depending on the quality of the relationship with the former opponent. This requires great flexibility in the frequency of interaction with various group members and with the same individual over time. Differential post-conflict anxiety could generate flexible responses to conflicts as shown in studies of longtailed and Japanese macaques. Rates of post-conflict scratching by the recipient of aggression were higher after conflicts between likely valuable partners (without being associated with higher rates of renewed attacks: Aureli, 1997; Kutsukake & Castles, 2001) and reunions occurred more often following conflicts between such partners (Aureli *et al.*, 1989; Kutsukake & Castles, 2001). Differential post-conflict anxiety may therefore reflect relationship assessment and mediate the effect of relationship quality on conflict resolution. The assessment could be based on the relative loss of benefits due to relationship disturbance: more valuable relationships provide greater benefits and their disturbance would produce greater loss and higher levels of anxiety. Anxiety may mediate not only post-conflict resolution, but could also play an important role in the flexible occurrence of other forms of conflict management (e.g., forms to prevent aggressive escalation: Aureli & Smucny, 2000).

A less detailed example derives from the study of the relationship between the establishment of dominance relations and the occurrence of affiliation (see Kummer, 1975, for pioneering work on this subject in gelada baboons). Schino and collaborators (1990) showed that when unfamiliar longtailed macaque females are paired, the lack of established dominance relations causes tension and anxiety in the form of high self-directed behavior. If dominance relations are not defined in the first minutes of interaction, affiliation is prevented and anxiety increases progressively. The increased anxiety may in turn be instrumental in promoting the definition of dominance relations thus allowing affiliative inter-actions (e.g., grooming) to take place. Grooming in turn reduces anxiety and tension (Schino *et al.*, 1988). Again, emotions appear to be part of a homeostatic mechanism that mediates macaque social interactions.

The above study can also be interpreted from the relationship assessment point of view. The nature of the relationship of each pair was reflected in the rate of self-directed behavior. When the unfamiliar females delayed the estab-lishment of dominance relationships, the rate of self-directed behavior was much higher than when dominance relationships were quickly established. The level of uncertainty about the relationship influences the degree of anxiety that individuals experienced. Thus, the assessment of dominance relationships with partners may be mediated by the level of anxiety experienced when in prox-imity with them and this in turn may affect the subsequent interactions. In fact, pairs with established dominance relationships engaged in social groom-ing sooner than pairs with undecided dominance relationships (Schino *et al.*, 1990).

A framework for future research

Their role in mediating social interactions and providing a means for relation-ship assessment makes emotions good candidates as proximate mechanisms underlying the variation in social organization across macaques. Differences in tolerance and kin bias (Chapais, Chapter 9; Thierry, Chapter 12), and possibly in other aspects of social life across macaque species could be related to potential interspecific variation in emotional experiences, which in turn trigger different social behavior. To test this hypothesis, systematic research on interspecific differences and similarities in emotional responses to a variety of social and non-social stimuli is needed. The degree of impulsivity, anxiety levels under various circumstances and the need for emotional support from others would be among the critical aspects that should be investigated with a comparative approach. The comparative research carried out so far on some of these aspects is limited but promising (Clarke & Boinski, 1995; Capitanio, Chapter 2). Future research needs especially to compare directly interspecific (and intergroup) differences in emotionality variables with interspecific (and intergroup) differences in social variables taking into account possible sex differences (i.e., males and females of the same species may differ in such variables) and examine whether variations in the two sets of variables are meaningfully interlinked. Interspecific (and intergroup) comparisons between emotional responses and multiple social variables could elucidate whether the diversity of social interactions across macaque species (and groups) is due to different causes or to one major underlying principle of covariation.

One way of systematically gathering information on the possible emotional causes of social variation is to expand the comparative study of brain neurochemistry and monoamine functioning. The first step could be genus-wide correlations between overall patterns of social interaction (e.g., tolerance, kin bias) or life history (e.g., dispersal, lifelong rank trajectory) and various measures of brain metabolism (e.g., cerebrospinal fluid monoamine metabolite concentrations, pharmacological challenges). Furthermore, interspecific (and intergroup) differences in emotional responses could be investigated through pharmacological manipulation. The development of comparative psychopharmacology could provide the opportunity, for example, to study the effects of anxiolitic and anxiogenic drugs or of opioid antagonists on the modulation of interspecific differences in tolerance or kin bias (see Schino & Troisi, 1992, for a successful case of intraspecific manipulation).

We would also like to suggest that the comparative study of emotional responses can provide a framework to identify which of the possible selective pressures (e.g., predation risk, between-group competition, infanticide risk) is at the basis of the original diversity in macaque dominance style. For example, experiments with predator models in different species can provide systematic evidence of differential behavioral and emotional responses to them and therefore indirect information on the relative attitude to such external threats and the relative need of forming tight bonds with group members for cohesiveness and possible cooperative actions. As previously proposed by Lott (1991), differential predation risk is likely to lead to differential anxiety, which would in turn promote differential group cohesiveness to cope with such a risk. The fact that a drastic reduction in the number of potential primate predators is among the first negative consequences of transformations imposed on the natural habitats of most macaque species by human actions leaves extrapolations from such experiments to be one of the few options we have to obtain insights into the role of predation as one of the original selective pressures. Similarly, the systematic investigation of behavioral and emotional responses to the exposure to same-sex or opposite-sex unfamiliar individuals (both experimentally in captive settings and opportunistically in the wild) in different macaque species can provide insights into the role of between-group competition and infanticide or sexual harassment in promoting differential intragroup tolerance and thus shaping social organization.

The comparative study of emotional responses to social and nonsocial factors can also be useful in order to acquire information on current patterns of social organization where data are difficult to gather in the wild (again often because there is no pristine natural macaque habitat left). An example of such investigations is the comparative study on responses to all-male group formation carried out by Clarke and collaborators (1995). Liontailed macaque males showed a much higher stress response than longtailed macaque males and the group formation had to be aborted. This outcome suggests high levels of male–male intolerance in liontailed macaques, a characteristic that certainly contributes to the higher proportion of one-male groups in wild populations of this species (Singh *et al.*, 2002).

Scientists need to make the most of the unique body of knowledge currently available on macaque societies and develop integrated approaches that combine socioecological variables with individual characteristics. The framework presented here is just one example of these approaches. Viewing emotions as causes and consequences of social life leads us to focus on their role in relationship assessment and mediation of social interactions. This is a bottom–up approach that integrates individual characteristics (e.g., emotional responses and personalities) with social interactions and relationships and that can reveal the building blocks of social organizations. We are still at the early stages of the integrative effort and more focused comparative research needs to be carried out to obtain a comprehensive picture of the development and evolution of macaque societies. The macaque model needs to be tested in other taxa so as to obtain unifying principles that could be at the root of human sociality.

Box 3   Power and communication

Signe Preuschoft

Primate facial expressions represent an evolutionary legacy. Old World monkeys use about seven facial displays in the context of bonding alone. The facial expressions used in the context of assertion are less well studied, but include approximately six additional displays (van Hooff, 1967; Redican, 1975; Preuschoft & van Hooff, 1995a). These displays are defined by their appearance, and fulfill different social functions in different species. While phylogeny cannot explain these differences in function, they coincide with differences in power asymmetry between the species.

Many facial displays are ancestral homologies of all the macaques, and also shared in common with other Old World monkeys, with apes and sometimes even New World monkeys (Preuschoft & van Hooff, 1995a,b, 1997). Among these widespread displays are the homologues to human laughter and smiling, the relaxed open-mouth display (ROM) and the silent bared-teeth display (SBT) (van Hooff, 1972; Preuschoft, 1995a; van Hooff & Preuschoft, 2003). However, the function of smiling (SBT) and laughter (ROM) displays varies between species, ranging from formal subordination, over submission and appeasement, reconciliation, to affiliation, reassurance, and even to playfulness (Preuschoft, 1992, 1995a; Preuschoft & van Hooff, 1997). Along with the functional differences, the extent to which the ROM and the SBT displays overlap morphologically differs as well (Fig. 3.6). When used in an overlapping range of functions the SBT and ROM displays tend to merge in appearance as well, producing an open-mouthed bared-teeth laughter face that is accompanied by the typical panting play vocalization.

Such functional variability is not limited to the genus *Macaca*. Importantly, even within the genus *Macaca*, the variation in the function of laughter and smile homologues is not predicted by phylogeny (but see Thierry, Chapter 12). Phylogeny would predict the functions of SBT and ROM in pigtailed macaques should resemble those in Tonkean and lion-tailed macaques. Barbary macaques, the most conservative macaque, should resemble gelada baboons – but this is not the case (Dücker 1996). The situation in common chimpanzees is similar to that in Barbary macaques, whereas bonobos and humans resemble Tonkean macaques with respect to the function of laughter and smile (van Hooff, 1972; Lockard *et al.*, 1977; Preuschoft & van Hooff, 1997; S. Preuschoft, personal observation). The differences within the great apes as well as the similarity between some great ape species and some macaque species is impossible to reconcile with the

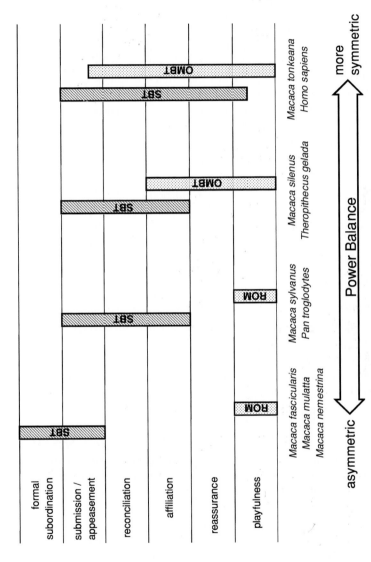

Fig. 3.6. The social functions of silent bared-teeth (SBT) and relaxed or open-mouth bared-teeth displays (ROM and OMBT respectively) coincide with characteristic differences in power balance but not with phylogeny (*M. fascicularis*: Preuschoft et al., 1995; *M. nemestrina*: S. Preuschoft, J. Flack, and M. L. Gong, unpub. data; *Theropithecus gelada*: Dücker, 1996; *M. silenus*: Preuschoft & Beckmann, 1995; Beckmann, 1997; *M. tonkeana*: Preuschoft 1995b; see reviews in Preuschoft & van Hooff, 1997; Preuschoft & van Schaik, 2000; Thierry, 2000).

idea that phylogeny predicts the function and form of smile and laughter homologues.

On the other hand, the functional differences between species of macaques do correspond to characteristic differences in power asymmetry (Fig. 3.6). Power asymmetries are evident in a set of covarying social traits, including both dominance and sociopositive interactions (Thierry, 1985a, Chapter 12; de Waal, 1989a; de Waal & Luttrell, 1989; Preuschoft & van Schaik, 2000). In obligatory social species the need for sociopositive interactions can profoundly alter the power dynamics in dyads, since it results in leverage power on the side of the needed individual (Hand, 1986; Lewis, 2002). Species-typical power asymmetries were established in independent studies (Thierry, 1985a, 2000; de Waal & Luttrell, 1989). On this basis macaque species can be arranged on a continuum from very asymmetric to more balanced power relations (Preuschoft & van Schaik, 2000; Thierry, 2000, Chapter 12).

The "Power Asymmetry Hypothesis" of functional changes predicts the greatest functional overlap of the SBT and the ROM display for species with the most symmetric power relations (Preuschoft & van Hooff, 1995b, 1997). The SBT functions as a genuine expression of friendly attraction, and may serve to appease or reassure an interaction partner and may even accompany a play invitation. Shading in with the SBT, the play face becomes a open-mouthed bared-teeth laughter face which not only expresses skittish playfulness but is also used in reconciliation and even appeasement (Thierry *et al.*, 1989; Preuschoft, 1995b). The species with the most asymmetric power relations, on the other hand, show the opposite pattern. As an indicator of formal subordination, silent-baring of the teeth is restricted to the communication of submission and appeasement. Confined to the play context, the morphologically wholly distinct relaxed open-mouth display earmarks playful transgression of rules.

The situation in pigtailed macaques is particularly interesting. While generally considered a member of the *silenus*-group (Delson, 1980; Fooden, 1980; Morales & Melnick, 1998) pigtailed power relations resemble those of the more distantly related longtailed macaques, and are clearly more asymmetric than those of their closer relatives, Tonkean or liontailed macaques (Castles *et al.*, 1996; Thierry, 2000; J. C. Flack, unpublished data). Pigtailed communication turns out to be exactly in line with the power asymmetry hypothesis: like other "despotic" macaque species, pigtailed macaques use only a ROM display, and not the open-mouth bared-teeth display that is typical of those with more balanced power relationships. The ROM is restricted to the context of play, and the SBT is used as an appeasement signal

Box 3 Power and communication 59

and expresses formal subordination (Fig. 8.1) (S. Preuschoft, J. C. Flack & M. L. Gong, unpublished data).

In species with strictly asymmetric power relations we find signals that are only performed by the subordinate member of a dyad and express a readiness to yield (de Waal & Luttrell, 1985; Preuschoft, 1999). Such indicators of formal subordination exist because conflicts of interest are consistently won by the dominant. Formal status indicators communicate a sender's perception of the power relationship between itself and the receiver (Preuschoft, 1999; Preuschoft & van Schaik, 2000). It is therefore expected that, contingent upon differences in power asymmetry, macaque species should also vary in the use of formal status indicators. This is indeed the case: species with rather balanced power relationships, such as Tonkean or lion-tailed macaques do not use any status indicators. Species with reduced power asymmetries, like Barbary or stumptailed macaques, do not use indicators of subordination but of dominance. Dominance indicators express a dominant's assertive tendency (Reichler, 1996). In these species the subordinate does not volunteer to yield without solicitation but waits until precedence is claimed by the dominant. Barbary macaques and geladas use a rounded-mouth stare as dominance indicator (Reichler, 1996; Preuschoft *et al.*, 1998). In stumptailed macaques formal biting is documented to function as a dominance indicator (Chevalier-Skolnikoff, 1974; Demaria & Thierry, 1990).

It also appears that the variation in power asymmetries is not limited to the genus *Macaca*. Similar variation also exists among great apes, and data indicate corresponding differences in communication patterns (van Schaik *et al.*, 2004). Among platyrhines, tufted capuchin monkeys have been likened to chimpanzees in their potential for food sharing, hunting and tool use, and there is research linking these behaviors with social tolerance and cooperation (Coussi-Korbel & Fragaszy, 1995; de Waal, 2000). It is therefore interesting that we found tufted capuchins to use a SBT in both courtship and play, but not as a response to aggression (D. R. Valenzano, E. Visalberghi & S. Preuschoft, unpublished data).

The selection pressures that influence a display's social function arise from the receivers. The social relationship between sender and receiver, that is whether emphasis is more on competition or on cooperation, is critical to how the receiver will respond to a display, and when it will suit a sender to publicize private motivational states (Dawkins & Krebs, 1978). When the relationship is mainly cooperative we expect displays to be rich in meaning, because misunderstandings are not dangerous, and receivers to exhibit little "sales-resistance", i.e., to respond to even subtle cues. On the other hand, when relationships are predominantly competitive, displays should be highly salient and unambiguous, while the skeptical receivers respond only where necessary (Krebs & Dawkins, 1984). Clearly, such differences in power balance will also produce differences in emotional dispositions (Aureli & Schino, this Chapter).

Such interpretations rely on a functional link between communication patterns and the power asymmetry characteristic for a macaque society. There is not always a close match between social variables established in field studies as opposed to those found in studies of provisioned or captive groups (e.g., Kawamura, 1958; Hill & Okayasu, 1996). The covariation between dominance styles and communication patterns, which are inherited, therefore provides strong evidence that dominance styles, or power asymmetries, should be interpreted as evolved, rather than ephemeral, context-induced variability.

4   *Reproductive life history*

FRED B. BERCOVITCH AND NANCY C. HARVEY

Introduction

Life-history strategies refer to developmental trajectories followed by individuals as a function of time, energy, and resources devoted to growth, maintenance, and reproduction (Gadgil & Bossert, 1970). Reproductive strategies are one component of life-history strategies that incorporate both mating and rearing effort (Low, 1978). The reproductive life history of an individual encompasses that period of life between onset and cessation of reproduction. In macaques, as in nearly all mammals, the two sexes follow alternative strategies characterized by two major distinctions. First, within a species, females mature more rapidly than males, a process known as sexual bimaturism (Bercovitch, 2000, 2001). Second, female reproductive success focuses on rearing, not producing, young, while male reproductive success is more dependent upon the opposite pattern (Trivers, 1972). In this contribution, we review and discuss the behavioral biology of three phases of reproduction: onset, maintenance, and cessation. A companion chapter (Soltis, Chapter 7) concentrates on describing behavioral strategies, while we emphasize links between age, physiological processes, and reproduction.

Onset of reproduction

The neurobiological cascade of stimuli initiating reproductive maturation in female and male macaques is identical (Plant, 1994; Terasawa & Fernandez, 2001). Nocturnal pulses of gonadotropin-releasing hormone (GnRH) from the hypothalamus prime the pituitary to produce bursts of polypeptides (luteinizing hormone (LH), follicle-stimulating hormone (FSH)) which spark the gonads into producing sex steroids (estrogen, testosterone) that stimulate gamete production (males) or release (females). When properly aroused, the follicle within

Macaque Societies: A Model for the Study of Social Organization, ed. B. Thierry, M. Singh and W. Kaumanns. Published by Cambridge University Press. © Cambridge University Press 2004.

the ovary will release an egg, but such an event occurs quite infrequently during the lifetime of female macaques because most of adulthood is spent either pregnant or lactating. In males, on the other hand, gametogenesis begins at puberty and the testes are constantly churning out sperm during the lifetime of male macaques. In male macaques, the initial sign of achievement of reproductive potential is descent of enlarged testes, whereas in female macaques, the initial sign of achievement of reproductive potential is a reddening of the sexual skin area in the anogenital region, which sometimes balloons in size.

Male reproductive maturation

The longer duration of the phase of reproductive maturation in males compared with females not only enables males to augment their body size, but also provides an opportunity for males to hone social skills regulating mating success (Bercovitch, 2000, 2001). On average, the interval between onset of reproductive capacity and attainment of complete adult body mass in male macaques requires about 4 to 6 years (from Bercovitch, 2000). During this lengthy adolescent phase, male macaques disperse from their natal troop and either directly join another troop or live a solitary existence (Gachot-Neveu & Ménard, Chapter 6), with troop integration possibly facilitated by social or aggressive interactions with females (Packer & Pusey, 1979; Jack & Pavelka, 1997). For example, female solicitation of non-troop males during the mating season sometimes precedes immigration, but movement into a new troop sometimes takes a few months (Bercovitch, personal observation). Male macaques may remain in a single troop for their entire lives or shift troops throughout their lives, with factors regulating troop tenure still unclear. A decrease in mating success is coupled to an increase in tenure in both rhesus (Berard, 1999) and Japanese macaques (Huffman, 1991), suggesting that female mate choice, and the number of sexually receptive females, might influence male tenure length. Dispersal has probable life-history costs in terms of increased likelihood of succumbing to predation, wounds, and disease (Dittus, Chapter 5), but such costs are probably offset by benefits associated with female mate preference for newcomer males (Huffman, 1991; Bercovitch, 1997; Berard, 1999).

Processes of merging into a new troop vary both within and across macaque species. In Taiwanese macaques, the highest-ranking resident male attempts to limit incoming males to the troop periphery (Hsu & Lin, 2001), whereas in rhesus macaques (Vessey & Meikle, 1987), males enter new troops at the lower echelons of the hierarchy. Bonnet macaques seem to have little difficulty immigrating (Simonds, 1973), and inter-male relationships in this species are fairly placid (Simonds, 1965; Koyama, 1973). Longtailed macaques often live alone after natal dispersal, a mating tactic designed to boost body size prior to immigration and challenge adult males for high-ranking positions (van Noordwijk & van Schaik, 1985). Rank and reproduction among males tend to be closely linked in this species (de Ruiter *et al.*, 1994). In Japanese macaques, males often enter troops as subordinate males when troop size is large, but topple, and supplant, the highest-ranking male when troop size is small (Sprague *et al.*, 1996, 1998).

Determining the onset of reproduction in male macaques is difficult to document due to dispersal patterns, combined with female tendencies to mate with multiple males. Male macaques acquire the ability to impregnate females before they generally have the ability to gain access to fertile females. Among rhesus macaques, males produce fertile sperm when about 3 and a half years of age, but only a fraction of males sires offspring before 5 years of age (Bercovitch *et al.*, 2003). At the time of their descent, testicles range in size from 7% (Matsubayashi & Mochuzuki, 1982) to 50% (Nieuwenhuijsen *et al.*, 1987) of adult dimensions, yet are producing fertile sperm (e.g., Dang & Meussy-DeSolle, 1984; Goy *et al.*, 1982). In both Barbary (Paul & Thommen, 1984) and rhesus macaques (Colvin, 1986), dominant males tend to disperse from their natal troop at older ages than subordinate peers, and in both species genetic analysis has revealed that the delayed dispersal is associated with production of offspring in the natal troop (*M. sylvanus*: Paul *et al.*, 1992a; *M. mulatta*: Bercovitch *et al.*, 2000).

Female reproductive maturation

Body composition, growth rates, and activity patterns influence metabolic state, which has strong repercussions on the tempo of reproductive maturation (Steiner, 1987; Bronson, 1989; I'Anson *et al.*, 1991). The hypothalamus, home to GnRH neurosecretory cells, contains the neural mechanisms responsible for regulating both food intake and energy expenditure (Frayn, 1986; Schmidt-Nielsen, 1997). The initial prepubertal pulsatile release of GnRH commences in a peculiar fashion because the neuroendocrine bursts arise when the inhibition that was dampening release during the juvenile stage is removed. Elevated prepubertal concentrations of gamma aminobutyric acid (GABA) inhibit neuropeptide Y, and free the hypothalamus from suppressed GnRH release (Terasawa & Fernandez, 2001). Leptin, a hormone discovered in 1994 that is synthesized in adipocyte tissue, also increases in concentration early in puberty, but rather than function as a "trigger" for the onset of puberty, leptin is one of a number of complex interacting factors that regulate reproductive maturation (Terasawa & Fernandez, 2001; Moschos *et al.*, 2002).

The coupling of body mass to age at menarche has given rise to the "critical weight/fat threshold" model (Frisch & McArthur, 1974; Frisch, 1984). According to this model, body fat provides a source of energy for maintaining pregnancy and lactation, despite the vagaries of the food supply, so surpassing a minimum fat level or weight-for-height threshold is required for triggering the onset and sustaining the maintenance of the ovarian cycle. Although the model has received widespread acceptance, the bulk of evidence conflicts with the model (see Bongaarts, 1980; Bronson & Manning, 1991; Ellison, 2001).

Diminished resource availability tends to correspond with reduction in rate of reproductive maturation, and comparisons between captive and wild animals indicates an association between enhanced nutrition and accelerated reproductive maturation (Sadleir, 1969), but the link between fertility and nutritional intake in primates is complex (Loy, 1988). Among stumptailed macaques, heavier females tend to initiate ovarian cyclicity at younger ages than their peers (Nieuwenhuijsen et al., 1987), but female rhesus macaques with first ovulation at 31 months had identical body weights to peers with first ovulation at 43 months (Wilson et al., 1986). Among captive longtailed macaques, 35% of females produce first progeny at 3 years of age, while only 12% bear first offspring by 4 years of age in wild populations (van Noordwijk & van Schaik, 1999). Age at first parturition was independent of food availability among longtailed macaques at Ketambe, Sumatra (van Noordwijk & van Schaik, 1999), but poor nutrition delays onset of reproductive maturation by up to 3 years among Japanese macaques (Mori et al., 1997; see also Mori, 1979a). Among the Japanese macaques resident on Koshima Island, those maturing when resources were plentiful had first offspring at an average age of 8.8 years, while those developing during lean years had first offspring at an average age of 10.9 years (Mori et al., 1997), both of which are substantially later than the 5.4 years among the provisioned Japanese macaques at Katsuyama (Itoigawa et al., 1992) and Arashiyama (Koyama et al., 1992). Among rhesus macaques living on Cayo Santiago, females are significantly more likely to generate first progeny at 3 years of age when population density is low (Bercovitch & Berard, 1993). In general, enhanced nutrition accelerates reproductive maturation in macaques.

One mysterious aspect of reproductive maturation among female primates is that first ovulation is not necessarily followed by regular ovulatory activity (see Table 4.1). Adolescent female rhesus macaques often have three or more anovulatory menstrual cycles between first and second ovulation (Bercovitch & Goy, 1990), while human primates may menstruate for five or more years prior to manifesting regular ovulatory cycles (Edwards, 1980; Worthman, 1993). Adolescent subfecundity terminates with first birth, but contrary to popular perception, motherhood, especially among primiparae, is quite a challenge (Hrdy, 1999). Adolescent subfecundity has been reasoned to be a nonadaptive

Table 4.1. *Timing of life-history reproductive events and physiological changes associated with reproduction in Macaca*

Species	First estrous, first swelling, first menses, first ovulation (years)	Estrous/menstrual cycle length (days)	Sex skin	Sex skin color changes	Post-conception estrous/swellings	First birth (years)
M. sylvanus	First swelling ~4[1]	NA	Large swellings[2]	NA	Estrous and swellings[3]	5[4]
M. silenus	First swelling 2.9[5] First menses 3.1[5] First conception 3[5]	Mean 32[5]	Large swellings[5]	Pregnancy[6]	Neither[6]	3.9[5]
M. nemestrina	First swelling 2.5–3.8[8]	Median 38[8]	Large swellings[9,10]	NA	NA	3.9[8]
M. nigra	First swelling 4[8]	Median 36[8,62] Mean 32.2[7]	Large swellings[7]	NA	NA	NA
M. nigrescens	NA	NA	Large swellings[11,a]	NA	NA	NA
M. hecki	NA	NA	Large swellings[11,a]	NA	NA	NA
M. tonkeana	First menses 4.4[12] First swelling 4.3[13] First consort 4.7[12]	Mean 37.4[12]	Large swellings[12]	NA	Swellings[12]	~5[12]
M. maurus	First swelling 4–6[14]	Mean 36.2[15]	Large swellings[14,15]	NA	No swellings[15]	6–7[14]
M. ochreata	NA	NA	Large swellings[11,a]	NA	NA	NA

(cont.)

Table 4.1. (*cont.*)

Species	First estrous, first swelling, first menses, first ovulation (years)	Estrous/menstrual cycle length (days)	Sex skin	Sex skin color changes	Post-conception estrous/swellings	First birth (years)
M. brunnescens	NA	NA	Large swellings[11,a]	NA	NA	NA
M. pagensis	NA	NA	Large swellings[16]	NA	NA	NA
M. sinica	First estrous ~4.5[65]	~29[19]	None[20]	Cycle[16]	NA	~4.5–5[17] ~6[18]
M. radiata	NA	Mean 27.8[21]	None[22]	Cycle[22]	NA	4[23]
M. assamensis	NA	Mean 32.0[24]	Adolescence only[25]	Pregnancy[19]	Swellings[25]	NA
M. thibetana	NA	Mean 26.4[26]	None[26]	NA	NA	~5.5[27]
M. arctoides	First menses ~2[28] First ovulation 3.7[29]	Mean 29.4[28] Mean 30.2[29] Mean 29.0[31]	None[28]	Pregnancy[28]	Estrous[30]	4.9[29]
M. fascicularis	First menses 2.4[32]	Mean 33.5[33]	Adolescence only[34]	Cycle and pregnancy[35]	Estrous and swellings[35]	5.2[36]
M. mulatta	First menses 2.5[37]	Mean 28.3[33]	Adolescence only[38,66]	Cycle and pregnancy[38,39,40]	Estrous[41]	4[42]
M. fuscata	First estrous 3.5[43]	Mean 26.5[44]	Adolescence only[43]	Cycle[43]	Estrous[43,45,46]	5[43]
M. cyclopis	First menses 3.3[47]	Mean 29.4[47]	Adolescence only[47]	NA	NA	4–6[48,49]

NA: data not available.
[a] Detailed sex skin descriptions or photographs of these species during estrous are lacking despite information by Fooden (1969, 1976) and Groves (2001).
For sources see Table 4.2.

byproduct of a lengthy immature stage, a "physiological atavism," and an adaptive mechanism promoting mate choice learning among females (Bercovitch & Ziegler, 2002).

Reproductive failure of primiparae is two to three times as high as among multiparae (e.g. Dazey & Erwin, 1976; Bercovitch *et al.*, 1998). Among captive pigtailed macaques, infant mortality resulting from dystocia (i.e., severe complications encountered during parturition) was much more pronounced among primiparous than multiparous mothers (Dazey & Erwin, 1976), possibly due to differences in body size. Primiparous female cercopithecines are significantly smaller than fully adult females (Bercovitch, 2000). Abortions and stillbirths account for a large fraction of the reduced reproductive success of primiparae (Dazey & Erwin, 1976; Wilson *et al.*, 1978; Altmann *et al.*, 1988; Paul & Kuester, 1996a). Developmental modifications in neuroendocrine feedback mechanisms participating in regulating adolescent female ovulatory activity might be adequate for conception, but inadequate for maintenance of pregnancy.

First-born infants tend to be smaller than subsequent infants, but larger relative to mother's size, which probably imposes additional burdens on females. Infants of primiparous captive rhesus macaques weigh an average of 6.5% of mother's weight, while infants of multiparae weigh an average of 5.6% of mother's weight (Bowman & Lee, 1995). Lactational obstacles involving both milk yield and milk delivery probably accompany first birth. Mammary gland development begins during gestation, but nipple development might sometimes be too meager to successfully nurture first born. Nipple growth depends upon estrogenic stimulation, but nipple enlargement depends upon suckling stimulation. For example, multiparous savanna baboon nipples are about 3 cm in length, while those of primiparae are about 1 cm long (Ransom & Rowell, 1972; Ransom, 1981). The small nipples of primiparae place them at a disadvantage in providing proper nourishment to progeny. In addition, milk yield is a function of maternal condition, and young primiparous females are leaner than older females (see Schwartz & Kemnitz, 1992). Neonatal mortality rates among primiparae are likely to arise from the imposition of large energetic demands on an organism striving to partition resources into both growth and reproductive functions (Lee & Bowman, 1995). An inadequate milk yield prompts frequent suckling, which drains additional resources from maternal reserves (Gomendio, 1989a).

Primiparous mothers are often clumsy and awkward when handling newborns, but maternal competence improves within a few days (Altmann, 1980; Hiraiwa, 1981; Chism, 1986). The most pronounced difference between primiparous and multiparous rhesus macaques is that the former tend to be more restrictive and less encouraging of independence (Berman, 1984), behaviors that are likely to increase, not decrease, survivorship prospects of first born. Maternal competence is probably built not so much on a foundation of experience acting as an infant caretaker as on experience received as an infant from caretakers, especially the mother. Infant mortality among orphaned, primiparous Japanese macaques was significantly greater than that among nonorphaned, primiparous mothers (Hiraiwa, 1981). Orphan females engaged in similar amounts of infant handling when immature as did nonorphan females, but received less maternal care.

Maintenance of reproduction

An average female rhesus macaque probably spends about 30 days during her entire life interacting sexually with males (i.e., 6-year reproductive lifespan (Bercovitch & Berard, 1993) × 5-day sexual consortship period (Loy, 1971) × 1-year interbirth interval (Rawlins & Kessler, 1986)), or a maximum of about 100 days (5-day sexual consortship × 20-year reproductive life (Tables 4.1 & 4.2)). Maestripieri (Box 10) describes maternal effort, and Singh and Sinha (Box 4) describe how reproductive life history is influenced by body size, phylogeny, and sex ratio of the group, so we concentrate on reproductive effort in males and females. About half of macaque species breed on a seasonal basis (Table 4.2), with the maintenance of male reproduction possibly depending upon the ability to enter a period of testicular quiescence during the birth season.

Male reproductive effort

Dominance rank influences progeny production in macaques (Gachot-Neveu & Ménard, Chapter 6; Paul, Box 6; Soltis, Chapter 7), and also mediates the relationship between hormones and male behavior. Testosterone concentrations correspond with ejaculatory rates among low-ranking males, but not among high-ranking males (Wallen, 2001). In general, testosterone facilitates sexual and aggressive behavior, rather than ignites them (Dixson, 1980; Bernstein *et al.*, 1983a; Bercovitch, 1999). Testosterone not only promotes sexual and aggressive behavior, but provides the substrate for conversion to estrogen, which probably has an impact on male feeding strategies and endurance rivalry, or the ability to sustain reproductive activity for lengthy periods of time (Bercovitch, 1992, 1997). In seasonally breeding macaque species, such as rhesus (Bercovitch, 1997) and Japanese (Nigi *et al.*, 1980) macaques, the annual reproductive cycle coincides with elevated testosterone concentrations, and enlarged testes.

   The annual pattern of regression and recrudescence of the testes suggests that it might not benefit males to maintain high testosterone levels all year long. In many avian species, the temporal pattern of testosterone concentration

Table 4.2. *Timing and duration of life history reproductive events and physiological changes associated with reproduction in* Macaca

Species	Birth seasonality discrete, seasonal peaks, year round	Gestation length (days)	Interbirth interval following infant survive (days)	Inter-birth interval following infant loss ≤ 202 days (days)
M. sylvanus	Discrete wild, Apr–June[1] Discrete semi-free, Mar–Apr[4]	Mean 164.7[3]	Mean 435.2[50]	Mean 408.4[50]
M. silenus	Year round captive[5]	Mean 171.8[51]	Mean 601[6]	Mean 393[6]
M. nemestrina	Year round in 2 groups. Nov peak in third group[10]	Median 171[8]	Median 405[8]	Median 270[8]
M. nigra	Year round[8]	Median 176[8]	NA	NA
M. nigrescens	NA	NA	NA	NA
M. hecki	NA	NA	NA	NA
M. tonkeana	Year round[52]	Mean 176[12,13]	NA	NA
M. maurus	Year round, March–June peak[9]	~175.5[15]	Mean 738[14]	Mean 475[14]
M. ochreata	NA	NA	NA	NA
M. brunnescens	NA	NA	NA	NA
M. pagensis	NA	NA	NA	NA
M. sinica	Peak, Dec–May[64]	~168[65]	~579[64]	NA
M. radiata	Discrete wild, Feb–Mar[22] Peak captive, Mar–May[53]	~168[50]	Mean 458.6[50]	Mean 382.5[50]

(cont.)

Table 4.2. (cont.)

Species	Birth seasonality discrete, seasonal peaks, year round	Gestation length (days)	Interbirth interval following infant survive (days)	Inter-birth interval following infant loss ≤ 202 days (days)
M. assamensis	Seasonal[55]	NA	NA	NA
M. thibetana	Discrete, Nov–June[56]	NA	~730[56]	~365[56]
M. arctoides	Year round[55]	Mean 172.4[54] Mean 176.6[55]	Mean 576.3[55] Mean 544.6[58]	Mean 291.9[55] Mean 308.7[58]
M. fascicularis	Birth peak, July–Nov[36]	Mean 162.7[54]	Median 390[8] ~461[61]	Median 315[8] ~320[61]
M. mulatta	Discrete, Dec–July[59]	Mean 166.5[60]	Mean 372[59]	Mean 336[59]
M. fuscata	Discrete, April–July[45]	Mean 173[46]	~722[46,60] ~847.9[63]	~403[46,60] ~603.9[63]
M. cyclopis	Discrete, Mar–June[48]	Mean 163.4[47]	Mean 415.4[48]	Mean 381.5[48]

NA: Data not available.

Sources: (1) Ménard & Vallet, 1993a; (2) Taub, 1980; (3) Küster & Paul, 1984; (4) Paul & Thommen, 1984; (5) Lindburg & Harvey, 1996; (6) D.G. Lindburg, pers. comm.; (7) Dixson, 1977; (8) Hadidian & Bernstein, 1979; (9) Tokuda et al., 1968; (10) Oi, 1996; (11) Fooden, 1969; (12) Thierry et al., 1996; (13) B. Thierry, pers. comm.; (14) Okamoto et al., 2000; (15) Matsumura, 1993; (16) C. Abegg, pers. comm; (17) Dittus, 1975; (18) Ratnayeke, 1994; (19) Hrdy & Whitten, 1987; (20) Melnick & Pearl, 1987; (21) Parker & Hendrickx, 1975; (22) Rahaman & Parthasarathy, 1969; (23) Silk, 1988; (24) Wehrenberg et al., 1980; (25) Fooden, 1971; (26) Flynn et al., 1989; (27) Zhao, 1996; (28) Trollope & Blurton Jones, 1975; (29) Nieuwenhuijsen et al., 1985; (30) Nieuwenhuijsen et al., 1986; (31) Steklis & Fox, 1988; (32) Dang, 1983; (33) Michael & Zumpe, 1993; (34) Nawar & Hafez, 1972; (35) van Noordwijk, 1985; (36) van Noordwijk, 1999; (37) Resko et al., 1982; (38) Baulu, 1976; (39) Lindburg, 1983; (40) Schwartz et al., 1988; (41) Bielert et al., 1976; (42) Bercovitch & Berard, 1993; (43) Hanby et al., 1971; (44) Hiraiwa, 1981; (45) Takahata, 1980; (46) Nigi, 1976; (47) Peng et al., 1973; (48) Hsu et al., 2001; (49) Wu & Lin, 1992; (50) Burton & Sawchuk, 1982; (51) Lindburg & Lasley, 1985; (52) Masataka & Thierry, 1993; (53) Silk, 1990; (54) MacDonald, 1971; (55) Bernstein & Cooper, 1999; (56) Zhao & Deng, 1988a; (57) Brüggemann & Dukelow, 1980; (58) Harvey & Rhine, 1983; (59) Rawlins et al., 1984; (60) Silk et al., 1993; (61) Nomura et al., 1972; (62) A. F. Dixson, pers. comm; (63) Takahashi, 2002; (64) Dittus, 1998; (65) W. P. J. Dittus, pers. comm.; (66) Rowell, 1967.

varies as a function of both male–male interactions and levels of paternal care (Wingfield *et al.*, 2000). Male birds that do not provide any form of paternal care have maximum testosterone concentrations over an extended breeding/hatching season, but the sustained high testosterone levels are associated with suppressed immune function and increased risks of injury and mortality. Adult male Barbary (Deag & Crook, 1971; Taub, 1980; Paul & Kuester, 1996a), Tibetan (Ogawa, 1995; Fitch-Snyder *et al.*, 1997), and stumptailed macaques (Estrada, 1984) regularly hold infants, but most male macaques simply tolerate the proximity of infants (Whitten, 1987). In a number of nonprimate species, prolactin concentrations among "helpers" are as high as among parents, and caretaking by male New World monkeys is also associated with elevated prolactin (Bercovitch & Ziegler, 2002). Detailed studies of the annual cycle of prolactin in male macaques, and potential effects on behavior, are unavailable for testing this hypothesis, but we propose that testicular regression is adaptive because the maintenance of steroidogenesis and spermatogenesis throughout the year would extract too high a cost from males. We suggest that the annual reproductive cycle in males might be adaptive because the reduction in testosterone production during the nonmating season reduces aggressive activity, facilitates tolerance by males toward females and infants, and avoids continued steroid-induced immunosuppressive effects.

Factors determining lifetime male reproductive success have not yet been thoroughly evaluated, but males depend upon a variety of reproductive tactics as a means of enhancing their reproductive success (Soltis, Chapter 7). The maintenance of reproduction among seasonally breeding male macaques is, therefore, dependent upon annual patterns of testicular regression and recrudescence that place a premium on telescoping spermatogenesis into the most probable period when females are receptive. Maintaining high testosterone levels throughout the year is probably not a good means of maintaining reproductive success across years.

Female reproductive effort

As noted by Soltis (Chapter 7), macaques are an unusual genus because, while all species live in multimale troops, not all species have pronounced sexual skin swellings during the follicular stage of the cycle (Table 4.1). The distribution of sexual skin swellings in macaques closely follows phylogenetic relationships (Thierry *et al.*, 2000). In some species, e.g., rhesus macaques, adolescent females are the only age class that has sexual skin swellings, with the swellings running along the sides of the hind limbs (see plate 2.1.D in Rowell, 1967), while in other species, e.g., Barbary and Sulawesi macaques,

a   b

Fig. 4.1. Female sexual skin swelling during the follicular stage of the cycle. (a) In adolescent rhesus macaques, sexual skin swellings are found in both the hindlimb area and the eyebrow region (Strasbourg Primate Center, France). (b) In crested macaques, estrous females display enlarged anogenital swellings (Tangkoko Nature Reserve, Sulawesi). (Photographs by B. Thierry.)

adult females have enlarged anogenital swellings (Fig. 4.1). Regardless of whether the periovulatory phase is accompanied by sex skin swellings, ovarian hormones exert a profound effect on female sexual behavior, with the connection between ovarian hormones and sexual activity mediated by social factors (Wallen, 1990; Ziegler & Bercovitch, 1990). For example, high-ranking female rhesus macaques are likely to initiate interactions with males at various times during their reproductive cycle, whereas low-ranking females are much more likely to initiate interactions with males only when estradiol concentrations are elevated (Wallen, 1990). The compressed allocation of time that subordinate females have for interacting with males probably reflects the subtleties of nonagonistic competition between females.

Some scientists suggest that female primate sexual activity is relatively disconnected from endocrine state, and that females are receptive throughout their cycles[1], but reproductive behavior tends to track estrogen levels, as do morphological cues indicating proximity to ovulation (Robinson & Goy, 1986). In Tonkean macaques peak sexual skin swelling sizes are obtained when estradiol concentrations reach their zenith (Aujard et al., 1998), while in rhesus macaques (Czaja et al., 1977), the sex skin obtains a bright red color in the anogenital region during the periovulatory period. Among Tibetan macaques, vaginal mucous flow becomes quite copious around the time of probable ovulation (Flynn et al., 1989). In some macaques, sexual skin swellings also occur during pregnancy, but whether this is a by-product of circulating endocrine concentrations or an adaptive feature remains to be determined (see Thierry, 1997).

The endocrine profile of all macaque species is reasonably consistent, despite differences in outward manifestations of proximity to ovulation (Robinson & Goy, 1986). The average ovarian cycle is about 1 month, and gestation lasts approximately 5 to 6 months (Tables 4.1 & 4.2). The largest variance among female macaques in reproductive life-history trajectories is the interbirth interval, which is generally condensed upon the loss of an infant. Because the loss of an infant can decrease the interbirth interval (Table 4.2), the use of interbirth interval as a measure of reproductive rate, and an indirect indicator of "fitness," is questionable. The shortest interbirth intervals will occur in those females who have no surviving infants (Bercovitch, 1997).

Cessation of reproduction

Termination of reproduction caps a female's reproductive career but the multimale mating structure of macaque societies hampers determining the lifetime reproductive success of males. Short-term variance in reproductive output among male macaques exceeds that of females (Paul & Kuester, 1996a; Bercovitch *et al.*, 2000), but whether the average lifetime reproductive success of males also exceeds that of females is unclear.

Reproductive senescence in male macaques

Investigations of reproductive senescence in male macaques are nearly absent from the literature. Although one component of the reproductive life history of women is menopause, men lack a comparable milestone. In both male and female macaques, maximum lifespan is nearly twice as long as average lifespan (e.g., Rawlins & Kessler, 1986; Tigges *et al.*, 1988; Fedigan & Zohar, 1997). The median age of death among male Japanese macaques at Arashiyama West was 12 years (Fedigan & Zohar, 1997), while average life expectancy among male rhesus macaques living in social groups at the Yerkes Regional Primate Research Center was about 15 years (Tigges *et al.*, 1988).

Among men, aging is accompanied by a reduction in circulating testosterone concentrations (Vermeulen, 1990), and 20-year-old male rhesus macaques on Cayo Santiago have testosterone concentrations much lower than prime-age males (Schwartz & Kemnitz, 1992). However, among captive male rhesus macaques, individuals over 20 years of age have similar androgen levels to those that are half as old (Robinson *et al.*, 1975; Chambers *et al.*, 1981). The most pronounced effect reported for aging among male macaques is that the latency to ejaculation is longer among very old males (≥25 years) when paired with females (Phoenix & Chambers, 1988). Despite the potential for reduced testosterone concentrations and lowered levels of sexual activity, older male macaques are fully capable of impregnating females. Using genetic analysis of paternity, the oldest known male rhesus (Bercovitch *et al.*, 2003) and Barbary (Kuester *et al.*, 1995) macaques who have sired offspring have been 24 years old.

In summary, the reproductive activity of male macaques probably declines with age, but whether the decline results from female rejection of familiar males or male competitive disability remains to be determined. Reduced prospects for reproductive success are most likely the result of limited access to females, rather than hormonal deficits, and very old males have sired offspring in semi free-ranging conditions.

Reproductive senescence in female macaques

Recently, it has been estimated that each ovary in a newborn pigtailed macaque has 71 483 primordial follicles (Miller *et al.*, 1999), which is similar to what has been reported in human females (O'Connor *et al.*, 2001). As the numbers of primordial follicles decrease at a log-linear rate after birth (Nozaki *et al.*, 1997; Miller *et al.*, 1999), the numbers of primary and tertiary follicles along with the size of the ovary continue to increase to around 10 years of age (Nozaki *et al.*, 1997). After 10 years of age, Japanese macaques experience a slow, but continuous decline in the number of primary and tertiary follicles available for ovulation until ages 26–28 years, when histological examination of ovaries revealed few or no remaining follicles for development (Nozaki *et al.*, 1997). Histological examination of ovaries from rhesus macaques in their late twenties also showed little or no evidence of follicular development (Walker, 1995; Gilardi *et al.*, 1997).

Follicular maturation is greatest during the sexually mature period and is reflected in fecundity rates from long-term studies in field and captive populations of macaques. In macaques, first birth typically occurs between 4 and 5 years of age and the birth rate increases rapidly over the next 2 to 3 years and remains at high levels until the late teens. In Japanese (Sugiyama & Ohsawa, 1982a; Takahata *et al.*, 1995; Pavelka & Fedigan, 1999), rhesus (Dyke *et al.*, 1986; Walker, 1995; Kemnitz *et al.*, 1998), longtailed (van Noordwijk & van Schaik, 1999), Barbary (Paul *et al.*, 1993a,b; Paul & Kuester, 1996a), and Tonkean macaques (Thierry *et al.*, 1996), female fecundity declines at about 20 years of age and usually ceases by the mid- to late twenties (Fig. 4.2).

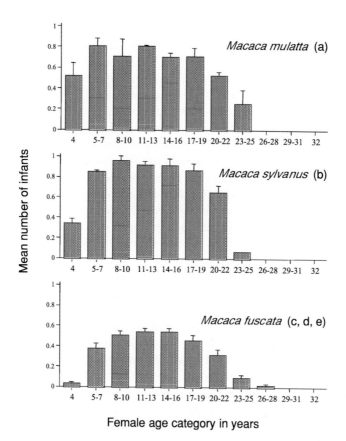

Fig. 4.2. Number of infants per number of females in each age category in three species of macaques (error bars: standard errors). (a) Johnson & Kapsalis, 1995: data from Key Louis and Racoon Key, Florida (1992–1994), adapted from table 1, p. 274; (b) Paul et al., 1993: data from Salem, Germany (1978–1998), adapted from table 1, p. 110; (c) Itoigawa, *et al.*, 1992: data from Katsuyama, Japan (1958–1986), adapted from table 5, p. 56; (d) Koyama *et al.*, 1992: data from Arashiyama, Japan (1954–1985) adapted from table 3, p. 37; (e) Watanabe *et al.*, 1992: data from Koshima, Japan (1952–1986), adapted from table 2, p. 17.

The transition from reproductive success to reproductive cessation in macaques is a gradual process possibly spanning up to 5 years. This transition phase is typically referred to as the perimenopause, and is characterized by episodes of increasing intervals between menses and/or perineal swelling (Hodgen *et al.*, 1977; Short *et al.*, 1989; Walker, 1995; Gilardi *et al.*, 1997) resulting in increasing intervals between births (Koyama *et al.*, 1992; Paul *et al.*, 1993a,b). In the post-reproductive phase, some of the long-term studies on macaques have described older females as displaying obvious morphological and physiological signs of reproductive senescence. Signs included poor body condition, complete absence of perineal swellings, menses, behavioral estrous, as well as a decline in social interactions (Nakamichi, 1988; Short et al., 1989; Thierry et al., 1996; Kemnitz et al., 1998; van Noordwijk & van Schaik, 1999).

The underlying endocrinological changes that occur during the transition from prime reproductive years through senescence have been documented for several species of macaques. Considerable individual differences in hormone profiles within a species have been documented – females enter the perimenopausal stage at a variety of ages. In Japanese macaques, a slight increase in LH, which typically indicates a decline in ovarian activity, was detected as early as 18 years of age. After 21 years of age, some female cycles became infertile, as determined by low levels of progesterone concomitant with slight increases in LH. By 27 to 28 years of age females had extremely high levels of LH, suggesting that they had transitioned from the perimenopausal to the postmenopausal stage (Nozaki et al., 1995, 1997).

Gilardi and collaborators (1997), using daily urinary hormone samples and records of menses over a 12-week period, found that female rhesus macaques exhibited typical perimenopausal endocrine characteristics when about 24 years of age. Some females had very long intervals between menstrual cycles, due to a lengthening of the follicular phase with very low levels of estrone conjugates (E_1C) persisting for a prolonged period of time until E_1C peaked. Other perimenopausal females had incidents of vaginal bleeding that did not correspond with either E_1C or pregnanedio-3-glucuronide (Hygeia [Hy]-Pdg) fluctuations. A few females never exhibited any signs of vaginal bleeding, but both E_1C and Pdg levels fluctuated in a manner indicative of normal follicular activity. Other studies have shown macaque females between the ages of 21 and 26 years exhibit signs of perimenopause, in that they are capable of showing both normal cyclical activity of fluctuating estradiol (E_2), FSH, LH, and progesterone levels, interspersed with anovulatory cycles (Hodgen et al., 1977; Short et al., 1989; Walker, 1995). By age 27–28 years females have been identified as postmenopausal as estrogen levels were consistently low while LH levels were consistently high (Walker, 1995).

From the endocrinological data it appears that ovarian activity begins to change around 20 years of age, with the mid-twenties showing the most variability between individuals in gonadal hormone patterns, menstrual, and/or swelling cycle length. During this time period, some females display signs of perimenopause or menopause. By the late twenties, many macaque females are postmenopausal as determined by the absence of developing ovarian follicles, menses, perineal swellings, and elevated levels of FSH and LH concomitant with exceedingly low and nonfluctuating levels of estrogen and progesterone. These are the same endocrine changes seen in human females in the transition from perimenopause to postmenopause stage (O' Connor et al., 2001).

Data on age of females at last parturition are few, but mean ages were 22 and 21.3 years for Japanese macaques (Takahata *et al.*, 1995; Pavelka & Fedigan, 1999) and 20.5 years estimated for Barbary macaques (Paul *et al.*, 1993a,b). The age at which female Japanese macaques truly became postmenopausal is unknown, but Barbary macaque females continue to have swelling cycles for 1–3 years after their last parturition, indicating some ovarian function (Paul *et al.*, 1993a,b). Post-reproductive lifespan of female macaques have been estimated, based on the interval from the female's last parturition to death or disappearance, as ranging from 1–11 years for Japanese macaques (Takahata *et al.*, 1995; Pavelka & Fedigan, 1999) and 3–8 years for Barbary macaques (Paul *et al.*, 1993a,b).

Only 7.9% of Japanese macaque females at Arashiyama West live to 20 years of age (Pavelka & Fedigan, 1999), which precedes the perimenopause stage identified hormonally in laboratory settings. In contrast, 60.0% of Barbary macaque females are estimated to survive to 20 years of age. After age 20 years, the percentage of females estimated as 21–25 years of age (Table 4.3) in non captive primates ranged from as low as 0.2% for rhesus macaques at Cayo Santiago (Rawlins *et al.*, 1984) to as high as 18.8% for rhesus macaques in the Florida Keys (Johnson & Kapsalis, 1995) and 18.6% for Japanese macaques at Arashiyama (Koyama *et al.*, 1992). For the two presumed postmenopausal age (26–30 years and 31–35 years) categories, the percentages of females still alive decreases dramatically, indicating that very few females live long enough to become truly post-reproductive.

There appears to be little difference in the individual case records on female longevity between free-ranging and laboratory macaques. Oldest longevity records for free-ranging Japanese macaques ranges from 22.3 to 33 years of age (Takahata *et al.*, 1995; Pavelka & Fedigan, 1999) and from around 26.5 to 32 years of age for Barbary macaques (Paul *et al.*, 1993a,b). The oldest rhesus macaque females reported in laboratory settings have been 29 and 34 years of age (Walker, 1995; Gilardi *et al.*, 1997), while the oldest captive pigtailed macaque lived to 31 years of age (Short *et al.*, 1989).

In summary, if female macaques live long enough they ultimately reach menopause. Although some controversy persists regarding whether or not menopause is a characteristic of macaques (Pavelka & Fedigan, 1999), long-term data reveal that some female macaques survive into their late twenties and early thirties without bearing offspring in their latter years. Since females become grandmothers long before menopause, a distinction between reproductive and post-reproductive grandmothers should be made when considering the validity of the "grandmother" hypothesis. The basic premise of the original grandmother hypothesis is that cessation of reproduction occurs a long time before the termination of life because it enables a female to maximize reproductive success by investing in her grandoffspring, but a variety of "grandmother"

Table 4.3. *Number and percentage of Macaca females that have been identified as living past 20 years of age at different free ranging sites*

Female age categories in years	M. fuscata fuscata Koshima 1972–1986[1]		M. fuscata Arashiyama 1954–1983[2]		M. fuscata Katsuyama 1958–1986[3]		M. fuscata A. West/Texas 1972–1993[4]		M. mulatta Cayo Santiago 1957–1983[5]		M. mulatta Florida Keys 1992–1993[6]	
	Total females	% (n/656)	Total females	% (n/253)	Total females	% (n/2086)	Total % females/No. years	% (mean)	Total females	% (n/571)	Total females	% (n/1427)
21–25	7	1.19	47	18.6	139	6.7	20.4/4	5.1	1	0.18	268	18.8
26–30	0	0.0	13	5.1	17	0.85	7.3/4	1.8	1	0.18	0	0.0
31–35	0	0.0	0	0.0	2	0.096	0.60/3	0.2	0	0.0	0	0.0

Sources: (1) Watanabe *et al.*, 1992: 17, table 2; (2) Koyama *et al.*, 1992: 37, table 3; (3) Itoigawa *et al.*, 1992: 56, table 5; (4) Pavelka & Fedigan, 1999: 460, table 2; (5) Rawlins *et al.*, 1984: 250, table I; (6) Johnson & Kapsalis, 1995: 274, table 1.

hypotheses have been proposed (see review in Peccei, 2001). Whether in Old World monkeys a post-reproductive lifespan is a unique evolutionary adaptation that allows females to invest in the survival of their children or their grandchildren or is simply a by-product of aging (Packer *et al.*, 1998; Sherman, 1998; Peccei, 2001) cannot be addressed at this time. Few female macaques live long enough to become post-reproductive (Table 4.3), so data to test the "grandmother" hypothesis are sparse. One report showed that when a post-reproductive grandmother was available in Japanese macaques, there was a significant increase in their grandchildren surviving to 1 year of age (Pavelka *et al.*, 2002). However, similar results were not found in survival rates for infant baboons, with post-reproductive grandmothers present in the group (Packer *et al.*, 1998). In general, female macaques do not survive long enough to become menopausal, but the few that survive to become post-reproductive show similar endocrine profiles to menopausal women.

Life-history theory proposes that a higher fitness will accrue to those who begin reproduction at younger ages (Stearns, 1992). However, fitness differentials in female primates originate from lengthy breeding lifespans, not age at first parturition. Age at first birth does not predict lifetime reproductive success in Japanese macaques (Fedigan *et al.*, 1986), rhesus macaques (Bercovitch & Berard, 1993), or savanna baboons (Packer *et al.*, 1995). Rhesus macaques bearing first offspring at 3 years of age tend to live until 9 years old, while those who delay first birth until 5 years of age tend to die at 11 years of age, indicating an average breeding lifespan of 6 years (Bercovitch & Berard, 1993). Among nonhuman primates, offspring survival rates and breeding lifespan, not age at first birth, are the principal contributors to differential reproduction among females.

Conclusion

The major difference in macaque reproductive life-history patterns across species is that some breed on a seasonal basis, while others reproduce throughout the year. However, the paradox is that the ancestral group, i.e., *silenus–sylvanus*, breeds in all months of the year, while the derived groups have re-incorporated the usual mammalian pattern of seasonal reproduction (Bronson, 1989). Hence, the key evolutionary change in macaque reproductive life history was the re-incorporation of a discrete breeding season from the ancestral mammalian condition into derived species groups of macaques, while the ancestral macaque species group expanded sexual activity into a year-round behavior. Furthermore, along with this change in reproductive cyclicity, females reduced the extent to which they displayed sexual skin swellings as indicators of the follicular stage of the cycle. And, concomitant with changes in female reproductive cyclicity, males developed seasonal reproductive patterns that incorporate testicular regression and recrudescence that might be adaptive in promoting year-long bonds between males and others in the group during the non-breeding season.

The consequences of aging in both female and male macaques remains understudied as one component of reproductive life history. Male lifespan tends to be less than female lifespan, but males have survived and reproduced long past their prime ages. The endocrinological changes underlying the transition from the reproductively active phase to the postmenopausal phase in laboratory macaque females are identical to the changes identified in women, but how these changes impact lifetime reproductive success is still unclear.

Promising research directions include collecting more specific data on patterns of troop integration following dispersal, documenting more precisely nutritional influences on reproductive performance, focusing greater attention on the development of female sexuality, incorporating more molecular genetic analysis into resolving issues about male reproductive success, identifying how social and demographic factors influence both male and female reproduction, and obtaining more longitudinal data on multiple species of macaques.

Footnote

1. Frank Beach is reported to have noted that any male who thinks that females are continually receptive is either a young male headed for a big disappointment or an old male with a faulty memory.

Box 4   Life-history traits: ecological adaptations or phylogenetic relics?

Mewa Singh and Anindya Sinha

The macaques constitute one of the largest genera of extant primates and represent arguably the most ecologically diverse species within the order. They also exhibit striking variation in their life-history traits, as summarized in Tables 4.1 and 4.2 and Table 4.4. A meta-analysis of the traits displayed by the 15 better-studied species was carried out, using these data, to determine the extent to which these traits have been shaped by the ecology of the various species and to discern whether such traits exhibit any phylogenetic inertia, given that these species share a recent common origin. For this, the respective values of each independent life-history variable were classified into either of two categories on the basis of the median value of each; these were assigned category labels of 1 and 2 respectively as follows: male body

Table 4.4. *Selected life-history traits of* Macaca

Species	Male body weight (kg)[a]	Female body weight (kg)[a]	Sexual dimorphism[b]	Mount pattern[c]	Female: male ratio	% of Category 2 values	Phylogenetic group[d]	Ecological group[d]
M. sylvanus	16.0	11.0	0.37	SM	1.1[1]	36.4	A	B
M. silenus	8.9	6.1	0.38	MM	9.9[2]	64.6	A	A
M. nigra	9.9	5.5	0.59	MM	3.5[3]	81.8	A	A
M. tonkeana	14.9	9.0	0.50	MM	NA	100	A	A
M. maurus	9.7	6.1	0.46	MM	4[4]	72.7	A	A
M. nemestrina	11.2	6.5	0.54	MM	6.3[5]	72.7	A	A
M. sinica	5.7	3.2	0.58	SM	2.4[6]	27.3	B	B
M. radiata	6.7	3.9	0.54	SM	1.7[7]	0	B	B
M. assamensis	11.3	6.9	0.49	SM	2.4[8]	50.0	B	A
M. thibetana	18.3	12.8	0.36	MM	1.9[9]	60.0	B	A
M. arctoides	12.2	8.4	0.37	SM	NA	40.0	B	A
M. fascicularis	5.4	3.6	0.41	SM	4.8[10]	36.4	B	B
M. mulatta	11.0	8.8	0.22	MM	2.9[11]	27.3	C	B
M. fuscata	11.0	8.0	0.32	MM	1.3[12]	45.4	C	B
M. cyclopis	6.0	4.9	0.20	MM	1.7[13]	10.0	C	B

[a] Adult body weight values are derived from Smith & Jungers (1997).

[b] Weight dimorphism was calculated as the natural logarithm of the male body weight minus that of the female body weight. This is equivalent to calculating sexual dimorphism as the natural logarithm of the male body weight divided by the female weight (Plavcan & van Schaik, 1997).

[c] Single-mount (SM) and multiple-mount (MM) copulatory patterns, derived from Thierry et al. (2000) and Soltis (Chapter 7) for M. radiata.

[d] Groups derived from Fooden (1976), Delson (1980) and Thierry et al. (Chapter 1).

NA: data not available.

Sources: (1) Ménard & Vallet, 1993a; (2) Singh et al., 2000; (3) O'Brien & Kinnaird, 1997; (4) Okamoto et al., 2000; (5) Oi, 1996; (6) Dittus, 1977; (7) Sinha, 2001; (8) Caldecott, 1986a; (9) Zhao, 1996; (10) Wheatley et al., 1996; (11) Lindburg, 1971; (12) Maruhashi, 1982; (13) Kawamura et al. (1991).

weight, less than vs. greater than 11.0 kg. Female body weight, less than vs. greater than 6.5 kg. Female age at sexual maturity, less than vs. greater than 43.8 months. Sexual swelling, absent vs. pronounced. Cycle length, less than vs. greater than 31 days. Mount pattern, single mount ejaculation (SME) vs. multiple mount ejaculation (MME). Gestation period, less than vs. greater than 171 days. Age at first birth, less than vs. greater than 60 months. Birth interval, less than vs. greater than 547 days. Birth seasonality, present vs. absent. Adult female: male ratio, less than vs. greater than 3.

An initial cross-correlational analysis among the life-history variables themselves revealed a significant association of certain traits with each other – a clear indication that certain coherent life-history strategies have been selected for among the macaques. Thus, among the morphological and physiological traits, male and female body weight are correlated with each other (Kendall's rank correlation, $\tau = 0.817$, $n = 15$, $P < 0.0001$), while a longer cycle length among females positively correlates with increasing sexual dimorphism ($\tau = 0.418$, $n = 15$, $P < 0.04$) and with pronounced female sexual swelling (point biserial correlation, $r_{PB} = 0.746$, $n = 15$, $P < 0.01$). Among the reproductive traits, a longer birth interval is correlated with delayed female sexual maturity (Kendall's rank correlation, $\tau = 0.421$, $n = 13$, $P < 0.05$) and to a longer gestation period ($\tau = 0.565$, $n = 12$, $P < 0.01$), while the latter, in turn, is correlated with the presence of pronounced sexual swelling (point biserial correlation, $r_{PB} = 0.514$, $n = 13$, $P < 0.05$). Finally, a reproductive trait that is responsive to the environment – the absence of birth seasonality – significantly characterizes species with relatively longer cycle lengths (point biserial correlation, $r_{PB} = 0.500$, $n = 15$, $P < 0.05$), while a demographic trait – a progressively female-biased adult sex ratio with relatively fewer males in the group – exhibits significant positive correlation with the reproductive trait of longer cycle lengths (Kendall's rank correlation, $\tau = 0.523$, $n = 12$, $P < 0.02$) as well as the environmentally-regulated absence of any birth seasonality (point biserial correlation, $r_{PB} = 0.625$, $n = 12$, $P < 0.05$).

The macaques were classified into three phylogenetic groups (Fooden, 1976; see Thierry, Singh & Kaumanns, Chapter 1), labeled A, B, and C (Table 4.4). For the 11 life-history variables under consideration, the percentage of category 2 values was calculated for each species (Table 4.4). It is noteworthy that, with the exception of cycle length, which varied significantly across the phylogenetic groups (Kruskal–Wallis test, $H = 7.78$, $n = 15$, $P < 0.02$), no phylogenetic bias could be observed among any of the displayed life-history traits. What is more striking, however, is that specific life-history strategies correlate most strongly with the ecology of the different species. Macaques with a relatively greater proportion of category 2

Box 4 Life-history traits 83

values thus primarily inhabit broadleaf evergreen forests (ecological category A) (Table 4.4) while those with a low proportion of category 2 values are inhabitants of non-broadleaf forests (ecological category B).

Species inhabiting broadleaf evergreen forests, in general, appear to follow a life-history strategy characterized by significantly prolonged gestation periods (ecological categories A and B: Mann–Whitney U-test; $U = 42$; $n = 7, 6$; $P < 0.003$) and longer interbirth intervals ($U = 36$; $n = 7, 6$; $P < 0.04$). Since these traits, in turn, correlate with several other characters including delayed female sexual maturity and longer cycle lengths, the life-history traits of these species seem to be characterized by an enhanced investment in individual growth and a correspondingly lower investment in reproductive effort – a strategy that would obviously result in relatively low rates of population turnover. A preponderance of category 1 values of these variables among the macaques that live in other, perhaps less stable, environments such as deciduous and secondary forests or human habitations (ecological category B), on the other hand, represents a contrasting life-history strategy – one that tends to maximize their reproductive output even while lowering the potential for individual development (see also Ross, 1988).

Wilson (1975) had earlier proposed a scheme for the evolution of social behavior wherein phylogenetic factors determine the extent and the speed of social evolution while ecological factors determine the actual direction of such processes. The present analysis reveals that although phylogenetic inertial forces may constrain the states that certain life-history variables can potentially achieve, the reproductive traits currently displayed by macaques appear to be selected for by the environment that these species inhabit. Given the ecological importance of such traits in the survival and propagation of the species, it is, in fact, not at all surprising that almost no vestige of any phylogenetic inertia can be discerned among these traits today.

Part II *Demography and reproductive systems*

Introduction

Groups of macaques have recognizable spatiotemporal boundaries. The groups are made up of individuals that interact regularly and more so with each other than with members of other groups (Struhsaker, 1969). Rates of birth, death, and dispersal underlie grouping patterns. Males regularly transfer between groups. Groups are cohesive units, though some may temporarily split up into foraging parties (e.g., Caldecott, 1986b; van Schaik & van Noordwijk, 1986). Growing groups eventually divide into smaller groups whereas shrinking ones may disappear or fuse with others (e.g., Dittus, 1987). The fates of individuals determine the demographic (Chapter 5) and genetic structures of populations (Chapter 6). Both structures reflect a double-layered hierarchy (Eldredge, 1985; Brooks & McLennan, 1991). On one side, the ecological hierarchy is an "energy-flow system" in which energy and matter are exchanged between individuals and their environment. Number and density, growth rate, access to resources, predation, and disease set the ecological stage. On the other side, the genealogical hierarchy is an "information-flow system" in which the perpetuation of generations is insured by the replication of DNA. Birth, history, and death of individuals punctuate the genealogical trail. Whereas the genealogical hierarchy supplies the players in the ecological game, it is the ecological hierarchy that establishes the dimensions of the playing field within the real world. Evolution results from the interaction between both systems. Changes in the genealogical hierarchy can redefine the boundaries of the playing field, and players from different genealogies influence one another in the playing field. Selective processes occur through differential reproduction of individuals. The contribution of individuals to the gene pool of future generations depends on their mating strategies (Chapter 7).

5 *Demography: a window to social evolution*

WOLFGANG DITTUS

Introduction

In this chapter, I examine the interplay between the basic parameters involved in social evolution: social structure, sexual dimorphism, competition, and environment. I use the demographic perspective as a tool in this inquiry. The chapter is not intended to be a literature review of primate demography, nor an exposition of demographic methods of analysis. Instead, I attempt to demonstrate that knowledge not only of demographic variables but also of their environmental determinants is critical to unraveling the nature of social evolution.

Present-day phenotypes evolved as adaptations to environmental settings and constraints that were far different from the contemporary ones in which most primate studies are carried out. Ideally, therefore, we might aim to investigate primate behavior in relation to environmental contexts that minimize known distortions. Given the paucity of such studies, particularly in the genus *Macaca*, and the rapid disappearance of natural habitats, the need for demographic and ecological studies of macaques in natural environments could hardly be greater.

In pursuing the question of "how societies are built" we should be clear on what constitutes a society: these are groups of individuals whose members show cohesiveness in their travel and other activities, stability in composition, and exclusivity in membership. One might add to that the long-term view of a group being self-sustaining, reproducing, and outbreeding – i.e., safeguarding Darwinian fitness by avoiding inbreeding. In mammals, the obligate and most basic social unit is the mother and her dependent offspring. Through natural selection this mother–family unit has been molded into various more complex configurations in different mammalian radiations (Eisenberg, 1981). Among primates, and especially the *Cercopithecinae*, it has mostly involved the retention of daughters which gives rise to both the stable matrilineal cores of social

Macaque Societies: A Model for the Study of Social Organization, ed. B. Thierry, M. Singh and W. Kaumanns. Published by Cambridge University Press. © Cambridge University Press 2004.

groups and the exchange of breeding-age males with other groups in kind. The structure of such groups, often labeled "female-bonded" (Wrangham, 1980), then comprises overlapping generations of matrilineally related females cohabiting with their immature offspring of either sex as well as with immigrant (not closely related) breeding males.

Parallel to this social structural development, by which females were spatially concentrated, was polygyny and its concomitant effects on sexual dimorphism through sexual selection. It has been argued that the relatively large size of cercopithecoid males was driven by an intensified level of male–male contests for reproductive opportunities under conditions where clusters of females, rather than single ones (monogamy), offered potentially high returns in male reproductive success (Darwin, 1871; Clutton-Brock *et al.*, 1977; Ralls, 1977).

This admittedly cursory phylogenetic sketch points to two major phenotypic attributes, social structure and dimorphism, which are shared by all species of *Macaca* as well as many other primates. No doubt, other major adaptations (e.g., ecological) and phenotypic embellishments (e.g., gestures of communication) have arisen among macaque species in response to local conditions, phylogenetic inertia and perhaps drift. From the point of view of social organization, however, the members of the genus *Macaca* are fairly consistent and conservative, the differences within and among species relating mainly to the size of groups, age and sex compositional variations, numbers of constituent matrilines, and numbers of outbreeding males. Many of these differences themselves might be related to local ecological conditions or recent histories. In terms of sexual dimorphism too, the genus *Macaca* is fairly consistent. Size differences among species (and locomotor specializations) appear to be ecologically related (e.g., terrestrial vs. arboreal) (Clutton-Brock *et al.*, 1977) but sexual dimorphism in body size is expressed by all species, though to varying degrees (see Singh & Sinha, Box 4).

Both phenotypic developments were honed over eons of time under a strict environmental regimen that set the limits to the numbers of survivors of different phenotypes. An understanding of the nature of these complex environmental constraints is fundamental to the appreciation of phenotypic adaptations. Yet, this important evolutionary dimension is perhaps the least well understood, especially in the genus *Macaca*, where the majority of studies have been done in captive or similar unnatural settings.

I have based most of this chapter on my own studies of toque macaques, for which a substantial database is available. Toque macaque are long-tailed and their locomotor anatomy allows them to travel equally well on the ground and in trees (Grand, 1972). They represent the typical macaque phenotype, particularly the more arboreal species.

What limits populations?

The answer to this question is important for understanding not only the environmental constraints under which phenotypes evolved, but also the phenotypic adaptations themselves. Three factors are generally cited as controlling the size or growth rate of primate populations: resource limitation, predation, and disease.

Resource limitation

Food availability – i.e., the carrying capacity of the habitat – commonly places an upper limit on the size of the population at least in ecologically balanced environments. Resource availability and its variation in space and time are notoriously difficult to measure. Cant (1980) pointed out that estimates of overall or annual food availability might lead to the wrong conclusion concerning food surpluses, as was the case for howler and spider monkeys at Tikal, Guatemala (Coelho et al., 1976, 1977). Given the seasonal and between-year fluctuations in climate and plant productivity, even in stable environments, the effect of resource limitation comes by way of fairly regular but brief "ecological crunches" or "resource bottlenecks" that impact population size (Cant, 1980; Emmons, 1984; Peres, 1994). Thus, it is during lean periods, such as the regular dry season, where a few critical or "keystone" resources, including water, may impact the population (McFarland Symington, 1988; Byrne et al., 1993). The evidence for resource limitation is often inferred or only grossly estimated. For example, the toque macaque population at Polonnaruwa in 1974 declined by 15% during a severe 11-month drought (7 months longer than the normal 2–4 months), the worst in 44 years up to that time (Dittus, 1977). Similarly, at Polonnaruwa in 1978, the worst cyclone in over 100 years destroyed nearly 40% of the tree canopy (Dittus, 1985). Presumably in response to this reduction in the food supply, toque macaque mortality increased, especially among infants and pregnant females, and several groups fissioned (Dittus, 1982, 1988) (see Okamoto, Box 5). When populations collapse during drought or famine, the vulnerable young, very old, pregnant, and socially low-ranking individuals most often bear the brunt of increased mortality in many primate taxa (Japanese macaques: Mori, 1979a; Barbary macaques: Ménard & Vallet, 1993a; vervets: Struhsaker, 1973; Cheney et al., 1981; Lee & Hauser, 1998; howler monkeys: Milton, 1982; yellow baboons: Altmann et al., 1985; chacma baboons: Hamilton, 1985; colobus monkeys: Decker, 1994; lemurs: Gould et al., 1999). Where a true food surplus exists, as when populations are artificially fed, populations predictably have grown (rhesus macaques at Cayo Santiago: Sade *et al.*, 1976, and feral populations in India: Malik *et al.*, 1984; Japanese macaques: Sugiyama & Ohsawa, 1982a; toque macaques: Dittus, 1977; and a number of other primates: Lyles & Dobson, 1988; see also Fa & Southwick, 1988). Taken together these observations indicate a close relation between food supply and primate population growth.

The food supply is thought to influence population growth by way of competition under a regimen of "density dependence." As the number of competitors increases, per capita food availability declines, leading to reduced birth rates, increased mortality – and group fission and emigration – and therefore slower population growth. These effects were observed, for example, when a dominant group of toque macaques at Polonnaruwa took over the home range of a neighboring group. Over a period of 8 years, the members of the dominant group, which had benefited from an expanded range and food supply, thrived numerically. The members of the subjugated group, on the other hand, were deprived of access to their normal food supply and were driven to extinction through increased mortality and lower birth rates (Dittus, 1986). Similarly, mortality increased and birth rates dropped in Japanese macaques when artificial feeding was reduced, resulting in a decline in population growth (Mori, 1979a; Ohsawa & Sugiyama, 1996).

Predation

The extent to which predation affects primate populations appears to vary depending on the species and the sites. Predators can have a devastating effect on primates, accounting for 50% of vervet monkey mortality at Amboseli, Kenya (Cheney *et al.*, 1988), and this increased to 70% in one year from a single leopard (Isbell, 1990). Predation by a single leopard similarly accounted for the crash of one population of Ugandan mountain gorillas (Butynski *et al.*, 1990). The dynamics between predator and prey can be complex; in the above examples a single leopard took up residency in the area and specialized on these primates for a short period. More commonly, leopards, lions, and hyenas are known to be occasional predators even of large-sized baboons at several sites in Africa (Altmann & Altmann, 1970; Busse, 1980; Cowlishaw, 1994). Predation by chimpanzees may eliminate 17–33% of red colobus monkeys annually in Gombe National Park (Tanzania) (Busse, 1977; Wrangham & Bergmann Riss, 1990; Stanford *et al.*, 1994), but at the Taï (Ivory Coast) this is only 4% (Wrangham & Bergmann Riss, 1990), and at Kahuzi-Biega National Park (Democratic Republic of Congo) chimpanzees kill 11–18% of *Cercopithecus* monkeys (Basabose & Yamagiwa, 1997). Raptors, too take a variable toll on primates. But this risk appears to fall disproportionately on species of small body size, like *Microcebus* in Madagascar (Goodman *et al.*, 1993) or squirrel monkeys and marmosets of Central and South America (Terborgh, 1983; Boinski *et al.*, 2000). In the Kibale Forest (Uganda), Struhsaker and Leakey (1990) estimated that although monkeys accounted for about 88% of prey items taken by eagles, this was distributed disproportionately over many species; only about 1% of the population of the relatively large-bodied red colobus fell victim and, in general, the adults of small sized species and the immatures of large bodied ones were more affected. But vulnerability by body size also depends on the predator; Cowlishaw (1994) for example noted that leopards are more likely to prey on adult baboons than on juveniles, and, among adults to take males rather than females. Predators, therefore may impact primate population size as well as age–sex structure, but these effects vary along with the primate species, location-specific properties of the environment and faunal assemblage, predator type, and variations in predator behavior. Cheney and Wrangham (1987) indicated that among 30 representative primate studies, half estimated annual predation rates of 0–5%, 12 sites lay in the 5–15% range, and three – all of small sized galagos and tamarins – estimated a >15% rate.

Although empirical evidence for predation on primates is slowly accumulating, rates of predation on primate populations are still today as difficult to quantify as they were some years ago (Boinski *et al.*, 2000). Sadly, this is particularly so for the genus *Macaca* because most species have been studied in habitats where humans have eliminated the large predators. Or, estimates at natural sites have been speculative: the relatively high estimate of <11% annual loss of longtailed macaques to predators at Ketambe (Indonesia), for example, was based on a single observation in 3 years, involving a python, plus another 46 events that were merely suspected (C. van Schaik & M. Van Noordwijk, personal communication, table 19–1 in Cheney & Wrangham, 1987). In a comparison among five natural sites in Asia, Seidensticker (1983) notes that the intensity of predation on macaques and langurs by tigers and leopards was inversely related to the availability and abundance of alternative prey. Predation by one or both cats was documented for rhesus macaques at the Royal Chitawan National Park (Nepal), and for longtailed macaques in Java (Indonesia) at Meru-Betiri, and (leopard only) at Udjung Kulon. Langurs *Semnopithecus entellus* too were prey at these sites as well as at Kanha National Park (India) and Wilpattu National Park (Sri Lanka). In all sites, as is common in Asia, langurs outnumbered macaques as prey, and leopards took primates far more often than did tigers. Tigers are absent from Sri Lanka, and, while leopards at the Wilpattu National Park commonly took langurs, wild boar and chittal deer, macaques were not preyed upon in the study sample (Muckenhirn & Eisenberg, 1973). This pattern was also noted at a second site, the Ruhunu National Park (A. Kittle & A. Watson, personal communication). In both of these arid zone parks, macaques were rare, however, being confined to alluvial forests, whereas langurs and other prey were more common and widespread. In contrast, a recent observation of a leopard killing a macaque in a secondary wet forest near Kandy, where the usual prey of langur and chittal deer were absent, confirms the threat of leopards to toque macaques under different ecological conditions (A. Watson & A. Kittle, personal communication). Leopards are absent from the study site at Polonnaruwa, but raptors, pythons, feral dogs, jackals and fishing cats *Felis viverrina* are present.

Disease

Disease, like predation, can markedly increase mortality in primate populations. Population crashes of South American howler monkeys have long been attributed to epidemics of yellow fever introduced from Africa (Collias & Southwick, 1952; James *et al.*, 1997). While not identifying the pathogen, Pope (1998) reports a 85% crash (during 1991–1996) of a population of Venezuelan red howler monkeys where entire social groups were wiped out. At a second site, 40% of red howlers were lost but the scourge affected mostly adult females. An outbreak of hemorrhagic diarrhea among chacma baboons similarly wiped out 85% over a period of 3 months, and the viral infection affected adult males more than females (Barrett & Henzi, 1998). The chimpanzees at Gombe have been reduced by a polio epidemic (Goodall, 1983).

Disease can also be more chronic in nature, of a type that, in the case of parasitism, fluctuates in severity between seasons and years in a number of primate species (Stuart & Strier, 1995). Stoner (1996) for example found that, in populations of howler monkeys in Costa Rica, nematode loads were higher under humid and wet conditions. Likewise, Milton (1996) reports that botfly and screwfly infestations and associated mortality for howler monkeys of Barro Colorado Island (Panama) peaked during the rainy seasons. The juveniles were more affected by botflies than adults or infants. In baboon populations, susceptibility to parasitic infestation differed depending on age, sex, nutrition as well as reproductive condition and dominance rank (Hausfater & Watson, 1976; Eley *et al.*, 1989).

The effects of disease are not necessarily fatal, however, as species adapt to their pathogens. Eley and collaborators (1989) found no mortality in Kenyan baboons that was associated with moderate infestations of helminths, including parasitic nematodes, such as ascarids. This is also true of the toque macaque population at Polonnaruwa (Sri Lanka) (Dewit *et al.*, 1991). Perhaps the best example of the evolution of resistance to disease comes from the genus *Macaca* in relation to malaria, where different species of macaques have developed resistance; but only to those strains of malaria normal in their own species' geographical range. Experimental infections with malarial strains typical to the ranges of other macaque species were fatal (Fooden, 1994). Similarly, toque macaques at Polonnaruwa were subject to an epizootic event of dengue virus type II infecting more than 90% of the population, but there was no morbidity or mortality associated with this epidemic (de Silva *et al.*, 1999). A related strain of dengue specific to humans – not found in macaques – causes illness and sometimes death in humans.

The relative importance of key factors in population fluctuations

It is clear from the above considerations that the food supply can affect rates of birth and death (and probably emigration), whereas predators and disease more directly impact mortality. Populations fluctuate in size in relation to variations in the impact of these factors that differ according to species and environment. Any attempt to weigh the relative importance of these factors on primate populations is difficult in light of sketchy empirical evidence, especially on predation rates. Notwithstanding, venturing some predictions may be useful. On an evolutionary time-scale one might expect a species either to succumb to disease or predation or to achieve a punctuated equilibrium with these threats. The cap set by resource availability, however, is actually or potentially constant measured on a generational (or multigenerational) time scale despite monthly or annual climate-generated fluctuations, and the additional threats of predation or disease. Barring rare catastrophic events, the role of disease or predation over generations would then pose additional or secondary limits to population growth.

The nature of these threats to survival also differs; predation and disease are exogenous factors requiring different adaptive responses from the ones they use in competition with conspecifics for food. Successful adaptations in immunology and behavior to ward off disease or predators might actually exacerbate the limiting effects through resource competition. In primates such as macaques, birth rates are high (usually between 0.5 and 0.8 infants per female per annum) (Dittus, 1975; Ménard & Vallet, 1996; Okamoto *et al.*, 2000). Therefore, in an average lifetime, a female produces far more offspring than can ever be sustained under density dependent population growth. In macaques and other cercopithecines, for example, it has been estimated that 50–80% of offspring die before reaching reproductive age (Dittus, 1980). This would be especially the case for populations that are close to the carrying capacity of the environment, where net population growth hovers near zero. The death of individuals through noncatastrophic or normal endemic disease or light predator pressure, therefore, would provide vacancies for the survival of resource competitors whose prospects for survival were otherwise poor. Or, under normal conditions, occasional losses to predators or disease have little or only a temporary effect in mitigating resource competition.

On the other hand, groping in an empirical vacuum, one cannot dismiss the possibility of a stronger predation effect or even that at least some primate species throughout their evolutionary history had been kept below resource carrying capacity by chronic erosion from predation and/or disease. Our knowledge base would suggest, however, that such an extreme would most likely have occurred, if at all, among primates of small body size. One would expect variations in the relative importance of these three limiting factors across the various taxa, habitats, and time frames. Primate social organization and behavior are the product of selective pressures from all of these quarters. Thus, propositions stating that predation is the primary driving force for group living in primates (e.g., van Schaik, 1983; van Schaik & van Hooff, 1983) have little empirical support or heuristic appeal.

Demography and the toque macaque example

Demography involves the statistical study of the vital processes of a population. It is concerned with the age and sex composition of social groups and populations, the rates at which the latter grow, and the frequencies with which individual animals give birth, die, and migrate. These measures call into play closely related considerations, such as how quickly animals grow and mature and how long they are capable of reproducing and living.

Estimates of these parameters are the direct consequence of, and therefore reflect, a complex and dynamic interplay among a variety of subsumed influences. These include a species' evolutionary history, its phylogenetic constraints and life-history strategy, but also the environmental and social milieu within which individual animals happen to live. The manner in which animals of a species behave in relation to each other in their local environment determines their individual survival, growth, and reproductive success, or, their Darwinian fitness. Demography measures these individual fitness outcomes averaged at the level of the population. In the absence of knowing individual fitness outcomes, therefore, demographic variables give an overview of the mean life and death processes of individuals of different age and sex for a particular population.

The methods for demographic measures and analyses are beyond the scope of this chapter. These have been well presented, especially by Caughley (1977), but also Wilson and Bossert (1971). For detailed practical applications of these methods to primates, see National Research Council (1981). A number of authors have applied life-table analyses (e.g., Dittus, 1975, 1980; Masui *et al.*, 1975; Sade *et al.*, 1976; Teleki *et al.*, 1976; Froehlich *et al.*, 1981; Dunbar, 1988; Dobson & Lyles, 1989).

The life table

The central tool for demographic analyses is the life table. As individuals grow older they are subject to shifting probabilities of reproducing and dying. These changes are commonly expressed as age-specific rates of birth and death in a life table. The raw data required for the construction of life tables can be either numbers of animals dying per age, or numbers living. Primate demographic studies have been "life-based" where census data from many groups or a population are expressed as numbers of animals observed by age or age-class for each sex. Given this information, vital statistics such as survivorship and mortality rate can be derived. The usual variables expressed in life tables are shown in Table 5.1 for toque macaques. It is important to keep in mind that the only independent variable in these analyses is the raw data: the frequency of animals by age and sex, on the basis of which survivorship l_x, is estimated. All other vital statistics d_x, q_x, e_x, are arithmetic derivatives of l_x, viewing the same data from a different perspective. This convertibility is handy, because if one value, say q_x, can be estimated, then the other life-table statistics can be derived from it. The second independent variable concerns age-specific fecundity rate m_x, which when combined with l_x can be used to estimate reproductive value v_x (Fig. 5.1) and other statistics such as the observed exponential rate of increase (r), net reproductive rate (R_o) and generation length (T).

In constructing and interpreting life tables and other derived statistics it is important to heed some constraints and basic ecological principles. Differences in the numbers of individuals observed by age are the product of past rates of birth, survival, immigration, and emigration, all of which are subject to environmental influences. Their combined effects would have underlain variations in the population growth rate over brief or prolonged periods.

The stable age distribution

Ecological projections indicate that when populations have been reproducing in steady environments they will approach a "stable age distribution" (assuming that the effects of immigration and emigration cancel one another). This means that the proportions of individuals belonging to different age–sex classes will maintain constant values generation after generation, regardless of whether the

Table 5.1. *Life table for toque macaques at Polonnaruwa, Sri Lanka, 1971*

Age and sex class	Age class (years)	Number of years per age class x	Number of individuals observed in each age class x	Mean number of individuals annually per age class x	Mean survival to age class x	Total mortality within age class x	Mean annual mortality rate within age class x	Mean life expectancy	Mean annual fecundity per age class x
	x	a_x	f_x	f_x/a_x	l_x	d_x	q_x	e_x	m_x
Females									
Infant	0	1	38	38	1.000	0.526	0.526	4.8	0.000
Yearling	1	1	18	18	0.474	0.158	0.333	8.6	0.000
Juvenile	2–4	3	36	12	0.316	0.163	0.172	10.7	0.000
Young adult	5–9	5	29	5.8	0.153	0.016	0.021	16.6	0.352
Prime	10–14	5	26	5.2	0.137	0.011	0.016	13.0	0.362
Middle-aged	15–19	5	24	4.8	0.126	0.021	0.033	9.3	0.406
Old	20–24	5	20	4.0	0.105	0.047	0.090	5.2	0.442
Very old	25–29	5	11	2.2	0.058	0.058	0.200	2.5	0.143
	>29	0	0	0	0				
Males									
Infant	0	1	38	38	1.000	0.395	0.395	4.5	0.000
Yearling	1	1	23	23	0.606	0.079	0.131	6.2	0.000
Juvenile	2–4	3	60	20	0.526	0.074	0.141	5.0	0.000
Subadult	5–6	2	23	11.5	0.303	0.100	0.328	5.1	0.022
Young adult	7–9	3	10	3.3	0.104	0.077	0.074	10.5	0.860
Prime	10–14	5	13	2.6	0.081	0.024	0.030	9.0	0.990
Middle-aged	15–19	5	11	2.2	0.069	0.076	0.110	5.2	0.623
Old	20–24	5	5	1.0	0.031	0.050	0.161	3.5	0.358
Very old	25–29	5	1	0.2	0.006	0.120	0.200	2.5	0.258
	>29	0	0	0	0				

(After Dittus, 1975.)

Mean survivorship, l_x: of those born, the mean proportion of individuals that survived to the onset of the age class x; or, the probability at birth of surviving to age class x.

Total mortality within age class, d_x: the mean probability at birth of dying between the onset and end of age class x.

Mortality rate, q_x: the mean rate (or probability) of death per annum for those that survived to the onset of age class x.

Mean life expectancy, e_x: the mean number of years of life remaining for individuals of age class x.

Female fecundity, m_x: one-half the birth rate; the mean number of female offspring born per year per female of age class x; or, the probability of a female of age class x giving birth to a female infant in any one year.

Male fecundity, m_x: one-half the mean number of offspring sired per year per male of age class x (estimated from the population mean annual birth rate and adjusted according to the adult sex ratio and differences in the probability of paternity among males of different age classes).

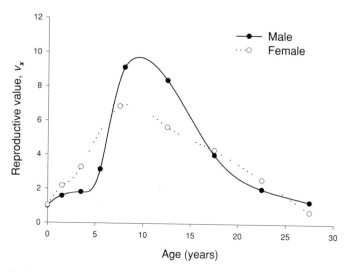

Fig. 5.1. The reproductive values of male and female toque macaques incorporating measures of survival and fecundity by sex (Table 5.1). Male fecundity was estimated, in part, from age differences in the probability of paternity as determined from DNA studies. (Keane *et al.*, 1997).

overall population is increasing in size, decreasing, or holding steady. Looked at another way, the effect of the steady environment is to stabilize age-specific rates of survivorship l_x and reproduction m_x through time and by this means the population structure converges towards a stable age distribution. It is unlikely, however, that a population of primates would increase or decrease indefinitely in a stable environment. At some point, population size would be expected to approach environmental carrying capacity and stop growing. The observed exponential rate of increase under this condition would be zero ($r = 0$) and the population also would be stationary. A stationary age distribution is a special case of the stable one, when $r = 0$. In terms of calculating a life table, if the observed distribution of ages is stable and stationery, then one can use these age frequencies to calculate the l_x schedule. If the population is merely stable then one would also need to know the population rate of increase (r) in order to estimate the survival schedules (see Caughley, 1977).

Methodological limitations and reducing sampling errors

Important sources of error in constructing life tables arise from small sample sizes of individuals enumerated (as is common in primate field studies) and from faulty age estimation. Judging a minimum required sample size is some-what case-specific and depends on the accuracy sought and the intended use for the life table. When a life table is drawn up from an age distribution census, Caughley (1977) had suggested that a table based on fewer than 150 deter-minations of age (per sex) is unlikely to be accurate enough for any purpose. The extent to which primate individuals' ages can be estimates with confi-dence varies between species (especially with their morphology) and studies (see National Research Council, 1981).

The method most appropriate for life-table analysis depends partly on the history of the population under consideration and what is known about it (see Caughley, 1977). The surest way to estimate survival schedules is through direct observation, following a cohort (all those born in a given year) through life and recording individual fates. Chance events and environmental fluctuations will introduce sampling error in estimates of mortality rates by age, but these might be averaged if more than one cohort (preferably many) are monitored. Not only do such studies require large cohort sizes, but ideally, they also involve an enormous investment of observation time lasting for as long as the longest lived individual in the sample cohort(s). Longevity in macaques can exceed 30 years (Hill, 1937; Jones, 1962; Bercovitch & Harvey, Chapter 4), and for this reason few cohort life tables have been attempted.

A "shorter-term" method is based on calculating mortality rates among age classes over brief 1–2 year periods. But this "instantaneous" or "time specific" method is most applicable to populations, such as captive ones, where the ages of all individuals are already known; otherwise long-term observation is required to estimate ages. Using such data, Sade and collaborators (1976), for example, estimated a life table for the managed colony of rhesus macaques on Cayo Santiago Island, Puerto Rico (Table 5.2). These authors combined data from two contiguous years for the advantages of producing a larger number of indi-viduals in each age class from which to estimate age-specific mortality (q_x), and of averaging random effects that might have occurred over short sampling peri-ods. On the downside, the method sacrificed independence of samples between successive ages. A similar approach, and combining short-term cohort-based observations with the time-specific method (but averaged across 3 years) has been applied to the provisioned colony of Japanese macaques of Takasakiyama (Masui et al., 1975). Dittus (1975) had applied a variable time span (see a_x in Table 5.1) for combining data from contiguous ages in order to reduce sampling error in age estimates.

Each of the aforementioned methods has its own set of underlying assump-tions and advantages and shortcomings when applied. All life-table methods suffer from the fact that environmental conditions fluctuate from year to year, so survival and birth rates obtained from one year may be different from those

Table 5.2. *Life table for female rhesus macaques at Cayo Santiago*

Age (years)	l_x	d_x	q_x	e_x	m_x
0	1.000	0.176	0.176	8.970	0.000
1	0.824	0.020	0.024	9.779	0.000
2	0.804	0.016	0.020	9.010	0.000
3	0.788	0.117	0.148	8.183	0.064
4	0.671	0.043	0.064	8.522	0.258
5	0.628	0.055	0.088	8.073	0.375
6–7	0.573	0.074	0.129	7.840	0.400
8–9	0.499	0.094	0.188	6.381	0.323
10–11	0.405	0.062	0.153	6.138	0.356
12–13	0.343	0.057	0.166	5.110	0.294
14–15	0.286	0.096	0.336	3.573	0.000
16–17	0.190	0.047	0.247	2.874	0.250
18–19	0.143	0.072	0.503	2.245	0.000
20–21	0.071	0.071	1.000	1.000	0.500
22–23	0.000			0	

(Modified after Sade *et al.*, 1976).
q_x measures the total mortality rate for the age (or age span) indicated.
See Table 5.1 for definitions of the other variables.

obtained in another. Faced with a set of data, the investigator needs to decide which model best fits the observed situation.

Estimating demographic stability from ecological information

In the absence of long-term demographic data to define population status and history, the focus for estimating "stability" – or its absence – might shift to secondary ecological indicators that are known to affect demography. For example, the following conditions are suggestive of long-term environmental stability (after National Research Council, 1981), that might lead to stable and stationary populations.

- A habitat in which the forest or other vegetation type has achieved an ecological climax condition and is undisturbed.
- Constant climatic regime. A climatic regime in which temperature and amount and seasonal distribution of rainfall follow the same pattern over many years would tend to induce predictable patterns of forest phenology and productivity. One assumes that primate populations are attuned to the predictable seasonal flux in resources.
- A balanced flora and fauna relative to environmental type.

Conditions indicative of unstable environments or populations are as follows:

- Forests that have been disturbed recently, whether through natural events, such as fire, hurricanes, or changes in soil salinity, or through felling by humans.
- Change in cultivation practices. The balance of populations in an area where they are partly or wholly dependent on agriculture would be changed by a shift in long-established cultivation practices, such as the introduction of a new crop, or of the use of pesticides.
- The introduction of unnatural or non-endemic disease.
- Change in the predator fauna and predation pressure (e.g., Isbell, 1990).
- Hunting, trapping, and poisoning by man.
- Highly irregular climatic regimes.
- The introduction of human settlements or domestic animals.
- An impoverished flora and fauna relative to environmental conditions.

The toque macaque population was assumed to have achieved a fairly long-term stationary condition by 1971, when the population census was carried out, and that provides the basis of the life-table estimate (Table 5.1). In the absence of observations of this population prior to 1968 there is, of course, no hard evidence to support this contention (a mistaken citation in Dittus (1975) of net replacement rate ($R_o = 1$) cannot be taken as evidence because its calculation assumed zero growth). Instead, the assumption of "stationary" was based on secondary estimates of long-term environmental stability as outlined above, as well as short-term population stability. Major cyclones that impact macaque demography are estimated to affect this site about once per century (seven macaque generations) (Dittus, 1985), droughts of 2–4 months occur annually, and severe ones with more than the usual demographic impact occurred at intervals of more than 20–40 years. Human influence in the last 800–900 years was minimal because an earlier civilization had abandoned the Polonnaruwa area from the thirteenth to twentieth centuries (>60 macaque generations). In the observed period 1968–2003, growth trajectories of individual groups differed greatly in relation to human disturbance. Groups with little access to human food have not grown in over >30 years (W. P. J. Dittus, unpublished data) and this ecological setting (excepting natural events such as droughts and cyclones) was the norm for most of the population 1968–1980 and probably before 1968. With the advent of tourism and increased settlement in the area in the early 1980s some groups gained access to human food and have grown. That is not to say that we can accept this population to have exhibited a perfectly stable age distribution and stationary status by 1971. However, this paradigm was accepted as a useful closest approximation.

Life-table differences in relation to resource availability

The colony of rhesus macaques on Cayo Santiago has a long history of management with known population growth, estimated by Sade and collaborators (1976) at $r = 0.0705$. The latter, "little r" or the observed exponential rate of increase r, can be estimated from a life table, but also from a comparison of population numbers at different times (see Caughley, 1977). The net reproductive rate R_o, is a related statistic that is also derived from a life table by $\sum l_x m_x$. This variable is equivalent to the number of female offspring born to each female during her lifetime and can be interpreted as the mean rate at which a female replaces herself. In expanding populations $R_o > 1$, in stable ones $R_o = 1$, and in declining populations $R_o < 1$. The demographic health of the rhesus population in 1973 and 1974 was good because, at $R_o = 1.801$, it was expanding rapidly.

It is instructive to compare the life-table statistics between toque macaques (Table 5.1), where I had assumed an approximately zero growth population status ($r = 0$, $R_o = 1$), with that of the expansion of rhesus macaques on Cayo Santiago (Table 5.2). Relative to female toque macaques, female rhesus macaques showed: (1) lower rates of mortality among the infants and juveniles; but (2) higher mortality among adults; (3) nearly twice the life expectancy, e_x, at birth; (4) an earlier age of onset of reproduction; (5) a shorter lifespan; and (6) a shorter mean generation length ($T = 8.35$ years in rhesus macaques and $T = 11.84$ years in toque macaques). In both populations fecundity, m_x, was lowest among the youngest and oldest females. These findings are consistent with differences in growth rates of these populations. The data also suggest that reproducing rhesus females "burn out" faster than those of wild toque macaques.

How representative are these different demographic profiles of the evolutionary norm in which social behavior evolved? One would expect all phases of population growth to have occurred in evolutionary time. Which phase predominated historically and had the greatest impact in molding phenotypes is unknown. It would seem, however, that rapid population growth, as is common under artificial feeding, was probably short-lived and rare. We should also keep in mind that behaviors and social structures differ under disparate demographic conditions (Altmann & Altmann, 1979).

Vital statistics in relation to social behavior

Typically, rates of mortality are high in mammals early in life, reach a minimum at the advent of adulthood, and remain low through most of adult life, and peak again in old age (Caughley, 1966). This pattern was also observed in toque macaques, but in addition there were sex differences; during the immature

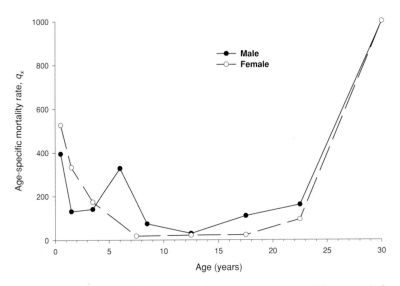

Fig. 5.2. Age-specific rate of mortality in toque macaques at Polonnaruwa, before 1972. (After Dittus, 1975.)

phase females were subject to greater mortality than males, and this pattern reversed in adulthood, beginning with a peak in mortality among subadult males (Fig. 5.2).

Mortality and resource competition

Age and sex specific rates of mortality have been linked to social behavior as at least a partial cause. Under normal resource competition within social groups juvenile females in general, and subordinate ones in particular, were supplanted from food resources more frequently than male peers. As a consequence, they ingested food items at slower rates (from less food-dense patches), obtained poorer quality foods, and spent more time (energy) in fulfilling their daily nutritional requirements (Dittus, 1977). I had suggested that these circumstances contributed to the higher death rates among immature females, especially under environmental conditions of food scarcity. Furthermore, when food is in surplus (e.g., garbage dump) the discrimination against juvenile females persisted, yet its effects on feeding ecology and survival were ameliorated, such that juvenile females survive as well or almost as well as juvenile males. A comparison of the demography of other primates, especially cercopithecines, suggested that, as in toque macaques, infant and juvenile females were subject to greater mortality than the males in natural populations, but less so in provisioned ones (Dittus, 1980).

These observations are consistent with the theoretical expectation, namely, that natal group kin should discriminate more against those resource competitors of their group that pose the greater threat to the individual and inclusive components of their fitness (Dittus, 1979). Females remain in their natal group and share the group's resources for life, whereas all males leave permanently at adolescence. Females, therefore, present a greater threat than males to the natal group's larder that, in large part, determines the survival and reproductive success of all its members.

Mortality, dispersal and mate competition

All males leave their natal group on average during adolescence, either solo or in small bachelor parties, and attempt to integrate themselves into the social fabric of other groups (see Gachot-Neveu & Ménard, Chapter 6). The immigration process is hindered by the xenophobic and competitive tendencies of resident group members. Subadult males, too, make their first serious attempts at competing with other males for mates. It is during this adolescent phase that males suffer a peak in mortality (Fig. 5.2). Male mortality reaches a low during their prime years as young adults. Nevertheless, the greater mortality sustained by males than females throughout adulthood has been linked to the greater frequency and severity of injuries that these males sustain in male–male contests for mates and to the risk of group transfers in the pursuit of reproductive opportunities throughout their adult careers (Dittus, 1977).

Census-based estimates of mortality in both subadult and adult males may be somewhat confounded by an imbalance in immigration and emigration of these age classes in the census population. In addition, where ages are estimated by morphological traits and body size, the growth spurt in adolescent males (e.g., Cheverud *et al.*, 1992) may seemingly deplete numerical representation of this class in a census, thus somewhat overestimating mortality of subadult males and underestimating it in the next age group, the "young adult males" (National Research Council, 1981). A measure of mortality based on numbers of carcasses of howler monkeys found on Barro Colorado Island, Panama, showed that, like toque macaques, male howler mortality peaked during adolescence and was greater in adult males than females (Otis *et al.*, 1981). Likewise, Fedigan and collaborators (1983) noted an increased rate of deaths among males of dispersal age in Japanese macaques. Demographic data from a number of macaques and other primate species suggested a pattern similar to that found in toque macaques (Dittus, 1980).

The avoidance of deleterious genetic effects and offspring mortality through inbreeding (e.g., Ralls & Ballou, 1982; Ralls *et al.*, 1988) has been suggested as the driving force for male dispersal in mammals (Greenwood, 1980; see

Gachot-Neveu & Ménard, Chapter 6). Indeed, in the toque macaque, where females are highly philopatric, no male ever returned to its natal group after dispersal, effectively eliminating the possibility of mating with maternal relatives, and this is common in other cercopithecines as well (Melnick *et al.*, 1984a; Gouzoules & Gouzoules, 1987; Pusey & Packer, 1987). Males often reproduce in more than one group in their lifetime, thus leaving paternally related female kin residing in several groups. Genetic studies showed, however, that none of the male toque macaques, whose paternity was known, dispersed into groups with reproductively active female paternal kin. There was no genetic evidence for inbreeding in the toque macaques (Keane *et al.*, 1997) and this was true also in rhesus macaques (Melnick *et al.*, 1984a). The rates of mortality associated with dispersal imply that the cost, though potentially high, is nevertheless not as great as that of inbreeding.

Reproductive success and effective population size

Male–male competition for mates is nearly universal in mammals, in particular polygynous ones, and has led to the evolution of sexual dimorphism and secondary sexual traits that are used in male contests and courtship (Darwin, 1871). Trivers (1972) pointed out that as females invest more heavily in the production of each offspring than males, they are the limiting resource for male paternity success. Males may increase the number of offspring that they produce by mating with additional fertile females. But, by females mating with many males does not increase the number of offspring produced. By this interpretation, large male size and fighting traits evolved because they enable well-endowed males to monopolize females at the expense of lesser-quality male competitors, and so to sire a greater number of offspring. These relationships imply a greater degree of variance in the reproductive success of males than females. DNA-based paternity data from toque macaques bear out that individual males indeed vary considerably in their success at siring offspring: in any one breeding season most males produced few or no offspring, and this also appears to be true for their lifetime reproductive success (Keane *et al.*, 1997). Furthermore, there was a marked imbalance between male and female contributions to succeeding generations, such that the effective population size for toque macaques is seemingly much smaller than indicated by demographic data alone.

Reproductive values

From a slightly different perspective, life-table data can portray the changing prospects for future reproductive success of males and females as they grow older. The reproductive value v_x, is defined as the expected number of female offspring that will be born to the average female during the remainder of her life, given that she has reached a specific age (x). It can be derived from a life table, following Wilson and Bossert (1971), v_x is calculated for each age x as:

$$v_x = \frac{e^{rx}}{l_x} \sum_{y=x}^{\infty} e^{-ry} l_y m_y$$

where $y =$ all the ages that a female has yet to pass through from age x to infinity (death).

The statistic incorporates changing probability of survival (l_x) from birth to old age, and changes in offspring production (m_x) with age. The statistic usually is applied to females, hence future reproduction is measured in terms female offspring only. However, the same concept can be applied to males and can be useful if, in addition to male survival schedules, we can estimate changes in the probability of paternity with age. This I have done in Fig. 5.1, using DNA-based paternity data of known-aged male toque macaques (e.g., Keane *et al.*, 1997). In both sexes, the reproductive values are low among juveniles and, in the total absence of reproduction, are determined mostly by survival prospects at different ages (life expectancy, e_x). The peak in v_x is reached with the advent of adulthood, where life expectancy is greatest in both sexes and reproduction begins (Table 5.1). Fairly low but constant mortality erodes v_x in both sexes with increasing age, but these trajectories are modified by age difference in female birth rates, or, in the case of males, by age related changes in the ability to sire offspring. Major sex differences stand out in these projections. Males begin reproducing at a later age than females, and are subject to a high but narrow peak in v_x. In other words, compared to females, the average male that survives to reproduce has a short but highly successful reproductive career. The curves in Fig. 5.1 represent mean trends, and from DNA paternity studies we also know (Keane *et al.*, 1997) that individual variance around these means is high, especially in males.

Life history and birth sex ratios

The above considerations encapsulate some of the major differences in male and female life-history strategies. One implication, supported empirically, is that large size in males contributes to their competitive ability (Dittus, 1977) and this in turn translates into greater reproductive success (W. P. J. Dittus, unpublished data). If we allow that individuals behave so as to maximize their survival and reproductive success (fitness), then we might expect females to raise the quality and sex of offspring that leads to the production of maximum

Fig. 5.3. A female toque macaque and her infant (Polonnaruwa, Sri Lanka). The black pigment of the ears and lips, and the strongly whorled cap (toque) are some of the traits that distinguish this species from the Indian bonnet macaque. (Photograph by W. Dittus.)

numbers of grandoffspring (Fig. 5.3). Females differ, however, in their ability to raise offspring of different qualities. The sex of offspring that might be expected to contribute most to maternal fitness has been linked to: (1) maternal body condition, the Trivers and Willard (1973) (TW) model; (2) local resource competition (LRC) among the non-dispersing sex (usually females) (Clark, 1978); and (3) dominance rank, which reflects individual differences in LRC among the philopatric sex (Altmann, 1980; Simpson & Simpson, 1982; Silk, 1983).

The TW and LRC hypotheses and their theoretical limitations

Briefly, the TW model suggests that mothers in good body condition are best able to produce sons that will eventually develop the physical prowess required for winning mates. The production of superior sons therefore would pay higher fitness dividends than a similar investment in daughters, which need not develop large body size to achieve reproductive success. On the other hand, faced with the high variance in male reproductive success, mothers in poor condition would experience better fitness returns from daughters than from poor quality (virtually non-reproducing) sons. The predictions based on LRC are opposite to those of TW, and rest on the observation that philopatric daughters, especially low-ranking ones, are more likely to die from the effects of resource competition than are dispersing sons. Therefore, according to LRC, high-ranking mothers, which are most able to protect and pass on their competitive advantage to their daughters, would be expected to produce more of them, whereas low-ranking mothers should produce sons. Empirical support exists for one or the other of these models, but it is marked all too often by inconsistency and contradictions (see reviews: Clutton-Brock & Iason, 1986; Hrdy, 1987; Altmann & Altmann, 1991; van Schaik & Hrdy, 1991; Dittus, 1998).

The roots of these inconsistencies are both empirical and theoretical. For example, difficult-to-measure differences in maternal body condition have relied almost exclusively on confounding proxy measures. In addition, however, the TW and LRC models have been taken as competing mutually exclusive alternatives, when in fact they are not. Each model is weighted according to only one aspect of life history and therefore is limited. The critical element of the TW model views differences in maternal fitness primarily as a function of maternal influence on variance in success among sons in mate competition, with only little consideration for additional potential effects owing to other aspects of female competition. The LRC hypotheses, on the other hand, focus on the effects of female resource competition on maternal fitness, and largely ignore those involving male–male competition for mates. In short, the TW model rests on male and the LRC on female life-history strategy.

The effects of TW and LRC occur concurrently not alternatively

Maximizing numbers of grandoffspring through sons taxes a different set of maternal abilities than achieving this through daughters because of sex differences in survival and reproductive strategy. Therefore, in allocating resources towards production of one sex or the other, a mother would be expected to produce that sex of offspring which has the highest fitness returns within the constraints imposed by her physical as well as social abilities in a particular environment. Given that male mate competition is universal in polygynous mammals we would expect the effects predicted by the TW model to be equally ubiquitous. The additional (not alternative) effects of LRC would be expected in only those polygynous mammalian societies (including most primates) where LRC is part of their lifestyle. In the phylogeny of species where male–male competition for mates and female ability to monopolize resources – for themselves, their offspring and kin – affect individual fitness, we would expect the effects of TW and LRC to co-occur. In other words, TW and LRC effects act concurrently and in concert at any one point in time in reference to the balance between a female's physical (body condition) and behavioral (competitive) qualities (Dittus, 1998). Using different arguments, a similar suggestion had been made by van Schaik and Hrdy (1991), but their conclusion that LRC and TW are the products of *counteracting* selection is not consistent with the phylogeny or operation of TW and LRC effects.

The interplay of these two effects was demonstrated in toque macaques, where maternal condition (ability to invest bodily resources in offspring) was dependent upon rank (ability to compete for resources) as well as environment (Dittus, 1998). By itself, rank had no effect on sex ratios. However, rank effects were strong when viewed in conjunction with those of condition: high-ranking mothers in robust body condition produced predominantly sons (as expected from TW), whereas those in lesser condition produced mostly daughters (in support of both TW and LRC predictions). The effects of LRC were strongest among low-ranking females in poor condition, which produced mostly sons. Measuring only the effects of body condition (irrespective of rank), robust mothers overproduced sons whereas those in poor condition produced more daughters (in support of TW). Following environmental events that led to food shortages, a greater proportion of females than normal was in poor condition, leading to an underproduction of females, which are most vulnerable to resource competition. On a broader interspecific comparative scale, we might expect a prevalence of robust females in growing food-provisioned colonies of macaques where the effects of LRC are ameliorated. At this level, van Schaik and Hrdy (1991) showed that the proportion of males produced by high-ranking females – compared to that of low-ranking ones – increased with population growth rate.

Resource allocation by sex per generation

The natural selection of facultative adjustment of birth sex ratios by macaque mothers in relation to their physical and social abilities underscores the importance to individual fitness of resource allocation between the sexes. It follows that such discrimination: (1) might continue after offspring birth; and (2) also involve close relatives whose inclusive fitness components may be affected differently by the growth and survival of male and female relatives (Dittus, 1979). A bias in the birth of offspring of one sex over another – or in the promotion of the survival to mean reproductive age of one sex over another among relatives already born – has important theoretical implications. Based on considerations of the evolution and genetics of sex, Fisher (1958) demonstrated that, in sexually producing organisms, total investment in male and female progeny or generations must be equal. I had explored (Dittus, 1979) this prediction in toque macaques using the following arguments. Where, as is the case for many macaques, social members in large part influence the production and survival of the young, the standing population can be used to reflect how life-supporting resources have been allocated (by social members in control) among individuals of different age and sex. This would be particularly true for a population at carrying capacity, whose size (i.e., fecundity and survival rates) is held in check largely by limited resource availability. Leaving aside the relatively unpredictable risks of unsual predation and disease, it is assumed that the survival of any individual is to a large extent the outcome of investment (e.g., tolerance or care) by its relatives in resource competition. Investment is used in the sense of Trivers (1972) as any behavior towards an offspring or other relative that increases its chances of surviving, at the cost of the parent's – or controlling relative's – ability to invest in others.

Male and female macaques grow at different rates (Cheverud *et al.*, 1992), and to take into account differences in the cost of producing individuals of different body size, I converted the known mean body weights by age and sex into metabolic weight ($kg^{0.75}$). In mammals, metabolic weight is proportional to the basic metabolic rate and food requirements for survival (Kleiber, 1975). Viewed at the level of the cohort, the cumulative product of the number of survivors and their metabolic weights (Fig. 5.4) can be taken as proportional to the energy that

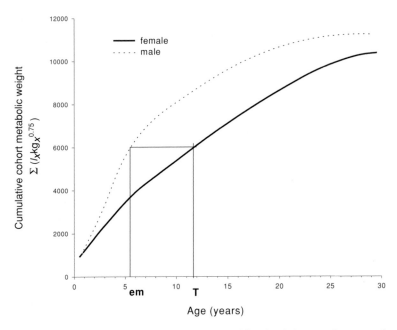

Fig. 5.4. Cumulative investment in male and female relatives or cohorts up to the end
of the period of parental (and family) care, measured in terms cumulative metabolic
weight with age. Total investment in males up to end of care, *em* (mean age of natal
dispersal) by natal group members is equal to that in the mean generation length of
females, $T = 11.84$ years (After Dittus, 1979.)

parents (and kin) must invest in the survival and growth needs of male versus
female progeny (or kin). The age-range and median age at which males permanently leave their natal group is demarcated and indicates the age at which all
investment in them ceases. The ages of females that correspond to an investment
equivalent to that in males can be read from these curves. They indicate that
the total investment in males up to the approximate median age of emigration
(5.5 years), at which point investment in males ceases, corresponds to the mean
generation length ($T = 11.84$ years) of the female cohort. In other words, total
investment in the males and females up to the end of parental care is equal.
This directly supports Fisher's (1958) theory. The mean generation length was
derived independently from life-table information, and may be understood as
the mean age that a female must attain in order to replace herself with one surviving daughter in her lifetime. (Compared to T in zero growth populations, T
is shorter in growing populations and longer in contracting ones.) The relationships in Fig. 5.4 indicate that total investment in the male and female cohorts is
equal per generation, but it is scheduled differently. Investment in males prior

to their emigration is greater than in their female peers, but ceases with their emigration. Females, on the other hand, benefit from the investment of their relatives throughout the mean life of their generation (Dittus, 1979).

Sexual dimorphism, investment, and demographic structure

The comparatively greater investment bestowed on young males rather than females (Fig. 5.4) can be seen as a by-product of intrasexual selection, where large body size or competitive ability in sons – or male relatives – are prerequisites for achieving fitness dividends through them. It follows that the degree of sexual dimorphism should influence investment patterns. For example, in Fig. 5.4, the vertical (ordinate) differences between the male and female curves reflect the degree of sexual dimorphism in morphology and behavior. Conceivably, in species where sexual dimorphism is less marked than in toque macaques, the vertical separation of the two curves would be less. And in almost monomorphic species the curves might nearly coincide. In the latter case, one might expect a reduction in the importance of body size to male–male competition for mates, a relaxation of confining dispersal to males only, and an attenuation of sex differences in mortality that are related to philopatry (LRC) as well as mate competition. On the other hand, in species such as baboons where sexual dimorphism is more strongly developed than in toque macaques, one would expect the vertical separation of the male and female curves (Fig. 5.4) to be greater. Keeping in mind the constraints of equal total investment in the sexes (Fisher, 1958), one would expect the costly production of oversized males to be compensated for by the reduction of investment in other ways, such as an earlier age of male natal dispersal or a reduction in the number of adult males admitted as resident group members – a prevalence of bachelor groups. Some of these life-history features have been observed, for example, in the highly dimorphic patas monkeys (Hall, 1965; Struhsaker & Gartlan, 1970). These arguments were more fully presented in Dittus (1979) and use an approach similar to that applied by Trivers and Hare (1976) in their investigations on the evolution of social insects. The relationships abstracted in Fig. 5.4 suggest that a demographically based analysis of this kind might be a useful tool in understanding the evolution of social organization in primates.

Macaque studies: a prospective hope

Temple sites, township parks, and managed colonies of macaques offer favorable conditions for observing the behavior of macaques, and no doubt for this reason such sites have also provided the vast majority of demographic data for different macaque species. Behaviors that we see today (or do not see because they are not manifested), however, were evolved as adaptive traits under very different social, demographic and ecological conditions than the ones where they have been in large part investigated. In the absence of the evolutionary contexts in which primate behavior evolved, our understanding of it will always be impoverished.

Given the rapidity with which natural habitats for macaque species are disappearing, there is a particular urgency to document macaque life histories in their original natural environments. Of the thousands of studies involving macaques, there is not a single one, known to me, where even natural predation rates have been reliably quantified. Small wonder that the role of predation in the evolution of social behavior is ripe for speculation rather than empirical science. Similarly, the vital statistics of natural macaque populations, even though time-consuming and costly to undertake, are grossly underrepresented. If we are to benefit from a true understanding of our fantastically rich primate heritage, as partly represented by the genus *Macaca*, these empirical lacunae cry for attention.

Box 5   Patterns of group fission

Kyoko Okamoto

Group fission refers to the permanent separation of the members of a social group into two or more new social groups (Struhsaker & Leland, 1988). It has been reported in various primate taxa, and there have been many detailed reports on the process and patterns of group fission in the genus *Macaca* (e.g., *M. fuscata*: Sugiyama, 1960; Furuya, 1968, 1969; Koyama, 1970; Nishimura, 1973; Maruhashi, 1982, 1992; Yamagiwa, 1985; Oi, 1988; *M. mulatta*: Missakian, 1973a; Chepko-Sade & Sade, 1979; Malik *et al.*, 1985; *M. sinica*: Dittus, 1988; *M. thibetana*: Li *et al.*, 1996; *M. sylvanus*: Prud'homme, 1991; Ménard & Vallet, 1993b; Kuester & Paul, 1997; *M. maurus*: Okamoto & Matsumura, 2001). Here, I outline what we know from patterns of group fission in macaques.

In many reported cases of group fission, it took up to a year or more from the subgroup formation to the completion of the fission (e.g., Missakian, 1973a; Ménard & Vallet, 1993b), whereas some group fissions appeared to be completed in a relatively brief period (e.g., Okamoto & Matsumura, 2001). The loss of cohesion has been described in large pre-fission groups (Malik *et al.*, 1985; Ménard & Vallet, 1993b).

Box 5 Patterns of group fission 113

An increase in group size prior to fission has been commonly reported both in provisioned groups (e.g., Furuya, 1969; Li *et al.*, 1996) and in wild groups (e.g., Malik *et al.*, 1985; Dittus, 1988; Ménard & Vallet, 1993b). Surplus males or a higher socionomic sex ratio (the number of adult males per adult female) before fission has been described by some authors (e.g., Furuya, 1969). However, no such tendency was found in a long-term study of wild toque macaques (Dittus, 1988), and an increased socionomic sex ratio in post-fission groups has also been reported (Maruhashi, 1982).

Mothers and offspring usually join the same groups, making matrilineal kin groups a unit of the division. In the cases of group fission in which the matrilineal lineage was known, the original groups had divided into two groups: one that contained basically higher-ranking kin groups, and the other that contained basically lower-ranking groups (Koyama, 1970; Chepko-Sade & Sade, 1979; Dittus, 1988; Oi, 1988; Maruhashi, 1992; Li *et al.*, 1996). In provisioned populations of Barbary macaques, the middle- to lower-ranking matrilines formed the new group, while the lowest-ranking ones remained in the main group along with the high-ranking ones (Prud'homme, 1991; Kuester & Paul, 1997). No such division patterns were found in moor macaques (Okamoto & Matsumura, 2001). Some authors have argued that the split patterns of adult females can be predicted by their association with one another, in such behaviors as grooming, prior to group fission (Oi, 1988). However, the furcation pattern in moor macaques did not agree with the grooming pattern observed during the previous year (Okamoto & Matsumura, 2001).

In seasonal breeding species, the distinction between the mating season and the birth season appeared to have a significant effect on the process of group fission. Different stages of the fission process were recognized in relation to this reproductive seasonality when it took more than 1 year to complete the fission (Missakian, 1973a; Ménard & Vallet, 1993b). Associations between adult males and estrous females seemed to promote the process of fission (Yamagiwa, 1985; Dittus, 1988; Prud'homme, 1991).

Several demographic and environmental factors, for example, sexual competition between males, competition among females for food resources, and loss of cohesion due to oversize groups, have been specified as the factors responsible for the occurrence of group fission. Intensified male–male competition has been observed in provisioned groups of Japanese (e.g., Furuya, 1969) and Barbary macaques (Prud'homme, 1991; Kuester & Paul, 1997), which included a large number of males. In a wild population of Japanese macaques, in which many visiting males from outside the group were seen during the mating season, nongroup males initiated the group fission by associating with estrous females, and by competing with resident males (Yamagiwa, 1985). Male aggression including infanticide might trigger off the fission process (Dittus, 1988). In most cases, however, the females appeared to play a bigger part than the males in the fission process. Reports from natural habitats have shown that the division of females occurred first (e.g., Oi, 1988; Ménard & Vallet, 1993b; Okamoto & Matsumura, 2001), and pre-fission female subgroups included no males or only a few (Dittus, 1988). Males, in particular lower-ranking males, seemed to play a role in promoting the fission process by joining initial subgroups formed by females.

Group members, and females in particular, would benefit from increasing group size in terms of anti-predator defense and between-group contest for food (Wrangham, 1980; van Schaik, 1983). However, a large group size would also lead to strong within-group competition. Increased within-group competition may incur significant costs to group members, and may become an incentive to fission for some individuals. Competition over food can be intensified not only by population growth, but also by a decrease in food availability caused by natural disasters such as draughts or cyclones (Dittus, 1988; Dittus, this Chapter), or by a cessation of provisioning (e.g., Sugiyama & Ohsawa, 1982b). A decline in female reproductive performance prior to fission (interbirth interval: Okamoto, 2004) or recovery after fission (birth rate: Maruhashi, 1982; Malik et al., 1985) also highlights the competition over food which underlies fission.

Different components of within-group competition may result in different patterns of group fission. Within-group contest is the dominance effect, while within-group scramble is the effect of group size with that of dominance removed (see van Schaik, 1989; Sterck et al., 1997). Within-group contest occurs when food patches are clumped, and when access to the patch can be monopolized. If within-group contest predominates, the food intake of lower-ranking females is much lower than that of higher-ranking ones. Lower-ranking females might also be the target of more direct aggression from higher-ranking females as the group grows. Thus, lower-ranking females would suffer from larger costs with increasing group size than would higher-ranking females. Group fission is expected to be caused by the desertion of subordinate females from the original group. In other words, original groups are expected to divide into two groups, one that contains higher-ranking females, and one that contains lower-ranking females (Dittus, 1988; van Schaik, 1989). This prediction agrees with the general tendency reported in macaques. The notable exception in moor macaques may suggest the lesser importance of within-group contest during the fission process. This is consistent with the results on proximity among group members during feeding (Fig. 5.5) (Matsumura & Okamoto, 1997), and interbirth intervals prior to the fission (Okamoto, 2004).

Box 5 Patterns of group fission   115

Fig. 5.5. Moor macaques at feeding in a fig tree (Karaenta Nature Reserve, Sulawesi). In this species, females may equally associate with relatives and non-relatives. (Photograph by R. Seitre.)

An alternative explanation, which is not mutually exclusive with the socioecological hypothesis described above, may be possible at a proximate level. If adult females of some species have less preference for relatives over nonrelatives as association partners (Thierry, 1990a), fewer clusters of matrilineal kin groups are likely to develop. This may result in an irregular sorting of kin groups during the fission process.

At present, most of the existing reports of group fission have been obtained from a few well-known species. Information from less well-known species could contribute greatly to our general understanding of how variations in macaque societies arise.

6 Gene flow, dispersal patterns, and social organization

HÉLÈNE GACHOT-NEVEU AND NELLY MÉNARD

Introduction

Many mammal species live in socially structured populations, which are subdivided into breeding groups maintained by sex-biased philopatry and dispersal (Greenwood, 1980; Pusey, 1987; Pusey & Packer, 1987). In mammals, females generally constitute the philopatric sex, staying in their natal group during their entire life. By contrast, males disperse from their natal group or range to another group or place on one or more occasions prior to or after breeding. Several explanations have been proposed to account for the predominance of male-biased dispersal in polygynous mammals: (1) adolescent males may be ejected from their natal group by resident adult males (Moore, 1984; Moore & Ali, 1984); (2) the potential benefits of dispersal are usually higher for males than females (Clutton-Brock & Harvey, 1976; Packer, 1979; Waser *et al.*, 1986); and (3) the costs of dispersal are generally higher for females because of their lower age at first breeding (Johnson, 1986), their more stringent energetic requirements (Budnitz & Dainis, 1975; Pusey, 1987), and their dependence on coalitions with relatives (Pusey & Packer, 1987). Berenstain and Wade (1983) suggest that females may solicit sexual activity with strange males to ensure that they and their younger female relatives will avoid intolerable levels of inbreeding. This behavior may have also been selected due to its beneficial effect on the genetic variability of females' offspring, although the importance of such variability may be minimal for long-lived ecological generalists (Kurland, 1977).

Groups are dynamic entities where social strategies lead to differences in reproductive success between individuals and to variable patterns of dispersion. Then, the social dynamics of groups condition the distribution of gene diversity within populations and determine a population's genetic structure.

Macaque Societies: A Model for the Study of Social Organization, ed. B. Thierry, M. Singh and W. Kaumanns. Published by Cambridge University Press. © Cambridge University Press 2004.

Chesser (1991a) theoretically evaluated the effect of female philopatry on gene diversity within a population. Female philopatry leads to social groups composed of one or several matrilines (nuclei of related females). In such groups, the distribution of gene diversity within and among matrilines differs from classically defined demic groups. Female philopatry and polygyny imply the formation of groups that represent a nonrandom sampling of the gene pool of adult females and adult males respectively. The subsequent gene distribution may result in an excess heterozygosity within lineages while population heterozygosity does not greatly change. The genetic divergence between matrilines depends on the lineage size and on the number of male breeders per lineage. One can expect the highest degree of relatedness within lineages or within cohorts of offspring and the highest genetic differentiation between groups when only few dominant males monopolize access to fertile females within lineages or groups (Smith & Small, 1987; Chesser, 1991a). Levels of relatedness among breeding females within a group will be much higher than those between randomly chosen females from different groups. Some of these theoretical developments received empirical confirmation in baboons where a high but short-term variance in reproductive success of dominant males led to a high relatedness among infants of the same age-cohort and social groups to be genetically substructured by age (Altmann *et al.*, 1996). Primates display various degrees of philopatry depending on the species. An extreme asymmetry of dispersal is found in macaques where females stay in their natal group during their entire life while males generally disperse before sexual maturity. Long-term studies of different species of macaques have found that social groups in these species are organized around stable female lineages (*M. fuscata*: Itani, 1975; *M. mulatta*: Sade *et al.*, 1976; *M. sylvanus*: Ménard & Vallet, 1993a), as is common in many mammals (Eisenberg, 1977).

In their social organization, macaque societies range from highly hierarchical to tolerant (Thierry *et al.*, 2000). Such differences are expected to induce differences in the variance of reproductive success of individuals and patterns of migration. Thus, they could be responsible for differences in the genetic diversity within and between groups and different species of macaques. This chapter explores the relations between different components of group social dynamics – such as reproductive and dispersal strategies – and the distribution of genes through populations in macaques. First, consequences of dominance rank and inbreeding avoidance on intragroup gene flow are explored. Then, intergroup gene flow is described through two mechanisms: male dispersal and group fission. A comparison will also be made between species, and the consequences of these microevolutionary processes on the speciation of macaques will be discussed.

Social organization and intragroup gene flow

While female philopatry could reduce the gene flow within social groups of primates, some social parameters, such as dominance relationships, mate choice, and inbreeding avoidance, would increase this intragroup gene flow.

Dominance rank and reproductive success

The variance in individual reproductive success is expected to influence the gene flow. The reproductive success of females largely depends on their ability to get sufficient food resources for gestation and to succeed in raising infants that also survive long enough to reproduce. Intragroup social relationships, leading to the hierarchical status of individuals, are potentially influential factors on female access to resources and, further, for their reproductive success. Among macaques, adult females are generally known to form linear dominance hierarchies (Flack & de Waal, Chapter 8; Chapais, Chapter 9). Dominance ranks remain relatively stable over time, with daughters assuming ranks similar to those of their mothers. However, whereas a number of studies have documented reproductive benefits to females of high rank, including reduced interbirth intervals, reduced mortality during times of food scarcity, and increased offspring survival (Manson, 1993), others found that fecundity and infant survival were not much affected by maternal rank (*M. fuscata*: Fedigan *et al.*, 1986; Takahata *et al.*, 1999; *M. sylvanus*: Paul & Kuester, 1996a). Most of the variance in female Barbary macaque reproductive success (61%) can be explained by differences in breeding lifespan, and fecundity was more important than infant survival (Paul & Kuester, 1996a). In fact, precise data on the relation between female reproductive success and dominance rank for larger macaque species are still unavailable.

By contrast, the reproductive success of males depends mainly on access to fertile females (Paul, Box 6; Soltis, Chapter 7). In macaques, group structure potentially allows females to mate with multiple males of their own group or with extra-group males, and males to mate with multiple females. However, male–male dominance relationships or female mate choice may prevent access to sexually mature females by some of the sexually mature males, thus limiting propagation of their alleles. Dominant males may monopolize fertile females while female mate choice might lead many females to mate with a given male over several years, or young females to mate preferentially with young males. Similarly, male mate choice could lead males to mate with females of a single matriline. As mating success is not always correlated with reproductive success

(Fedigan, 1983; Bercovitch, 1986; Ménard et al., 2001), paternity of males can only be known through genetic analyses.

Few studies in macaques provide data on male reproductive success and its determinants, and they rarely give the actual proportion of offspring a male sires relative to the other males living in the same group. Existing results are contradictory regarding the correlation between dominance rank and male reproductive success (Paul, Box 6). Such correlation was found in wild groups of longtailed macaques, where the first-ranking males were responsible for 62 to 81% of paternities (de Ruiter et al., 1994); whereas in captive groups of the same species no such correlation was found (Shively & Smith, 1985). In a captive group of Tonkean macaques, the mean reproductive success of three successive first-ranking males was about 94% (Gachot-Neveu et al., 2004). In rhesus macaques, data on the relation between rank and reproductive success has been contradictory (Berard et al., 1993; Smith, 1993). When the relation was found, it was not known whether this advantage persisted after dispersal (Smith & Smith, 1988). High-born Barbary macaque males reproduced earlier and sired more infants than low-born males (Paul et al., 1992b). However, in none of the four mating seasons examined did the correlation reach statistical significance when the low-ranking subadult males with poor reproductive success were excluded (Paul et al., 1993). Both the first and second higher-ranking males sired about 26% of four infant cohorts (drawn from Paul & Kuester, 1996a). Variance in adult male reproductive success was about three- to fourfold higher than that of females (Kuester et al., 1995). In Japanese macaques, male dominance rank was not correlated with reproductive success (Inoue et al., 1991; Takahata et al., 1999). Possible bias due to studying captive groups or including subadult males in analyses has been suggested to explain differences in results between studies or species (Bercovitch, 1986). However, as subadults are able to fertilize females, excluding them from analyses remains controversial (Colishaw & Dunbar, 1991). To summarize, the relationship between dominance rank and reproductive success seems to depend on group size, demographic conditions, or sampling bias. In addition, for most species, limited paternity tests do not allow any definitive conclusions on the relationship between dominance and reproductive success. Nevertheless, the few available results suggest that variance in male reproductive success is higher in longtailed and Tonkean macaques than in Barbary macaques.

The relationship between reproductive strategies of males and the genetic structure of groups or populations remains poorly explored. In longtailed macaques, a high variance in male reproductive success associated with the preferential mating of first-ranking males with females of higher-ranking matrilines induced a higher relatedness within high-ranking matrilines than within low-ranking matrilines (de Ruiter & Geffen, 1998). It remains to be verified whether similar tendencies exist in Tonkean macaques, which also show a high variance in male reproductive success. By contrast, in Barbary macaques, a lower variance in male reproductive success is associated with low kinship coefficients among infants of the same cohort (Lathuillière, 2002). Thus, the significant intergroup genetic differentiation observed in this species (Segesser *et al.*, 1999) cannot be explained by male reproductive strategies. Noticeable genetic differentiation between matrilines was explained: (1) by male tendencies to mate with females from particular matrilines in rhesus macaques at Cayo Santiago (McMillan & Duggleby, 1981; Melnick & Pearl, 1987); or (2) by preferential mating of matrilineal members with particular patrilines, at least when a large number of matrilines and relatively few patrilines, which includes natal males, are represented in the group (three captive groups of rhesus macaques, Smith & Small, 1987). This occurred despite high rates of male emigration, and is consistent with the maintenance of high levels of within-matriline kinship, which might enhance lineal solidarity and make fission less likely (Chepko-Sade & Olivier, 1979; Olivier *et al.*, 1981). In rhesus groups in Pakistan, on the other hand, small group size and the ability of dominant males to monopolize the majority of fertile females may result in high kinship coefficients across matrilines and among individuals of similar age (Melnick & Kidd, 1983). Such mating patterns may be responsible for higher intergroup differentiation than if matings were less restricted.

Inbreeding avoidance

It has been shown in captive primates that inbreeding can lead to increased homozygosity and to the expression of deleterious alleles, which increases rates of infant mortality and decreases fertility and fitness (Ralls & Ballou, 1982; Flesness, 1986). Inbreeding has also been reported in wild populations of mammals (Ralls *et al.*, 1986; Wildt *et al.*, 1987), but not in primates, or at least not in macaques.

In a semi free-ranging group of Barbary macaques, although males can reproduce in their natal group, only two of 62 potentially inbred infants were actually inbred via the paternal line and mating between close matrilineal relatives almost never occurred (Paul & Kuester, 1985; Kuester *et al.*, 1994). Wild groups of rhesus macaques were also found not to be inbred (Melnick *et al.*, 1984a). A high gene flow among groups and avoidance of consanguineous matings throughout the population produced an excess of heterozygotes, and the five groups studied were genetically similar. The time a male spends in a troop is positively correlated with social rank and probably with reproductive success (Meikle & Vessey, 1981), which limits inbreeding in patrilines.

Studies on several captive macaque groups have shown that mating with close relatives or with familiar individuals is generally avoided (*M. mulatta*: Sade, 1968; *M. fuscata*: Inoue *et al.*, 1992; *M. sylvanus*, Paul & Kuester, 1985; Smith, 1995), and sex-biased dispersal in the field would further limit the potential for inbreeding. Some matings between maternal half-siblings have been reported in free-ranging groups (Loy, 1971; Smith, 1986a). However, it may be established that close inbreeding is usually avoided (Bixler, 1981; *M. mulatta*: Sade, 1968; Missakian, 1973b; *M. fuscata*: Itoigawa *et al.*, 1981; *M. arctoides*: Murray & Smith, 1982; *M. sylvanus*: Paul & Kuester, 1985; *M. sinica*: Keane *et al.*, 1997; Dittus, Chapter 5).

In short, mechanisms preventing inbreeding in macaques are sex-biased dispersal, relatively short non-natal group male tenure, seemingly random direction of male migration, natal migration at the age of puberty that separates males and females of the same matriline, intergroup transfer, and active behavioral avoidance of mating among co-residing close kin (Melnick *et al.*, 1984a; Moore & Ali, 1984; Paul & Kuester, 1985; Kuester *et al.*, 1994).

Intergroup gene flow

Macaque populations are commonly described as sets of social groups occupying adjacent home ranges and exchanging individuals and genes and thus exhibiting marked genetic similarities (Melnick, 1988). Gene spreading flow, due to group fission and male transfers, can influence the differentiation of allele frequencies among, and distribution of genotypes within, social groups (Chesser, 1991b). Gene flow is generally estimated as the "effective number of migrants per generation," namely the product of local effective population size and migration or dispersal rate (Wright, 1978; Slatkin, 1987; Beerli & Felsenstein, 1999). Ober and collaborators (1980) described the proportional contributions of demographic components to total allele frequency change. Male migration has the greatest effect on gene allele frequency change, reaching at least 50%.

Male dispersal

When linked to subsequent reproductive success, dispersal, including transfer among subpopulations in a subdivided system, is a mechanism for gene flow (Crow & Kimura, 1970; Melnick, 1987; Melnick & Pearl, 1987). While intergroup transfer by males is routine in macaques (Kawamoto *et al.*, 1981; Bercovitch & Harvey, Chapter 4), it is rare in females (Moore, 1984; Dittus, 1986). It has been reported, however, that females may occasionally disperse in rhesus and Japanese macaques, and also in bonnet macaques (Moore, 1984).

Fig. 6.1. A bachelor male liontailed macaque (Anaimalai Hills, India). Dispersal is a risky life event, which remains poorly documented in macaques. (Photograph by B. Thierry.)

Dispersing is a potentially risky event. Indeed, in bonnet macaques male mortality reaches a peak in adolescence, which coincides with the age of migration (Dittus, 1977, Chapter 5). Thus, dispersal is expected to increase individual fitness. Population level dispersal patterns probably reflect results in the balance between the advantages of philopatry and the costs of intrasexual competition (Moore & Ali, 1984). Subordinate males may maximize their access to matings by transferring to groups in which more matings are available (Paul & Kuester, 1985) or in which they have a better chance to reach a higher rank than in their previous group (Harcourt, 1978; Pusey & Packer, 1987). Thus, social rank can influence the patterns of male dispersal among groups.

Relatively little is known about the dispersal patterns of wild macaques (Fig. 6.1). Studies on different macaque species have mentioned that after the natal dispersal, males often migrate among non-natal groups several times during their lives (*M. mulatta*, Melnick & Pearl, 1987; *M. sinica*: Dittus, 1975; *M. sylvanus*: Paul & Kuester, 1985). They can migrate in the company of maternal or paternal half-brothers (Paul & Kuester, 1985) or transfer into the same social groups as their older maternally-related brothers (*M. mulatta*: Meikle & Vessey, 1981; *M. sylvanus*: Paul & Kuester, 1985; Ménard & Vallet, 1996) as is common in many primate species (Pusey & Packer, 1987). The main difference between species is variation in age at first transfer. In some species, most males leave their natal group before reaching sexual maturity (Melnick, 1988), whereas in other species up to 50% of males stay in their natal group after reaching sexual maturity and then participate in matings in their natal group (*M. sylvanus*: captive groups, Paul & Kuester, 1985; wild groups, Ménard & Vallet, 1996). They often migrate to neighboring groups (Melnick, 1987; Ménard & Vallet, 1996).

In accordance with global female philopatry and male dispersal, a lower genetic relatedness among males than females was reported in wild groups of longtailed (de Ruiter *et al.*, 1994) and Barbary macaques (Lathuillière *et al.*, 2004). Migration between groups is the most important factor contributing to local changes in allele frequencies, as shown in a study conducted over 4 years in free-ranging rhesus macaques (Ober *et al.*, 1984). Consequently, the frequency of inbreeding is probably small in macaques and social groups are largely outbred (Melnick, 1987), even in Barbary macaques that show delayed natal dispersal. In this latter species, maximal avoidance of consanguineous matings within social groups is well developed (Paul & Kuester, 1985). Gene flow between longtailed macaque groups likely explained the high degree of relatedness observed between matrilines of adjacent groups (de Ruiter *et al.*, 1998). Male migration is generally uniformly distributed among adjacent social groups (Melnick *et al.*, 1984a). However, preferential migration into immediate neighboring groups explained correlations between geographical distance and genetic distance (*M. sylvanus*: von Segesser *et al.*, 1999). Migration of closely-related males into the same group where they reproduce could increase the degree of relatedness within groups. However, the presence of many unrelated males in the new group appears sufficient to dilute their average relatedness, as found in longtailed macaques (de Ruiter & Geffen, 1998).

Group fission

While male transfer tends to homogenize the genetic structure of macaque populations, group fission tends to segregate close maternal relatives, which increases the degree of relatedness within groups (Moore, 1984). But group fission gives rise to founders of new populations and may accelerate genetic differentiation within a population (Melnick & Kidd, 1983). Group fission and fusion then, not only produce the most dramatic short-term changes in allele frequencies due to lineal effects, but also cause changes in social relationships that result in increasing rates of male migration (Ober *et al.*, 1980).

Some studies have documented group fission processes (Okamoto, Box 5). In most cases, macaque groups were under observation during too short a time to know genealogies and to have any opportunity to describe this relatively rare event. Group fission was observed mainly in captive colonies (Chepko-Sade & Sade, 1979), sometimes in wild provisioned groups (Prud'homme, 1991;

Kuester & Paul, 1997) and rarely in wild nonprovisioned groups (Ménard & Vallet, 1993b). It is a process by which a parent group divides into two or more distinct daughter groups (Melnick & Pearl, 1987). Although group fission is rare, it is one of the most important dynamic events within a population because it is the main way by which females disperse. When group fission occurs, it commonly results in the separation of matrilines, with maternally related females staying together. By contrast, patrilineal relatives can be distributed by chance into the new groups, and fission may stimulate the dispersal of resident males and an influx of new males (captive *M. mulatta*: Chepko-Sade & Sade, 1979; Ober *et al.*, 1984; wild and provisioned *M. sylvanus*: Ménard & Vallet, 1993b; Kuester & Paul, 1997).

The lineal effect of group fission has been well documented in rhesus groups living on Cayo Santiago Island. It was shown that group fission was responsible for more than 10% of total gene allele frequencies in groups and explained intergroup genetic differentiation, while male migration and genetic drift counteracted this effect, being responsible for at least 50% of change in allele frequencies in groups (Cheverud *et al.*, 1978; Ober *et al.*, 1984). One of the effects of group fission is to raise the average degree of relatedness within daughter groups compared with the original group when members of this latter group fall below the cousin level (*M. mulatta*: Chepko-Sade & Olivier, 1979; Melnick & Kidd, 1983; wild *M. sylvanus*: Lathuillière, 2002; Lathuillière *et al.*, 2004). It has been documented that after group fission in Barbary macaques, mean relatedness among females within daughter groups reached cousin or half-sibling levels depending on the group (Lathuillière, 2002; Lathuillière *et al.*, 2004). Interestingly, the effects of fission on variation in highly heritable morphological traits among eight social groups of rhesus macaques indicated strongly negative autocorrelation coefficients, as groups most recently formed by fission were most dissimilar (Cheverud & Dow, 1985).

In groups that underwent fission and fusion, the timing of these processes was related to the largest short-term changes in allele frequencies (*M. mulatta*: Ober *et al.*, 1984, Smith, 1986a). However, immigration and emigration had the greatest long-term effect (Kawamoto *et al.*, 1981).

Consequences for genetic diversity

Diversity among macaque species

The greater the level of gene flow and the more random and widespread its distribution, the more likely it is that every regional population, local population, and social group will have a complete representation of the species gene diversity (Nei, 1977).

Lewontin (1974) indicated that the average heterozygosity in species of animals is about 10%, but this number does not illustrate with exactitude the fact that heterozygosity, both expected and observed, is vastly different for different sorts of genetic markers. Low levels of heterozygosity have been observed in large mammals such as pongids, elephant seals, moose and elk (Ryman *et al.*, 1980). In the Asian macaques, the level of genetic variability varies among species; the rhesus and longtailed macaques maintain genetic variability more than other macaque species (Prychodko *et al.*, 1969; Ishimoto, 1973). Population structure in the ancestral lineage as well as evolutionary events (drift and selection) may influence the level of genetic variability in descendant populations.

Wright (1951) introduced a quantitative framework for investigating the hierarchical population structure of genetic variability through the use of fixation indices or F-statistics: F_{IS} indicates the deviation from Hardy–Weinberg equilibrium averaged within local populations; F_{ST} indicates the differentiation in allele frequency between local populations; and F_{IT} indicates deviation from Hardy–Weinberg equilibrium over the total population.

Nei's (1973, 1975) gene-diversity analysis extended this approach to loci with multiple alleles and provided one way to apportion the genetic variation within a species among the levels at which a species is subdivided. Among macaque species, one can distinguish at least five hierarchical levels: the species, the "regional population," the "local population," the social group and the individual. Gene flow tends to increase genetic homogeneity within local and regional populations (Melnick *et al.*, 1986).

For five macaques species (*M. mulatta, M. fuscata, M. fascicularis, M. sinica* and *M. nemestrina*: Melnick, 1988), allozyme data indicate very little genetic difference among regional populations and among social groups in the same local population. In contrast, the differences between local populations within the same region constitute almost one-quarter of the genetic diversity in the species. And, for example in rhesus macaques, 59.1% of the total species gene diversity can be apportioned to genetic differences between members of the same social group (Melnick *et al.*, 1984b).

Table 6.1 compares the genetic variability for different macaque species in calculating average heterozygosities H_T and the fixation index F_{ST}, on the basis of comparable protein markers. Kawamoto (1982), Nozawa *et al.* (1982), and Hayasaka *et al.* (1987) used the same blood protein loci, as well as most of the other studies reported by Melnick *et al.* (1984b) and Melnick (1988). There is no evident relation between the social grade, phylogenetic lineage and heterozygosity, but we distinguish three major clusters.

Table 6.1. *Range of genetic variability measured by average heterozygosity (H_T) and inter-population differentiation (F_{ST}) among different Macaca species. Social grades refer to the classification proposed by Thierry (2000), ranging along a gradient from highly hierarchical and nepotistic species (grade 1) to tolerant species (grade 4)*

Species and lineage	Social grade	Country	Number of loci	H_T	F_{ST}	Sources
silenus–sylvanus lineage						
M. sylvanus	3	Algeria	23	0.031–0.083	0.113	1
M. nigra	4	Sulawesi	31	0.033–0.107	–	2, 3
M. nigrescens	4	Sulawesi	31	0.040	–	3
M. hecki	4	Sulawesi	31	0.085–0.109	–	2, 3
M. tonkeana	4	Sulawesi	31	0.052–0.053	0.250	2, 3, 4
M. maurus	4	Sulawesi	31	0.052–0.054	0.000–0.067	3, 4
M. ochreata	4	Sulawesi	31	0.066	–	3
M. brunnescens	4	Sulawesi	31	0.055	–	3
M. nemestrina	2	Thailand, Malaysia, Sumatra	25	0.076	–	2
sinica–arctoides lineage						
M. sinica	3	Sri Lanka	32	0.071–0.087	0.029–0.069	2, 5, 6
M. radiata	3	India	–	0.078	–	2
M. assamensis	3	India	–	0.092	–	2
fascicularis lineage						
M. fascicularis	2	Thailand, Philippines, Indonesia, Malaysia	22–33	0.012–0.096	0.083–0.293	2, 7, 8, 9, 10
M. mulatta	1	China, Thailand, India, Pakistan	29	0.053–0.080	–	9, 11
M. fuscata	1	Japan	29–33	0.001–0.044	0.006–0.086	2, 9, 10, 12, 13, 14
M. cyclopis	1	Taiwan	23	0.044	–	2

Sources: (1) Scheffrahn et al., 1993; (2) Melnick, 1988; (3) Kawamoto, 1996; (4) Evans et al., 2001; (5) Shotake & Santiapilai, 1982; (6) Shotake et al., 1991; (7) Kawamoto et al., 1984; (8) Kawamoto et al., 1989; (9) Kawamoto et al., 1982; (10) Kawamoto et al., 1981; (11) Melnick et al., 1984b; (12) Hayasaka et al., 1987; (13) Nozawa et al., 1982; (14) Nozawa et al., 1991.

Only one macaque species exhibits a very low heterozygosity: between 0.001 and 0.044 for Japanese macaques (Table 6.1). Nozawa and collaborators (1982) conclude that geographical distribution of the genetic variation in Japanese macaques was not uniform, but variants occurred only in limited areas, and the genetic diversity is remarkably lower than those estimated for other species. The proportion of polymorphic loci is 9.1% on average, and the average heterozygosity per individual is 1.3%. Selander (1970) tabulated the results of quantification of genetic variability in populations of several animal species examined by different biologists. Genetic variability in the Japanese macaque troops is about the same as in some animal species inhabiting narrow, closed environments. The Japanese macaque population has a structure capable of being influenced by random genetic drift, which is considered responsible for the low variability within each troop and the marked genetic differentiation between troops, as well as in other macaque species. Neighboring groups of Japanese (Nozawa *et al.*, 1982) and rhesus macaques (Melnick, 1983) within local populations are found to be genetically similar compared with groups that were geographically very distant, which confirms that dispersal happens mainly among neighboring groups. The tendency to split may be considered essentially as a product of the troop-making social behavior of this species, and is perhaps also due to the fact that the natural habitats have contracted because of growing human activity. A characteristic distribution pattern of variant alleles of blood proteins is that most of the variable loci showed remarkable polymorphism in most of the troops, which is different from the distribution of variant alleles of the Japanese macaque.

The highest heterozygosities are found in longtailed, toque, bonnet, Assamese, and Hecks macaques. The influence of social organization on the genetic structure of a population can be strong; group size, reproductive and dispersal patterns all influence the genetic structure of the population. In long-tailed macaques, migrating males reproductively connect groups of resident females. F_{IS} is consistently negative because genetic drift causes differentiation in allele frequencies between groups, and due to this differentiation, allele frequencies differ between resident females and immigrant males, leading to offspring with an excess of heterozygotes (de Jong *et al.*, 1994). Low values of H_T are found in islands where bottleneck effects are strong (Kawamoto *et al.*, 1981).

The genetic variability of toque macaques is high, and is about the same as that of longtailed macaques in Thailand, a continental macaque (Kawamoto *et al.*, 1989). The F_{ST} values of toque macaques are smaller than those obtained in Japanese macaques (Table 6.1), which are both endemic to islands. Genetic differentiation between troops of toque macaques is not very marked and the alleles in the populations of toque macaques have been well mixed throughout the whole island (Shotake *et al.*, 1991).

The other species present mean heterozygosities ranging from 0.038 to 0.080 (Table 6.1). Although the mean heterozygosity of Barbary macaques was higher than for Japanese macaques, the former present a low genetic variability compared to other macaque species, which could be explained by successive bottlenecks (Scheffrahn *et al.*, 1993). F_{ST} is also relatively high (11%), which illustrates a differentiation among the Algerian populations (Table 6.1). Although the mean heterozygosity was comparable within most Sulawesi macaques, some great differences were observed among F_{ST} values. Although highly structuring in mtDNA, nuclear DNA analyses demonstrated that subdivisions among moor macaque populations are very low compared to those of Tonkean macaques (Evans *et al.*, 2001).

Macaques speciation

While species of macaques can be distinguished from each other by their morphological traits, some of them are sympatric and interbreed (Fooden, 1964; Bernstein & Gordon, 1980a; Eudey, 1980), they may be in the process of homogenizing. The number of chromosomes of all macaque species is 42. This genetic similarity within the genus facilitates hybridization between different species. Interspecific hybrids, usually viable and fertile, have been reported for a number of macaque species (Bernstein & Gordon, 1980a; Fa, 1989; Wong & Ni, 2000; Evans *et al.*, 2001). This phenomenon may have substantial consequences for macaque evolution.

Social organization influences microevolutionary change (Storz, 1999). Correlation between rank and reproductive success is consistent with strong intrasexual selection and may accelerate the rates of some evolutionary processes by reducing the effective population size (N_e) (Hoelzer *et al.*, 1998). The social subdivision, inbreeding, and reduced N_e of nonhuman mammalian populations (Kawamoto *et al.*, 1981; Lucotte *et al.*, 1984) have been suggested as an explanation for rapid rates of chromosomal evolution and speciation among mammals (Melnick *et al.*, 1984a). Especially rapid rates of chromosomal change and species formation among primates have been singled out as excellent examples of the presumed genetic effects of sociality (Wilson *et al.*, 1975; Bush *et al.*, 1977).

Local genetic differentiation may be an important evolutionary factor in mammalian populations. According to the Bush–Wilson hypothesis, joint effects of drift and inbreeding in socially structured populations facilitate fixation of underdominant chromosomal rearrangements, thereby establishing a postmating isolation mechanism that promotes stasipatric speciation (White, 1978; Bush, 1981; Sites & Moritz, 1987). In this model, reproductive isolation is enforced by reduced fecundity of chromosomal heterozygotes due to meiotic malassortment and impaired gametogenesis (Storz, 1999).

Since lineal fission is important in the production of significant amounts of intergroup variation, it is a major factor in the evolutionary dynamics of primate populations (Cheverud & Dow, 1985). Thus populations may diverge genetically even when exposed to similar selection pressures if the genetic variance/covariance matrix is different among related populations due to founder effects. Although genetic consequences of group fission are not predictable, models should be able to predict that genetic divergences between matrilines might play an important role in facilitating evolutionary events.

Conclusion

As stressed by Melnick and Pearl (1987) for cercopithecines, several behavioral and ecological factors may have opposite consequences on gene flow and genetic structure in macaques: (1) Nuclear genetic diversity is relatively evenly distributed throughout any local population of groups, apparently as a result of male migration. Male migration tends to homogenize nuclear genetic diversity throughout local populations. (2) Macaque groups are highly socially structured, so they tend to be genetically differentiated and inbred. Because they contain clusters of relatives, social groups are distinctive in their specific frequencies of alleles. In many cases, the degree of genetic differentiation among groups is greater than might be expected if migration patterns were completely random. (3) Sex-biased dispersal prevents inbreeding both at the group level and population level. (4) Group fission appears to contribute to genetic differentiation, but it is also responsible for the genetic similarity among neighboring groups in mtDNA.

The genetic consequences of a large range of social organization in macaques cannot easily be studied in natural conditions. Chesser (1991a) and Hoelzer and collaborators (1998) proposed different models showing the effects of a highly structured social organization, gene flow and breeding tactics on gene diversity within groups. Both studies confirm the important role of female philopatry and inheritance of female dominance on the genetic differentiation within groups, rather than the gene flow among groups.

We may predict that species with a socially structured population in nature are particularly vulnerable to severe inbreeding depression upon disruption of the

Box 6 Dominance and paternity 131

social organization, as when brought together into one population in a single zoo (de Jong *et al.*, 1994). This should have considerable consequences for genetic management of animals both in captivity and in the wild (Wong & Ni, 2000).

Box 6 Dominance and paternity

Andreas Paul

Why did hierarchies evolve? There are two possible answers to this question. According to the first hypothesis, which was championed by early etholo-gists (e.g., Lorenz, 1963), complex social groups could not exist without a hierarchical order because in the long run members of such groups would ruin themselves and their society in endless fights over access to scarce resources. In this view, all individuals, even the lowest-ranking ones, ben-efit from a hierarchical organization because it guarantees a peaceful life. But while it is true that in established dominance relationships the risk of escalated aggression is often greatly reduced (e.g., Preuschoft & van Schaik, 2000), it has also long been suspected that the relative benefits and costs of being at the top or the bottom of a hierarchy may be highly asymmetrically distributed. For example, in his classic study on sexual behavior in rhesus macaques, Carpenter (1942) found a highly significant correlation between male dominance and sexual activity. Moreover, when only one female was in estrus, he noticed that "she is usually possessed by the male with highest dominance status" (p. 158). "A low dominance status," Carpenter concluded, "is one in which the expression of heterosexual motivation is to some degree frustrated, restricted or prohibited" (p. 160).

Although Carpenter's observations led to the formulation of dominance-based priority-of-access models (Altmann, 1962; Suarez & Ackerman, 1971), the evolutionary significance of his findings was not immediately obvious, however, and Darwin's initial note that sexual selection is "less rigourous than natural selection" since "the result is not death to the unsuc-cessful competitor, but few or no offspring" (Darwin, 1999 [1859]: 74) clearly contributed to the long neglect of his theory of sexual selection. But in fact, restricted access to mates, or "few or no offspring," do not only affect an individual's emotional state, but also, and more importantly, its fit-ness: differential reproduction is the only way natural (and sexual) selection works. This insight neatly led to the second hypothesis, after which hierar-chies evolved because dominance might enable individuals to produce more offspring than their competitors (e.g., Clutton-Brock & Harvey, 1976).

Despite several decades of research on the relationship between dominance and paternity, the issue remains a controversial one (Gachot-Neveu & Ménard, this Chapter). In a thorough review on the subject, Ellis (1995) concluded that "the accumulated evidence supports the view that dominance hierarchies have evolved because they have reproductive consequences" (p. 305), but he also acknowledged that the situation "is particularly complex and equivocal in the case of primates" (p. 304). Exactly this led other authors to maintain that "social stratification probably evolved as a mechanism for reducing the probability of escalated aggressive encounters by providing a reasonable degree of predictability and stability to social relationships" (Bercovitch, 1991: 315).

Because of their diversity in hierarchical relationships and sexual and reproductive behavior, macaques represent a particularly interesting taxon for testing the association between dominance and paternity. Moreover, much more data on dominance and paternity have been accumulated for macaques than for any other genus of nonhuman primates, and most nonassociations have been reported from this taxon (Table 6.2). In fact, almost all other recent studies combining genetic and behavioral data revealed that dominance is tightly linked to paternity (Hanuman langurs: Launhardt et al., 2001; sooty mangabeys: Gust et al., 1998; savanna baboons: Altmann et al., 1996; bonobos: Gerloff et al., 1999; common chimpanzees: Constable et al., 2001). So, do macaques violate the general rule, and if so, why? Given that some macaque species appear to be much more egalitarian than others (e.g., Thierry, 2000), it might be argued, for example, that the association between paternity and dominance may be also more relaxed among these more egalitarian species. Alternatively, a variety of other factors such as local ecological conditions, numbers of male competitors, alternative male mating tactics, effects of female choice, extended periods of female receptivity, polyandrous mating, and/or the degree of breeding seasonality may offset conventional priority-of-access models (e.g., Cowlishaw & Dunbar, 1991; Altmann et al., 1996; Takahata et al., 1999; Heistermann et al., 2001; Soltis, Chapter 7).

Cross-species comparisons indeed suggest that breeding seasonality strongly affects male monopolization potential (Oi, 1996; Paul, 1997), although the crucial variable is not breeding seasonality per se, but the probability that there will be two or more females simultaneously in estrus on any given day (e.g., Dunbar, 2000). The currently available data provide further support for this hypothesis (Table 6.2): almost all studies on nonseasonal species found a strong relationship between dominance and paternity, while among seasonal species the association was much more variable. In fact, only two studies on nonseasonal species reported negative results: in

Table 6.2. *Summary of genetic studies on paternity and dominance in Macaca*

Species	Social grade[a]	Breeding seasonality[b]	Context of the study population[c]	Number of paternities	Number of potential fathers	Correlation rank-RS[d]	Sources
M. sylvanus	3	2	Captive: 1 group, 5 years	32	2–3	(+)[e]	1
			Free: 1 group, 4 years	75	16–33	(+)[f]	2
M. tonkeana	4	0	Captive: 1 group, 4 years	8	2	+	3
M. nemestrina	2	0	Captive: 1 group, 1 year	13	6	0	4
M. arctoides	3	1	Captive: 1 group, 8 years	21	2–7	+	5
M. fascicularis	2	1	Captive: 1 group, 3 years	35	6–7	–	6
			Wild: 3 groups, 3 years	40	2–10	+	7
M. mulatta	1	2	Captive: 1 group, 2 years	25	7–8	(+)[g]	8
			Captive: 6 groups, 3 years	202	2–4	+	9
			Captive: 1 group, 8 years	48	5–10	0	10
			Captive: 3 groups, 1 year	47	5–13	0	11
			Captive: 2 groups, 8 years	282	~10 per group	(+)[h]	12
			Captive: 1 group, 15 years	268	4–18	(+)[i]	13
			Free: 1 group, 1 year	11	11	0	14
			Wild: 2 groups, 2 years	?	2–5	+	15
M. fuscata	1	2	Captive: 1 group, 1 year	26	21	+	16
			Captive: 1 group, 4 years	35	9–10	0	17
			Captive: 1 group, 1 year	13	8	0	18
			Wild: 1 group, 1 year	9	15	+	19

[a] Based on Thierry's (2000) four-grade scale of social organization ranging from highly hierarchical and nepotistic societies (grade 1) to more tolerant/egalitarian ones (grade 4).

[b] Based on van Schaik et al. (1999); 0: < 33% of births in peak 3-month period; 1: between 33% and 67%; 2: > 67% in peak 3-month period.

[c] Captive: captive population; Free: captive, but free-ranging population; Wild: unprovisioned, wild population.

[d] RS: reproductive success; +: positive relationship; 0: no relationship; –: negative relationship.

[e] First-ranking male in 4 out of 5 years more successful than second-ranking male, but due to frequent rank changes both adult males sired similar numbers of offspring.

[f] Significant correlation between rank and RS in 3 out of 4 years, but no significant correlation if analysis was confined to adult males only.

[g] Statistically significant correlation only during second year.

[h] Significant correlation between the males' average rank and RS, but during most single years no significant correlation between their actual ranks and RS.

[i] Highly significant correlation between the males' average rank and RS, but during most single years no significant correlation between their average rank and RS, but during most single years no significant correlation between actual ranks and RS.

Sources: (1) Witt et al., 1981; (2) Paul et al., 1993; Paul & Kuester, 1996; (3) Thierry et al., 1994; (4) Gust et al., 1996; (5) Shively & Smith, 1985; (7) de Ruiter et al., 1992, 1994; de Ruiter & van Hooff, 1993; (8) Duvall et al., 1976; (9) Smith 1981; (10) Curie-Cohen et al., 1983; (11) Stern & Smith, 1984; (12) Smith & Smith, 1988; (13) Smith, 1993, 1994; (14) Berard et al., 1993, 1994; (15) Melnick, 1987; (16) Bercovitch & Nürnberg, 1997; (17) Inoue et al., 1992, 1993; (18) Soltis et al., 1997b; (19) Soltis et al., 2001.

one group of captive longtailed macaques, two low-ranking males sired the majority of offspring (Shively & Smith, 1985; the only negative correlation between male rank and reproductive success ever reported from any nonhuman primate species to date), and in one group of captive pigtailed macaques the fourth-ranking male sired the majority of offspring. However, apart from the fact that in this latter case the highest-ranking male was apparently sterile, this study rather supports the hypothesis because all 13 offspring were born within a very limited period of only four months so that the situation was effectively a seasonal one (Gust *et al.*, 1996).

Within-species comparisons also indicate that among seasonal species the dominant males' monopoly is stronger when few females are mating simultaneously (rhesus macaques: Melnick, 1987; Japanese macaques: Soltis *et al.*, 2001). Taken together, these data strongly suggest that dominance is tightly linked to paternity, but that the number of co-cycling females affects male monopolization potential.

The hypothesis that among the more egalitarian macaque societies the association between dominance and paternity is more relaxed than in despotic societies does not seem to be well supported. Both among Tonkean and stumptailed macaques (grades 4 and 3; see Thierry, Chapter 12) the limited available data indicate that dominant males are at least as able to effectively monopolize paternity as dominant males from more despotic species (see Table 6.2). At least in sexual matters, therefore, males from "egalitarian" species appear to be just as dominance-oriented as males from "despotic" species, although small power asymmetries and/or conditions favoring scramble competition over access to co-cycling females with prolonged mating periods may sometimes result in effectively more egalitarian relationships (Preuschoft & Paul, 2000, for male Barbary macaques).

It seems reasonable to conclude from these data that among macaques, as well as among other species, dominance is the single most important factor affecting male reproductive success. Partly because the reproductive interests of the sexes are not identical and females tend to mate polyandrously (e.g., van Schaik *et al.*, 2000; Soltis *et al.*, 2001), dominance does not, however, explain all the variance in male reproductive success. Moreover, short-term differences in reproductive success do not necessarily translate into differences in lifetime reproductive success (Altmann *et al.*, 1996), although among macaques male dominance hierarchies tend to be much more stable over time than among baboons (e.g., Bercovitch & Nürnberg, 1997).

7 *Mating systems*

JOSEPH SOLTIS

Introduction

The multimale, multifemale social organization of macaques provides for the complex interaction of male and female reproductive strategies observed in this genus. It is not surprising, therefore, that the mating system of macaques has been also described as polygynous, polygamous and promiscuous (Dixson, 1998: 37). Male reproductive strategies may include direct contest competition over access to females, endurance rivalry, sperm competition, and sneak copulation. Female reproductive strategy may incorporate both selective mate choice and mating with multiple males, and females mate both when conception is likely and when it is unlikely or not possible.

Here I review male and female reproductive strategies and their interaction. I explore outstanding controversies, such as the strength of male dominance rank and female mate choice in determining male reproductive success, the function of multiple-mount versus single-mount copulatory patterns, and the function of female sexual swellings. Additionally, I attempt to relate male and female reproductive strategies to the variable dominance styles observed across macaque species. Much of the discussion that follows relies on incomplete evidence, but conclusions based on current evidence are offered, and areas for further investigation are suggested. I approach male and female reproductive strategies from a behavioral perspective, emphasizing general patterns in adults (for reproductive life history, see Bercovitch & Harvey, Chapter 4).

Male reproductive strategy

Male dominance rank and reproductive success

Historically, the study of reproductive strategy in macaques and other primates has focused on competition among males over access to females (Altmann,

Macaque Societies: A Model for the Study of Social Organization, ed. B. Thierry, M. Singh and W. Kaumanns. Published by Cambridge University Press. © Cambridge University Press 2004.

1962). Typically, males in a group are ordered into a dominance hierarchy based on the direction of aggressive and submissive behaviors, and correlations between male dominance rank and various measures of reproductive success are obtained (Fedigan, 1983). Although it has become clear that dominant males cannot completely monopolize females, male dominance rank in macaques has an overall positive influence on mating success, just as it does for primates generally. Male dominance rank is usually positively correlated with mating success, with various levels of statistical significance, but rarely negatively correlated with mating success (Cowlishaw & Dunbar, 1991).

In recent decades, observations of mating behavior have been supplemented with DNA paternity studies that reveal the actual number of offspring that males sire. Early DNA studies on captive macaques yielded surprising results, suggesting that male rank has little or no effect on actual reproductive success, and in some cases, a demonstrable negative effect (e.g., *M. fascicularis*: Shively & Smith, 1985). In light of the accumulating evidence, however, it is clear that male dominance rank has an overall positive effect on paternity in macaques (see Gachot-Neveu & Ménard, Chapter 6; Paul, Box 6). In nonwild contexts, most correlations between male dominance rank and number of offspring sired are positive, with variable levels of statistical significance (e.g., *M. fuscata*: Inoue *et al.*, 1990, 1991, 1993; Soltis *et al.*, 1997b; *M. mulatta*: Curie-Cohen *et al.*, 1983; Smith, 1993, 1994; Bercovitch & Nürnberg, 1997; *M. arctoides*: Bauers & Hearn, 1994; *M. sylvanus*: Witt *et al.*, 1981; Paul & Kuester, 1996a), and rarely negative (e.g., *M. fascicularis*: Shively & Smith, 1985; *M. nemestrina*: Gust *et al.*, 1996). The few paternity studies on macaques in wild habitats have yielded positive associations between male rank and number of offspring (*M. fuscata*: Soltis *et al.*, 2001; *M. mulatta*: Melnick, 1987; *M. fascicularis*: de Ruiter *et al.*, 1994).

Figure 7.1 shows the ranges of correlations between male rank and two measures of reproductive success (mating and number of offspring) for several macaque species. Both mating and paternity are positively associated with male rank. This suggests that overall mating success may be a good proxy for actual paternity, although it should be noted that observed mating does not reflect the sire of any particular offspring (Bercovitch & Nürnberg, 1996; Soltis *et al.*, 1997b). In the three macaque species for which there are data, correlations between male rank and mating success are higher than those between male rank and actual paternity. This suggests that observations of mating behavior overestimate the reproductive success of dominant males, and implies additional mechanisms of competition among males, such as alternative mating tactics and sperm competition, as well as the potential counter-influence of female choice (discussed below).

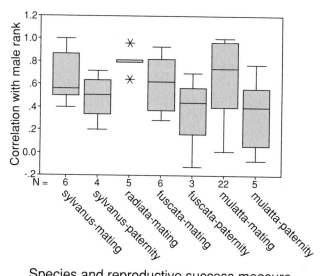

Species and reproductive success measure

Fig. 7.1. Spearman's rank correlations between male dominance rank and two measures of reproductive success (mating and paternity) for four macaque species. Horizontal lines are medians, boxes are inter-quartile ranges, and bars indicate entire range of correlations, unless there are extreme outliers, indicated by asterisks (SPSS, version 10.01). Mating data are from Cowlishaw & Dunbar, 1991. Paternity data: *Mucaca fuscata*: Inoue *et al.*, 1991; Soltis *et al.*, 1997b; Soltis *et al.*, 2001; *M. mulatta*: Smith, 1994; *M. sylvanus*: Paul & Kuester, 1996a. There are no paternity data for *M. radiata*. Number of correlations (*n*) for each category (species-measure of reproductive success) is indicated. Paternity correlations with male rank are weaker than mating correlations, but no within species differences are significant. Taken together, however, *M. fuscata*, *M. mulatta* and *M. sylvanus* mating correlations are significantly higher than paternity correlations for the same species (Mann–Whitney *U* test, $U = 118.5$, $n_1 = 34$, $n_2 = 12$, $P = 0.032$).

A fundamental influence on the ability of dominant males to monopolize paternity of offspring is the number of females mating simultaneously. The priority-of-access model (Altmann, 1962) predicts that when one female is mating, the first-ranking male will have priority of access, and that when two females are mating, the first and second-ranking males will have access, and so on. Several studies support this expectation. In seasonally breeding species, more females are expected to mate simultaneously compared to nonseasonal breeders. As predicted, correlations between male rank and reproductive success are weaker in seasonally breeding primates (Paul, 1997), and this relationship is the same for macaques (Oi, 1996; Aujard *et al.*, 1998; Paul, Box 6). Also, dominant males are more likely to monopolize female mating on days when fewer females mate (*M. fuscata*: Soltis *et al.*, 2001; *M. nemestrina*: Oi, 1996), and dominant males gain more paternity of offspring in mating seasons in which fewer females mate at the same time (*M. fuscata*: Soltis *et al.*, 2001).

Mechanisms of male–male competition

Behaviorally, relatively high-ranking males sometimes monopolize females by aggressively interrupting the mating attempts of lower-ranking males, forcing subordinate males to engage in sneak copulations. In various macaque species, spanning all the major phylogenetic branches, higher-ranking males are observed to interrupt the mating attempts of lower-ranking males (*M. fuscata*: Tokuda, 1961; Huffman, 1987; Perloe, 1992; *M. mulatta*; Manson, 1996; *M. fascicularis*: Dixson, 1998: 72; *M. arctoides*: Niemeyer & Anderson, 1983; Bruce & Estep, 1992; Dixson, 1998: 71; *M. thibetana*: Zhao, 1993; *M. nemestrina*: Shively *et al.*, 1982; Oi, 1996; *M. sylvanus*: Witt *et al.*, 1981; Kuester & Paul, 1992; but see Taub, 1980). When subordinate males are excluded from mating due to such direct contest competition, they engage in sneak copulations outside of the presence of higher-ranking males (*M. fuscata*: Soltis *et al.*, 2001; *M. mulatta*: Berard *et al.*, 1994; Manson, 1996; *M. fascicularis*: Gygax, 1995; *M. arctoides*: Nieuwenhuijsen *et al.*, 1988; *M. thibetana*: Zhao, 1993; *M. sylvanus*: Paul & Kuester, 1996a; *M. maurus*: Matsumura, 1993).

Dominant males also often guard females from rival males. The best evidence for such mate-guarding is in rhesus and Japanese macaques, although the behavior has also been observed in other macaques (Fig. 7.2) (e.g., *M. thibetana*: Zhao, 1993; *M. tonkeana*: Thierry *et al.*, 1996). In rhesus macaques, dominant males engage in longer mount series than do subordinates, and females are less likely to leave males during such mount series (Manson, 1996). In Japanese macaques, dominant males also engage in longer mounting series than do subordinates and they enjoy the highest mating success (Soltis, 1999; Soltis *et al.*, 2001).

Multiple mounting before ejaculation is not necessarily a component of mate-guarding, since males could guard females without mounting. In rhesus macaques, however, females are less likely to leave males during mount series compared to when no mounting is taking place (Manson, 1996). In any case, multiple mounts before ejaculation may satisfy functions in addition to mate-guarding such as increasing sperm output per ejaculate (Bercovitch & Nürnberg, 1996). Increased sperm output would benefit all males, but disruption by rival males results in shorter mount series in subordinates.

Male macaques do not solely engage in direct contest competition, but may also compete in endurance rivalries and sperm competitions. For example, rhesus macaque males competing for mates lose weight across the mating season,

Fig. 7.2. Mate-guarding in Tonkean macaques (Strasbourg Primate Center, France). Consortships last several days in this species. The highest-ranking male (left) closely follows the female in estrus, excluding other males (right) from mating. (Photograph by B. Thierry.)

and those with highest weight at the beginning of the mating season enjoyed the highest reproductive success (Bercovitch & Nürnberg, 1996, 1997). The authors argue that large size in males confers greater reproductive output not so much through direct contest competition, but because fatted males have more endurance to reduce feeding and increase mate-guarding throughout the mating season. For other macaques showing such a concentrated breeding season, similar endurance rivalries may also take place.

Female macaques often mate with multiple partners, which increases sperm competition. Consistent with such competition, primates forming multimale groups, such as macaques, have large testes for body size (Harcourt *et al.*, 1981; Dixson, 1998: chapter 8; see also Bercovitch & Rodriguez, 1993). In rhesus macaques, Bercovitch and Nürnberg (1996, 1997) showed that males with larger testes, independent of body weight, produced more offspring than their less well-endowed rivals. In a phylogenetic analysis, Thomsen and collaborators (2003) demonstrated that primate species with high levels of sperm competition (e.g., macaques) were the most likely to exhibit male masturbation, and argued that masturbation increases the quality of subsequent ejaculations in sperm competitions (Baker & Bellis, 1995: chapter 9).

In some baboons and chimpanzees, males form coalitions to secure females from other males (Dixson, 1998: 57). Such coalitions have been observed in two macaque species: the Tonkean (B. Thierry, personal communication) and Barbary macaques (Taub, 1980; Witt *et al.*, 1981).

In conclusion, macaque males compete with one another over access to females and high relative rank confers an advantage in such contests. Male rank, however, does not explain all of the variance in male reproductive success. Even without consideration of female mate choice, male competitive strategies that are not based solely on dyadic fighting (i.e., endurance rivalries, sperm competitions, sneak copulation tactics and coalition formation) may also influence male reproductive success.

Male–male competition and dominance style

Across macaques, male dominance style ranges from intolerance of other males to relatively tolerant inter-male relationships (e.g., Shively *et al.*, 1982; de Waal & Luttrell, 1989; Hill, 1994; Matsumura, 1999; Thierry, 2000, Chapter 12). Characteristics of the tolerant male dominance style include less severe aggression, more affiliation, and more reconciliation after fights. This variation in dominance style may be related to inter-male competitive reproductive strategies.

One expectation is that those species with a tolerant dominance style should show weaker correlations between male dominance rank and reproductive success. While there are no such data for most macaque species, the best data available do not support this prediction. Bonnet and Barbary macaques are two species commonly cited as having relatively tolerant male dominance styles (*M. radiata*: Glick, 1980; Shively *et al.*, 1982; *M. sylvanus*: Taub, 1980; Shively *et al.*, 1982; Preuschoft *et al.*, 1998; but see Kuester & Paul, 1992, 1996), while Japanese and rhesus macaques are commonly cited as having intolerant male dominance styles (*M. fuscata*: Takahata, 1982; Shively *et al.*, 1982; Soltis, 1999; *M. mulatta*: Shively *et al.*, 1982; de Waal & Luttrell, 1989). Figure 7.1 shows, however, that the correlations between male rank and reproductive success are just as high in the two tolerant species as they are in the two intolerant species.

Dominant males in tolerant species may gain a reproductive advantage by mechanisms other than direct contest competition, such as endurance rivalries and sperm competitions, as described earlier for rhesus macaques. According to this argument, dominance rank will be positively correlated with reproductive success when high-ranking males perform best in such endurance rivalries and sperm competitions. Thus, a tolerant male dominance style does not necessarily imply a lack of male–male competition (see also Kuester & Paul, 1992), but an attenuation of overt physical fighting, which may evolve when the costs of direct fighting are too high (Preuschoft *et al.*, 1998). This is not to say that all mechanisms cannot operate in a single species (e.g., *M. mulatta*: Bercovitch & Nürnberg, 1996, 1997), only that different species may rely more heavily on some mechanisms of competition compared to others. It will be possible to test this proposal once the male dominance styles of macaques have been resolved (see below).

Multiple versus single mount ejaculations

Macaque species can be divided into those that exhibit multi-mount ejaculatory patterns and those that exhibit a single mount with ejaculation (Table 7.1) (Shively *et al.*, 1982; Dixson, 1998: 121). Caldecott (1986a) explained multiple mounting as a product of female choice. In species in which females are promiscuous, according to the argument, males take advantage of all mating opportunities with quick single-mount ejaculations. In species in which females are selective, on the other hand, males engage in multi-mount ejaculations, in order to be chosen by the female (for unspecified reasons, but see Troisi & Carosi, 1998).

The model proposed by Caldecott, however, is not cogent. First, there is no evidence that females of macaque species characterized by multi-mount ejaculations are more selective than females of species characterized by single-mount ejaculations. Females in species with the multiple mount pattern mate with multiple partners (e.g., *M. fuscata*: Tokuda, 1961; Soltis *et al.*, 1999; *M. mulatta*: Manson, 1992; Bercovitch, 1997), just as do females in species with the single-mount pattern (e.g., *M. radiata*: Glick, 1980; *M. sylvanus*: Small, 1990a; Kuester & Paul, 1992). Although systematic, quantitative data may allow for a classification of macaques according to the degree of "female promiscuity," at present it appears that female macaques of both mounting patterns can be labeled as such.

Second, when females attempt to mate with multiple males, it is not clear why males should mate quickly and in succession, rather than monopolize such females by mate-guarding. Of course, when *multiple* females mate at the same time, dominant males must choose between monopolizing one female or attempting to mate with more than one in succession (e.g., Bercovitch, 1987; Kuester & Paul, 1992). The explanation put forth by Caldecott is not specifically about simultaneously mating females, however, but the degree to which

Table 7.1. *Male dominance style and mounting pattern in* Macaca *spp.*[a]

Species	Thierry et al., 2000[b]	Shively et al., 1982[c]	Other sources	Mounting style[d]
M. sylvanus	Tolerant	Tolerant		SM
M. nemestrina	Intolerant	Intolerant	Intolerant: Oi, 1996	MM
M. radiata	Tolerant	Tolerant		SM
M. arctoides	Tolerant	Tolerant	Tolerant: de Waal & Luttrell, 1989	SM[e]
M. fascicularis	Intolerant	Intolerant		SM, MM
M. mulatta	Intolerant	Intolerant	Intolerant: de Waal & Luttrell, 1989	MM
M. fuscata	Intolerant	Intolerant	Intolerant: Soltis et al., 2001	MM
M. silenus	Tolerant	Intolerant	Intolerant: Lindburg & Harvey, 1996	MM
M. nigra	Tolerant	Intolerant	Intolerant: Reed et al., 1997	MM
M. tonkeana	Tolerant			MM
M. maurus	Tolerant			MM
M. ochreata	Tolerant			
M. sinica	Tolerant	Intolerant		SM
M. assamensis	Tolerant			SM
M. thibetana	Tolerant		Intolerant: Zhao, 1993	MM
M. cyclopis	Intolerant			

[a] The top seven species show apparent consensus in male dominance style (tolerant vs. intolerant) across authors, while the bottom nine do not.
[b] Systematic classification of dominance styles at the species level (both sexes). Intolerant refers to dominance grades 1–2, and tolerant refers to grades 3–4.
[c] Classification of male dominance styles based on literature review.
[d] SM and MM: single-mount (S-M) and multiple-mount (M-M) copulatory patterns. Sources: Shively et al., 1982; Aujard et al., 1998; Dixson, 1998: 121; Cooper & Bernstein, 2000.
[e] *M. arctoides* engage in many more thrusts than typical single mounters (1–170 thrusts per mount; Hrdy & Whitten, 1987).

females attempt to mate with multiple males. Males can benefit from mate-guarding whenever females attempt to mate with rival males, regardless of female preferences for mating with one versus multiple males.

Another explanation by Shively and collaborators (1982) is that macaques with intolerant dominance styles engage in multiple mounting as a mate-guarding tactic, while species with more tolerant dominance styles engage in single mount ejaculations. Current evidence concerning this argument is mixed. First, there is no systematic classification of male dominance styles for macaque species. Shively and collaborators (1982) classified male dominance styles based on an evaluation of non-comparative and often anecdotal reports of male–male interaction. Thierry and collaborators (2000) constructed a comprehensive and systematic classification of macaque dominance styles, but at the species level (i.e., for both males and females). Table 7.1 shows the putative male macaque dominance styles as proposed by different authors. In seven species for which there is apparent consensus on the male dominance style, intolerant species show the multi-mount copulatory pattern, and tolerant species show the single-mount pattern, consistent with the contention of Shively and collaborators (1982). In the nine species in which there is no consensus or in which there is only data for the dominance style of the species as a whole, however, the relationship breaks down. In particular, Sulawesi macaques uniformly engage in the multi-mount pattern but appear to have relaxed male dominance styles (according to the species level analysis).

More comparative work on inter-male relationships in macaques is needed to determine the relationship between male reproductive tactics and inter-male tolerance. There is no consensus that demarcates the tolerant from the intolerant, and different researchers are likely to come to different conclusions. In Barbary macaques, for example, Taub (1980) observed that dominant males tolerated mating by subordinate males, but contest competition over females was observed in other studies (Kuester & Paul, 1992; Paul & Kuester, 1996a). Similarly, the male dominance style of the crested macaque has been described as both intolerant (Shively *et al.*, 1982; Reed *et al.*, 1997) and tolerant (Thierry *et al.*, 2000, Chapter 12). Additionally, inter-male tolerance may vary as a function of demographic differences between groups within the same species (Hill, 1994) or the presence of estrous females (e.g., males may be tolerant in the birth season compared to the mating season). A systematic, quantitative measure of male dominance styles across macaque species is required before any firm conclusions can be drawn.

Male mate choice

Most discussion of male mating tactics in macaques implies indiscriminate mating on the part of males, but there are circumstances in which males may benefit from exercising mate choice. When more than one female is mating but only one can be monopolized, for example, it may benefit males to choose. In the moor and crested macaques, males preferentially associate and copulate with one female when two are showing sexual swellings (Dixson, 1977; Matsumura, 1993). When such opportunity for choice arises, males could benefit by choosing the most fertile female, and male macaques do sometimes reject young females who have low fertility (e.g., *M. fuscata* and *M. radiata*: Smuts, 1987). Also, males may benefit from mating with high-ranking females who enjoy higher reproductive success than lower-ranking females (Keddy-Hector, 1992). In bonnet and Japanese macaques, for example, dominance ranks of heterosexual mates are correlated (Glick, 1980; Perloe, 1992), suggesting that both females and males choose high-ranking mates. Additionally, in Japanese macaques, dominant males disproportionately interrupt the matings of high-ranking females (Perloe, 1992). The best evidence is in Barbary macaques, in which high-ranking females enjoy highest reproductive success, males compete for mating opportunity with high-ranking females, and high-ranking males tend to win these contests (Kuester & Paul, 1996).

Female reproductive strategy

Female choice

Female mate choice occurs when female behavior increases the likelihood of fertile mating with particular males (Halliday, 1983; see Paul, 2002). Female macaques often express mate choice behaviors toward particular males, such as proximity maintenance and copulation vocalizations, and they often are able to reject the mounting attempts of disfavored males (e.g., *M. fuscata*: Huffman, 1987, 1991; Perloe, 1992; Soltis *et al.*, 1997a,b; *M. mulatta*: Wallen *et al.*, 1984; Manson, 1992; *M. nemestrina*: Eaton, 1973; *M. sylvanus*: Semple, 1998). Despite some selectivity, female macaques are also known to mate with multiple males (e.g., *M. fuscata*: Soltis *et al.*, 1999; *M. mulatta*: Manson, 1992; Bercovitch, 1997; *M. nemestrina*: Oi, 1996; *M. radiata*: Glick, 1980; *M. sylvanus*: Small, 1990a), and to mate when conception is unlikely or impossible (*M. arctoides*: Murray *et al.*, 1985; *M. fuscata*: Soltis *et al.*, 1999; *M. mulatta*: Wilson *et al.*, 1982; *M. silenus*: Clarke *et al.*, 1993; *M. sylvanus*: Small, 1990a; *M. tonkeana*: Aujard *et al.*, 1998; see also Vasey, Box 7).

Genetic benefits that females might receive from pre- or post-copulatory choice include avoiding inbreeding depression, acquiring high quality genotypes, and avoiding genetic incompatibility (Waser *et al.*, 1986; Andersson, 1994: chapter 3; Zeh & Zeh, 1996, 1997). Genetic studies show that matrilineal inbreeding is avoided in captive groups of rhesus (Smith, 1986b) and Japanese macaques (Inoue *et al.*, 1990) (see Gachot-Neveu & Ménard, Chapter 6). In Barbary macaques, matrilineal relatives and paternal relatives of the same age-cohort avoided inbreeding (Kuester *et al.*, 1994). While inbreeding is deleterious for both sexes, it is more costly for females (Waser *et al.*, 1986). Consistent with theory, it has been shown in two macaque species that females are responsible for avoiding mating with matrilineal relatives (*M. fuscata*: Soltis *et al.*, 1999; *M. mulatta*: Manson & Perry, 1993).

Females of non-macaque primate species often choose large, dominant, or fully mature males (e.g., *Cebus apella*: Janson, 1984; *Chlorocebus aethiops*: Keddy, 1986; *Pongo pygmaeus*: van Schaik & van Hooff, 1996; *Saimiri oerstedi*: Boinski, 1987). Among macaques, such choice for males of potentially superior quality has been observed only in Japanese macaques, in which females expressed peri-ovulatory mate choice toward those males who performed the most displays (conspicuous, repetitive behaviors including shaking a substrate, leaping vertically, or tossing the torso rapidly back and forth; Soltis *et al.*, 1999). But the benefits of choosing males who display most, if any, are not known.

Female macaques sometimes choose novel males and grow to disfavor familiar males after several years (*M. fuscata*: Huffman, 1991; Takahata *et al.*, 1999; *M. mulatta*: Bercovitch, 1997, but see Manson, 1995). Possible benefits of mating with novel males include increasing heterozygosity (Brown, 1997), or non-genetic benefits such as recruiting males for group defense (Wrangham, 1980) and inhibiting potentially infanticidal males (Hrdy, 1979). While female macaques are clearly active participants in the mating process, it is usually unclear, with the exception of inbreeding avoidance, what genetic benefits they receive by doing so.

Non-procreative benefits to female mating

A large proportion of female mating in macaques is non-procreative, often with multiple males (sources above). Mating is known to be costly in terms of energy, and risk of disease and injury from males (e.g., Reynolds & Gross, 1990; Manson, 1994). Commonly proposed compensating benefits for primates are: recruiting males for group defense, gaining extra paternal investment and inhibiting infanticide (Soltis, 2002).

Females may use mating to recruit males to their group for defensive purposes (Wrangham, 1980). While male primates sometimes do defend groups (e.g., Baldellou & Henzi, 1992), there is no evidence that females can actively influence male group transfer by mating with them. Sprague (1991a), for example, found that Japanese macaque males who entered troops were no more likely to have mated with resident females than those who visited but did not join troops.

Female multiple mating may recruit males to participate in infant care (Taub, 1980). Theory suggests that mating to receive paternal care can account for mating with only a few males, not, however, for mating with many group and non-group males, which is often observed in macaques (Soltis & McElreath, 2001). First, the benefits of mating with multiple males are limited because females must share paternal investment with other promiscuous females. Second, paternal investment is likely to experience diminishing returns. The first male helper may contribute substantially to offspring fitness, but the tenth such helper is unlikely to further benefit offspring even with the same amount of care.

Females may benefit from male care of offspring in some species of macaques, but the evidence is limited. Male–infant interactions in macaques range from never to extensive (Whitten, 1987; Maestripieri, Box 10). Barbary macaque males interact extensively with infants, by grooming, protecting, and playing with them (Taub, 1980), and longtailed macaque males may protect infants from infanticide (de Ruiter *et al.*, 1994), but there is no evidence that such interactions result in higher offspring fitness, although the benefits of infanticide protection could be substantial.

The final explanation for female multiple mating is infanticide inhibition (Hrdy, 1979). Male infanticide, generally consistent with the sexual selection hypothesis, has been observed in three macaque species (*M. fuscata*: Soltis *et al.*, 2000; *M. fascicularis*: de Ruiter *et al.*, 1994; *M. thibetana*: H. Ogawa, personal communication). Inhibition of infanticide and infant harassment has the potential to explain very pronounced female multiple mating, if males refrain from attacking infants of former mates (Soltis, 2002). Females can gain from mating with increasing numbers of males as long as the cost of mating with a male (in terms of losing offspring) is less than the cost of that male attempting to harass or kill her offspring. Supportive evidence is available for Japanese macaques, in which males who did not mate with mothers were eight times more likely to physically attack their infants (including a fatal attack) compared to males who did mate with the mothers (Soltis *et al.*, 2000). It remains to be seen if inhibition of infanticide or infant harassment is a robust explanation for female multiple mating across macaque species.

Sexual swellings and female mating strategy

Female sexual swellings have arisen independently three to five times in the evolution of primates, they tend to bracket ovulation, and have sexually arousing effects on conspecific males (Dixson, 1983, 1998: chapter 8; Sillen-Tullberg & Moller, 1993; Stallman & Froehlich, 2000). Potential functions of sexual swellings include long-distance signaling of fertility, signaling supernormal attractivity, honest signaling of quality, incitement of male-male competition, facilitating paternity confidence, and facilitating paternity confusion (Dixson, 1983, 1998: chapter 8; Stallman & Froehlich, 2000). Nunn (1999) attempted to reconcile these sometimes conflicting functions by arguing that sexual swellings are graded signals that advertise ovulation in a probabilistic manner such that females can gain both the procreative and nonprocreative benefits of mating.

There is some evidence that female sexual swellings are supernormal stimuli evolved to overwhelm male mate preferences. Males often prefer high-ranking, fully adult females (e.g., Kuester & Paul, 1996), and the sexual skin of many macaques is most exaggerated during puberty, when such rejection would be highest (Dixson, 1998: 210). In baboons, there is evidence that sexual swellings are honest signals of female quality (Domb & Pagel, 2001). Similarly, sexual swellings in Japanese macaques were manifest outside the normal season and earlier developmentally, when nutritional conditions were artificially improved (Mori *et al.*, 1997).

While there is little additional evidence as to the function of sexual swellings in macaques, some patterns in the data are consistent with the garnering of both procreative and non-procreative benefits. First, dominant males tend to monopolize females at the peak of sexual swelling in primates (Dixson, 1983; Nunn, 1999). This pattern is consistent with receiving good genes or direct benefits from high quality males. Second, while sex skin coloration and swelling in macaques do bracket ovulation, they do so in a less than perfect manner (e.g., *M. nemestrina*: Blakley *et al.*, 1981; *M. nigra*: Thomson *et al.*, 1992; *M. silenus*: Lindburg & Harvey, 1996; *M. tonkeana*: Thierry *et al.*, 1996; Aujard *et al.*, 1998), and sometimes appear to obscure ovulation altogether (e.g., *M. fuscata*: Nigi, 1975, 1976). These observations suggest that females promote mating with multiple males outside of the peri-ovulatory period, a pattern of mating consistent with gaining nonprocreative benefits such as infanticide inhibition or increasing male parental care.

The genus *Macaca* has the most variability in sexual skin and swellings of any primate genus, from pronounced swellings to almost no swelling or coloration of sex skin (Bercovitch & Harvey, Chapter 4). Pronounced sexual swelling is the ancestral condition for macaques, but it has been lost to various degrees in some groups, in particular the *sinica* and *arctoides* groups. Dixson (1983, 1998: 211) argued that sexual swellings were lost because they reduced hybridization in those species that evolved sympatrically with other macaques, and that they have been retained in those species that evolved in relative isolation. The *sylvanus–silenus* macaque group retains the most pronounced sexual swellings. This group dispersed earliest and individual species subsequently became geographically separated (Fooden, 1980), with modern populations in North Africa (*M. sylvanus*), Southern India (*M. silenus*) and Southeast Asia (*M. nemestrina* and the Sulawesi macaques). The Sulawesi species evolved in partial isolation on separate islands, which only later formed a single-land mass (Dixson, 1998: 211). The other macaque groups, whose sexual swellings are greatly attenuated or lost altogether, dispersed later than the *sylvanus–silenus* group and experienced less geographic disjunction (Fooden, 1980).

Do males know when females ovulate?

If males evolved mechanisms that allowed for the determination of female fertile periods, then they could not be easily deceived in ways implied by the "parental investment" and "infanticide inhibition" hypotheses. Males may be able to determine female fertile periods through a variety of cues. First, visually conspicuous sexual swellings often bracket ovulation, as described above. Second, males may be influenced by olfactory cues. In the single macaque species with no sexual swellings or sex skin coloration, *M. arctoides*, ejaculations also peak at mid-cycle (Murray *et al.*, 1985), suggesting olfactory cues, but evidence for olfactory cues is mixed in the rhesus macaque (Goldfoot, 1981; Dixson, 1998: 355–356). Third, males may use changes in female behavior itself as a cue to ovulation. In rhesus macaques, for example, female proceptive behaviors increase during the period of most likely conception (e.g., Wallen, 1995), and in Japanese macaques the criteria for choosing mates change during periods of likely conception (Soltis *et al.*, 1999).

Other evidence suggests that males cannot, however, determine female fertile periods perfectly. Despite ejaculatory peaks at ovulation in many macaque species (*M. fascicularis, M. fuscata, M. maurus, M. mulatta, M. nemestrina, M. nigra, M. silenus, M. sylvanus*: Dixson, 1998: 338; *M. tonkeana*: Aujard *et al.*, 1998), males also mate with females outside of the peri-ovulatory period (sources above). In Japanese macaques, for example, there is an ejaculatory peak around ovulation, but more than half of ejaculations occur after females are already pregnant, continuing up to 10 weeks after conception (Soltis *et al.*, 1999). Additionally, mating over the entire mating season, regardless of the actual time of conception, influenced male behavior toward offspring in the following birth season (Soltis *et al.*, 2000).

Thus, males may possess mechanisms to estimate female fertility based on behavioral, visual, and olfactory cues, but such estimations are not perfect. To the extent that males and females are in conflict with regard to the transparency of fertile periods, one would expect an arms race in which males evolve mechanisms to determine better the timing of ovulation, while females evolve mechanisms to obscure it further.

Female–female competition and dominance style

Female dominance style in macaques is thought to be influenced by the patterns of competition over food (e.g., van Schaik, 1989; Sterck *et al.*, 1997) and by phylogenetic inertia (Thierry *et al.*, 2000). The nature of competition over male mates may also impact female dominance styles. Experimental evidence suggests that female macaques do compete over males when male mates are scarce. In tests where single males were introduced to all-female groups, dominant females successfully monopolized males by inhibiting subordinates from sexual interaction (*M. mulatta*: Zumpe & Michael, 1987). Also, in single-male, multifemale groups of rhesus macaques, dominant females mated more often outside of the peri-ovulatory period than subordinates, but in multimale, multifemale groups all females mated with males outside the peri-ovulatory period (Wallen, 1990). In multimale, multifemale groups, female–female competition may exist in subtle form, but it has not been detected even when specifically investigated (e.g., Niemeyer & Anderson, 1983; Kuester & Paul, 1996).

The interaction of male and female reproductive strategies

Males and females will be in conflict whenever female choice rules out a male as a potential mate. Even when females attempt to mate with multiple males, there can still be conflict between the sexes because males may attempt to monopolize them. In macaques, males are larger than females, so conflict between the sexes could be resolved through male coercion of females, which is defined as using force or the threat of force to increase the probability of mating (Smuts & Smuts, 1993).

There is conflicting evidence as to whether male sexual coercion or female choice will prevail when male and female reproductive interests conflict. Some evidence suggests that females are not susceptible to coercion. In several macaque species, for example, females can reject the immediate mounting attempts of males by refusing to assume the proper mounting position or by simply walking away (*M. fuscata*: Huffman, 1987, 1991; *M. mulatta*: Manson, 1992; *M. nemestrino*: Oi, 1996). Additionally, in Japanese and rhesus macaques, mating harassment by dominant males does not usually result in copulation with the interrupted females (*M. fuscata*: Huffman, 1987, 1991; *M. mulatta*: Wilson, 1981), and male-to-female aggression in pair tests did not result in mating (*M. mulatta*: Bercovitch *et al.*, 1987). Finally, in Japanese macaques, female mate choice was a stronger predictor of mating success and paternity than male dominance rank, even though females expressed preferences toward males of high and low dominance ranks (Soltis *et al.*, 1997b).

Other evidence suggests that macaque males, however, can coerce females into mating with them. First, in several macaque species, aggression toward females increases in the mating season and is directed toward estrous females (*M. fuscata*: Enomoto, 1981; Smuts & Smuts, 1993; Soltis, 1999;

M. mulatta: Smuts & Smuts, 1993; *M. nigra*: Reed *et al.*, 1997; *M. thibetana*: Zhao, 1993). Second, when dominant males attack mating pairs involving lower-ranking males, they often (sometimes most often) attack the female of the pair, not the male (*M. fuscata*: Huffman, 1987; see also Vasey, 1998; *M. mulatta*: Bercovitch, 1997; *M. nemestrina*: Oi, 1996). Third, although females often avoid immediate mating with aggressive males, persistent aggression may increase the probability of future mating (e.g., *Papio*: Bercovitch, 1995). In Japanese and rhesus macaques, aggression against females is positively correlated with mating success (Smuts & Smuts, 1993; Soltis *et al.*, 1997a; Barrett *et al.*, 2002), and during mount series in Japanese macaques, male aggression toward females is associated with the female recipient of aggression subsequently approaching her former aggressor (Soltis, 1999). Current evidence suggests that male sexual coercion is one of many factors influencing mating and reproductive outcome in macaques.

Despite the potential influence of male coercion of females, female choice probably explains much of the variance in male reproductive success unexplained by male dominance rank (for a multivariate analysis, see Soltis *et al.*, 1997a,b). A major factor influencing the outcome of conflict between the sexes is the number of simultaneously mating females. When few females mate, dominant males are more likely to monopolize them, and the strategies of dominant males may prevail. When more females mate at the same time, there is more opportunity for female choice of subordinate males, and the strategies of females may prevail (e.g., *M. fuscata*: Soltis *et al.*, 2001; *M. nemestrina*: Oi, 1996).

Conclusion

Macaques live in bisexual social groups, in which various reproductive strategies of both sexes operate simultaneously. Males compete by a variety of mechanisms including direct contest competition, endurance rivalry, sperm competition and sneak ejaculations. Females exert selective mate choice toward and mate with multiple males, both when conception is likely and unlikely. Because of reproductive conflicts among males, among females, and between the sexes, no single factor, such as male dominance rank or female mate choice, is likely to explain all the variation in mating and reproductive outcome.

Topics in particular need of further inquiry include the function of multiple mounting, the function of nonprocreative mating by females, and the function of female sexual swellings. Additionally, because so many mechanisms of sexual selection operate simultaneously (e.g., female choice, male coercion, etc.), experimental studies of sexual selection would be useful in macaques. For

Box 7 Homosexual behavior 151

heuristic purposes, I attempted to relate intra-sexual dominance style to reproductive strategy, and future work on dominance styles may help to elucidate
the relationship between the two.

Box 7 Homosexual behavior

Paul L. Vasey

Courtship displays, mounting and genital contact/stimulation between same-
sex individuals have been observed in a variety of macaque species including
liontailed, Tonkean, moor, crested, pigtailed, bonnet, stumptailed, longtailed,
rhesus, and Japanese macaques (reviewed in Nadler, 1990; Vasey, 1995;
Bagemihl, 1999). These same-sex interactions are frequently referred to as
"homosexual behavior" in the macaque literature, although some researchers
(e.g., Wallen & Parsons, 1997; Dixson, 1998) argue that this label is mislead-
ing because homosexuality is a uniquely human phenomenon. This argument
is predicated, in part, on the assumption that *homosexual behavior* is synonymous with *homosexuality* when, in reality, these two terms are not conceptually equivalent. Homosexual behavior refers to discrete acts or interactions,
whereas homosexuality refers to a specific type of sexual orientation (i.e., an
overall pattern of sexual attraction/arousal over time as measured by multiple parameters such as behavior, fantasy, feelings, attraction, genital blood
flow, etc.). Care should be taken not to make conceptual leaps from terms
that denote behavior, to those that denote sexual orientation, sexual orientation identity (i.e., the sexual orientation that individuals perceive themselves
to have, see Cass, 1983–84; Troiden, 1989), categories of sexual beings
(i.e., homosexuals, heterosexuals, etc., see Foucault, 1990; Katz, 1995; for
non-Western examples, see Nanda, 2000), or sexual preference (see Vasey,
2002a,b).

  Homosexual behavior in macaques is not a uniform phenomenon. This
makes generalizations difficult and arguably, inappropriate. Indeed, it is
probably more accurate to speak in terms of macaque homosexual *behaviors* that vary both within species (i.e., inter-sexually) and between species
in terms of their form, frequency, motivation, cause, development, function,
and evolutionary history.

  One macaque species in which homosexual behavior has been particularly
well studied is the Japanese macaque, probably because female homosexual
behavior is commonly manifested in *some* free-ranging (e.g., Jigokudani:
Enomoto, 1974; Arashiyama-East (Texas): Fedigan & Gouzoules, 1978;
Wolfe, 1979, 1984; Gouzoules & Goy, 1983; Arashiyama East: Takahata,
1982; Wolfe, 1984; Minoo: Perloe, 1989) and captive populations (Oregon

Fig. 7.3. Female homosexual behavior in free-ranging Japanese macaques (Arashiyama, Japan). The mounter performs a "sitting" mount with pelvic thrusts, while rubbing her clitoris against the back of the mountee. The mountee turns to gaze into the eyes of the mounter and reaches back to grasp the mounter while she rubs her clitoris with her tail. (Photograph by S. Kovacovsky.)

troop: Eaton, 1978; Rome Zoo: Lunardini, 1989; Corradino, 1990; Université de Montréal colony: Chapais & Mignault, 1991; Vasey, 1996, 1998, 2002a, 2004; Chapais *et al.*, 1997; Vasey & Gauthier, 2000; Vasey *et al.*, 1998; Calgary Zoo: Rendall & Taylor, 1991) (Fig. 7.3). For example, Vasey (2002a) reported that during a typical mating season (October 1993–February 1994) every sexually mature female in the University of Montréal colony ($n = 14$ sexually active females; $n = 5$ sexually active males) engaged in female homosexual behavior. Moreover, the majority of sexual relationships formed ($n = 95$) were homosexual ones (55%), not heterosexual ones (45%). During homosexual consortships, female solicited their partners for sex on average, 28 times per hour of observation and they mounted each other, on average, 31 times per hour of observation ($n = 129$ observation hours).

Key proximate (i.e., organizational theory of brain and behavioral development; Phoenix *et al.*, 1959) and ultimate theories (e.g., sexual selection and

Box 7 Homosexual behavior 153

parental investment theories; Darwin, 1871; Trivers, 1972) in the biological sciences predict that sexual behavior should be oriented toward opposite-sex sexual partners (see Soltis, this Chapter). Thus, it seems reasonable to ask why female Japanese macaques in certain populations engage in such high levels of homosexual behavior?

On a proximate level, observational (Wolfe, 1984) and experimental (Vasey & Gauthier, 2000) research suggest that female homosexual behavior in Japanese macaques increases in the context of female-biased operational sex ratios (i.e., the number of sexually active males to receptive females). This pattern of female sexual activity is typically explained in terms of a lack of male mates. However, some males are invariably available *and* motivated to mate with females under these demographic conditions (Vasey, 1998, 2002a; Vasey & Gauthier, 2000). Many females, nevertheless, rebuff solicitations by such males in favor of same-sex sexual partners. As such, heterosexual deprivation is not an adequate explanation for the increase in female homosexual behavior observed under these demographic conditions. Wolfe (1984) has argued that the expression of female homosexual behavior in the context of female-skewed operational sex ratios reflects a quest on the part of the females for sexual novelty. Vasey and Gauthier (2000) have argued that higher levels of female homosexual activity observed under these demographic conditions is not caused by a lack of males per se, but rather results from the paucity of preferred male mates coupled with an abundance of preferred female sexual partners. In addition, they note that male sexual competitors are scarce under these demographic conditions and, as such, females are more able to access and maintain preferred, same-sex sexual partners. This particular perspective takes into account the fact that female Japanese macaques that engage in homosexual behavior are bisexual in orientation, not preferentially heterosexual, as is commonly assumed, and they must often compete inter-sexually for access to female sexual partners (Vasey, 1998, 2002a; Vasey & Gauthier, 2000).

At an ultimate level, it has been suggested that female homosexual behavior in Japanese macaques might be a socio-sexual behavior (*sensu* Wickler, 1967), that is, sexual in terms of its outward form, but enacted to facilitate some type of adaptive social goal or breeding strategy. However, research indicates female Japanese macaques do not use homosexual behavior to attract male mates (Gouzoules & Goy, 1983; Vasey, 1995), impede reproduction by same-sex competitors (Vasey, 1995), form alliances (Vasey, 1996), reduce social tension, communicate about dominance relationships (Vasey *et al.*, 1998), obtain alloparental care, or reconcile conflict (Vasey, 1998, 2004). It is noteworthy that despite over 40 years of intensive research on this species, there is not a single study demonstrating that this behavior serves any adaptive function. If female homosexual behavior in Japanese macaques is not adaptive, as the data seem to suggest, then how might we account for these phenomena within the context of an evolutionary framework?

Answers to this question may require a fuller appreciation of the evolutionary history of female mounting in this species. Sexually proceptive female Japanese macaques frequently mount-prompt adult males and, often, they follow through with full mounts involving pelvic thrusting (Gouzoules & Goy, 1983). Males, in turn, often respond by mounting the females that have mounted, or mount-prompted, them (Vasey, unpublished data). As such, female–male mounting in Japanese macaques may be an adaptation that proceptive females use to prompt sexually sluggish, or disinterested, males to copulate with them. But why would an ability to mount opposite-sex sexual partners translate into a desire to mount same-sex sexual partners in the absence of any fitness-enhancing pay-off? To understand why, it is important to note that female Japanese macaques frequently engage in clitoral stimulation during female–male mounts. They do so by rubbing their clitorises against the backs of male mountees or by masturbating with their tails during female–male mounts (Vasey, unpublished data). Once females evolved the capacity to mount, and the capacity to derive sexual reward during mounts through clitoral stimulation, they could do so just as easily by mounting females as males. Within the context of this evolutionary scenario, female–female mounting and occasional same-sex sexual partner preference can thus be seen as the by-product of an adaptation, namely, female–male mounting. By-products of adaptations are characteristics that do not evolve to solve adaptive problems, and thus, do not have a function and are not products of natural selection. Instead, by-products are "carried along" with particular adaptations because they happened to be coupled with those adaptations (Gould & Vrba, 1982; Futuyma & Risch, 1984; Buss *et al.*, 1998).

Part III *Social relationships and networks*

Introduction

The several behavioral dimensions of animal societies have been successfully tackled using the three-level scheme delineated by Hinde (1976b): (1) "social interactions" are defined by the current behavior of participants; (2) "social relationships" involve a succession of social interactions between participants known to each other; and (3) "social structures" are networks of social relationships. As an example, two individuals learn their respective strengths through a number of agonistic interactions. They then establish a dominance-subordination relationship, which makes their behavior toward each other mutually predictable. The dominance ranks of group members may be ordered into a social hierarchy. The gradient of the hierarchy varies according to multiple factors (Chapter 8). According to partner choice and the proportion of affiliative interactions, we may likewise recognize preferential relationships between relatives and kin networks, the members of which provide mutual support in conflicts. Alliance patterns feed back on dominance structures and rank inheritance (Chapter 9). One generation shapes the next one, influencing not only the behavior of individuals but also the networks of their relationships (Chapter 10). We must add that the three levels of Hinde's scheme do not have the same epistemological status. Whereas interactions are events that occur in the physical world, relationships and networks are inferred rather than observed (Altmann, 1981). If social networks are only explanatory concepts, they cannot have causal effects in the lives of individuals. On the other hand, experimental data indicate that macaques can distinguish different kinds of social relationships among their mates (Dasser, 1988a). The recognition of relational properties and use of such knowledge by individuals introduce a further degree of complexity into the social organization. The formal acknowledgement of social relationships can change the state of power at the group level (Chapter 8). To understand the interplay between social behaviors and the other components of the social realm, we must be ready to deconstruct the three-level scheme whenever necessary.

155

8 Dominance style, social power, and conflict management: a conceptual framework

JESSICA C. FLACK AND FRANS B. M. DE WAAL

Introduction

Following Kroeber and Parsons (1958), "society" refers to the relational system of interaction among individuals and subgroups, taking into account the statuses and roles of all individuals within the larger group. Studying how societies are built, the topic of this volume requires systematic, comparative study to identify the general laws or principles that underlie their diversity. However, as noted by Radcliffe-Brown in the preface to *African Political Systems* (Fortes & Evans-Pritchard, 1940), comparative empirical observation does not by itself lead to the identification of underlying general principles; the diversity of types must be classified to make abstraction possible. And, in its preliminary phases, this process requires focus on particular aspects of society, such as the political system or kinship system, even though these subsystems are likely to be highly interrelated. Having said this, the goal of this chapter is to provide a preliminary taxonomy of the political systems, defined here in the Parsonian sense as the interplay between power structure and conflict management, of macaque societies.

The macaque genus is an ideal starting point for investigating how societies arise from aggregates of individuals. There are three reasons for this, the details of which are reviewed in Thierry (2000): (1) the kinship and demographic structures across macaque species are relatively similar, which decreases the number of variables that need to be taken into account when testing hypotheses about how societies arise; (2) there does not seem to be ecological variability of significant importance across macaque habitats; and (3) there is, nonetheless, interesting variation in conflict and conflict management patterns across

---

Macaque Societies: A Model for the Study of Social Organization, ed. B. Thierry, M. Singh and W. Kaumanns. Published by Cambridge University Press. © Cambridge University Press 2004.

species that does not always fall out along phylogenetic lines. The confluence of these factors has produced a natural experiment that is particularly ripe for comparative investigation of the social causes and consequences of variation in power distributions and conflict management mechanisms across macaque species.

For example, despite similar kinship and demographic structures, the social relationships of rhesus macaques are characterized by large agonistic asymmetries, whereas social relationships of Tonkean macaques appear to be characterized by agonistic *symmetries* (Preuschoft & van Schaik, 2000; Thierry, Chapter 12). Stumptailed macaques are reported to exhibit more reciprocity in interventions and higher rates of post-conflict affiliative behavior than rhesus macaques (de Waal & Luttrell, 1989). Reviews of this variation by a number of authors have revealed covariation in conflict and conflict management traits. In macaque species in which there is less of an emphasis on kinship, conflict is frequent, low intensity, often bidirectional, and post-conflict affiliative behavior is relatively high. In species with more of an emphasis on kinship, conflict is frequent but severe, unidirectional, and not typically followed by affiliative post-conflict behavior (Butovskaya, Box 8; Thierry, Chapter 12).

This covariation in conflict and conflict management mechanisms, first documented by Thierry (1985a), has been used to characterize macaque societies, and various explanations from phylogenetic inertia (Thierry, 1990a, 2000) to differences in aggression levels (Hemelrijk, 1999a, Chapter 13), have been offered to explain why one social organization arises rather than another. Here we propose a possible framework for how macaque societies arise that differs from previous accounts primarily in its attention to the mechanisms and processes by which a particular system emerges from social interactions at the relationship level. The framework we present links the degree of agonistic asymmetry (or symmetry) at the relationship level, dominance style, the distribution of social power at the system level, and the resulting conflict management system.

Dominance style: a relationship-level concept

The term "dominance style" was originally introduced to the animal behavior literature as a relationship-level concept that referred to species-typical patterns of expressed asymmetry in agonistic relationships (de Waal & Luttrell, 1989). However, it is often used to describe the observed covariation in conflict and conflict management traits across macaque species. Although this usage has provided a starting point for comparative studies of animal social organizations, it has also confounded the relationship and system levels thereby making it difficult to study the processes by which particular power distributions or conflict management systems emerge from individual interactions. We favor a slightly modified version of the term's original usage, for reasons we make clear in the following paragraphs.

When discussing species-typical dominance styles, researchers have relied upon four terms: "despotic," "tolerant," "relaxed," and "egalitarian." Before discussing how we use these terms, we provide a brief history of their meaning. "Despotic" was originally a sociological term applied to human societies in which power is concentrated in a single individual, and in which there are no limitations on rule by that individual. Its meaning in the sociological literature is, for our purposes, more or less synonymous with "absolutism," "authoritarianism," and "totalitarianism." "Egalitarian" was originally used by anthropologists to describe variation in the degree to which individuals in human societies divided benefits in proportion to investment (e.g., Fortes & Evans-Pritchard, 1940; Sahlins, 1958; Flanagan, 1989).

Building on the ideas of Alexander (1974) and others, Vehrencamp (1983) formally introduced the terms "despotic" and "egalitarian" to the study of animal behavior by developing an optimization model that specified the conditions under which despotic and egalitarian societies might evolve. In Vehrencamp's treatment, despotic and egalitarian dominance relationships describe the degree of asymmetry in agonistic relationships, and the term that best summarizes the dyad typical degree of agonistic asymmetry, designates whether the society is described as egalitarian or despotic. It is important to note that whereas sociologists and anthropologists used the terms despotic and egalitarian to qualitatively describe a particular distribution of social power in society, Vehrencamp used them to refer to the degree of agonistic asymmetry in relationships, and then extrapolated to society. The implicit assumption was that societal types, a system-level characteristic, corresponded directly and linearly to degree of agonistic asymmetry at the relationship level.

Wrangham (1980), van Schaik (1989), and Sterk and collaborators (1997) have provided potential ecological explanations for why dyadic relationships in some species are best characterized as egalitarian whereas in others, despotic is the more appropriate classification. At about the same time, van Rhijn and Vodegel (1980) and Hand (1986) discussed despotic (dominance–subordination) and egalitarian relationships in terms of conflict resolution, which complemented a view de Waal (1986a) was developing of dominance–subordination relationships as the product of resource competition and social bonding. Thus emerged the view that the degree of agonistic asymmetry in relationships should covary with other conflict and conflict management traits, such as conflict severity, rate of post-conflict affiliation, and diversity of grooming partners.

Thierry's (1985a) study of Tonkean, rhesus and longtailed macaques, and de Waal and Luttrell's (1989) study of stumptailed and rhesus macaques were the first cross-species comparisons of conflict management patterns that attempted to empirically verify covariation of conflict management related traits in primate social organization. De Waal (1989a) suggested that this covariation is due to the different dominance styles of each species, a view perhaps first expressed by Maslow (1940). This initial usage of "dominance style" was meant to capture the discrepancy in some species between the degree of inherent agonistic asymmetry in relationships and the degree of expressed or manifest asymmetry. In de Waal's usage, despotic dominance relationships are those in which there is a high degree of inherent agonistic asymmetry and this is expressed in behavior. "Tolerant" relationships, in de Waal's usage, are those with high to moderate inherent agonistic asymmetry, but moderate to weak expression of that asymmetry during social interaction. Since those studies, the classification of dominance style has become increasingly fine-grained. Thierry (2000), for example, attempted to classify the 20 species in the macaque genus into "grades" along the despotic – egalitarian continuum. Based on a review of the data, it was argued that traits relating to dominance, conflict management, and nepotism cluster around four social organization attractors resulting from phylogenetic inertia.

In the course of validating the observed covariation in traits across macaque societies, and in testing whether that covariation might be due to dominance style, the dominance style concept and the observed covariation converged in meaning and are now at risk of being collapsed into one phenomenon. For example, despotic dominance style has been used to describe systems in which there is unidirectional aggression that is directed at subordinate individuals, frequent severe aggression, strong emphasis on kinship, and infrequent post-conflict affiliation (e.g., Preuschoft & van Schaik, 2000; see also Sterk *et al.*, 1997). Rhesus, Japanese, and possibly longtailed macaques are thought to fall into this category. Tolerant, relaxed, and egalitarian are applied by different (and occasionally the same) authors (B. Thierry, personal communication) to systems that are characterized by higher levels of bidirectional aggression, less emphasis on kinship, more frequent post-conflict affiliation, and more frequent but less intense conflicts. For example, stumptailed macaques have been described as more tolerant than rhesus macaques because of their lower levels of severe aggression, higher rates of post-conflict affiliative behavior and bidirectional aggression. Moor macaques have been described as having a relaxed dominance style (Matsumura, 1998) because of their relatively high rates of bidirectional aggression, infrequent intense aggression, and relatively high rates of post-conflict affiliation.

There are several reasons for restricting the dominance style concept to the relationship level. The first of these is that dominance style allows for a discrepancy between the degree of inherent agonistic asymmetry between two individuals and the expression of that asymmetry, as originally noted by de Waal. Consider the following example: individual A has won all of its agonistic encounters with individual B and B acknowledges that it is subordinate. A can take from B whatever it wants. However, A nonetheless exercises restraint toward B, and is tolerant of B's "transgressions," making the relationship seem less asymmetrical than it actually is. This is the case for pigtailed macaques. The degree of agonistic asymmetry that characterizes pigtailed dominance relationships is variable: in some cases there is substantial asymmetry, in others, very little. However, regardless of the degree of asymmetry, almost all relationships are weakly expressed in that they are reinforced through weak to moderate, rather than severe, aggression (Flack & de Waal, unpublished data). Variation of this sort has been reported for other macaque species as well, including moor macaques (Matsumura, 1998) and Barbary macaques (Deag, 1974).

This distinction facilitates disentangling those contextually stable and temporally invariant proximate factors (like the presence of many long-term allies or individual fighting ability) that influence the degree of *inherent* asymmetry in relationships from those socioecological and demographic factors (such as within-group scramble competition as opposed to within-group contest competition) that influence through evolutionary processes and social learning how tolerant dominants generally are of subordinates, and thus the degree of *manifest* or *expressed* asymmetry in relationships. Dominance style is a useful concept because it describes this interface by summarizing whether the inherent asymmetry in the relationship tends to match its expression or is discrepant.

To summarize, we favor distinguishing between relationship-level concepts, such as dominance style, and system-level concepts, such as political system type. Restricting the dominance style concept to the relationship level in no way devalues the observation that sets of conflict and conflict management traits covary. What it does do is facilitate disentangling causes and consequences of this covariation. For example, is the higher-rate of post-conflict affiliation observed in stumptailed macaques (compared to rhesus) a direct consequence: (1) of the degree of inherent agonistic asymmetry that characterizes relationships; (2) of dominance style; or (3) is it more directly a consequence of a system-level characteristic, such as societal power structure, that is itself influenced by dominance style?

Another advantage of restricting the dominance style concept to the relationship level is that it enables the study of the processes by which a particular political system emerges from dominance style. This allows for the possibility that, under certain conditions (we will discuss these in later sections of this chapter), different political systems can result from the same species-typical dominance style.

Determining dominance style

In the first column of Table 8.1 we present four dominance style types, which include: despotic, tolerant, relaxed, and egalitarian, and clearly distinguish the criteria of each. Thierry (Chapter 12) points out that the term "egalitarian dominance style" is somewhat of a misnomer. This is a reasonable objection from a semantic perspective. However, because we define dominance style as the discrepancy between inherent and expressed agonistic asymmetry, an egalitarian dominance style is not an inappropriate usage of terms. This is because there are two possible ways of producing an egalitarian dominance style: (1) when no or little inherent agonistic asymmetry is matched with no or little expressed agonistic asymmetry; and (2) when moderate or high inherent agonistic asymmetry is matched with no or little expressed agonistic asymmetry (this former case is a less interesting example of egalitarian relationships because individuals typically have overlapping interests).

Specified in the first and second columns of Table 8.1 are the defining characteristics of each dominance style, as well as those macaque species that tentatively fit into each one. This dominance style taxonomy is identical to Thierry's (2000) four grades, except that the defining criteria of each dominance style includes only the direction of aggression, the level of aggression used to reinforce the relationship, and the type of signal used to indicate roles in the relationship. We favor this classification system because it allows for distinguishing those traits that are directly related to dominance style from those traits, such as level of post-conflict affiliation and grooming partner diversity, which might be mediated by additional factors at higher levels of analysis.

One of the criteria we use to determine dominance style is the type of signal used to indicate role or status in the agonistic relationship. Status signals were first observed by Angst (1975) and de Waal (1977) in longtailed macaques, and by Bygott (1979), Nishida (1979), Noë and collaborators (1980) and de Waal (1982) in chimpanzees. They are discussed in detail by Preuschoft and van Schaik (2000). De Waal and Luttrell (1985) called these signals formal dominance signals because they have the following properties: (1) they are nearly 100% unidirectional signals in peaceful contexts, regardless of the presence of third parties or other factors that might influence the expression of the relationship; (2) the individual emitting the signal in peaceful situations is typically the one to yield in agonistic ones; and (3) these signals are found in species that have tolerant as well as despotic relationships, suggesting that they indicate the presence of a perceived agonistic asymmetry, rather than whether the agonistic relationship is characterized by a despotic or tolerant dominance style.

Preuschoft (1999) divided formal dominance signals into two classes: formal signals of subordination (Table 8.1), which includes those signals that are emitted in peaceful situations by the individual that typically yields in agonistic situations, and formal signals of dominance (Table 8.1), which includes those signals that are emitted in peaceful situations by the individual that typically aggresses in agonistic situations. Preuschoft introduced this division because the original signals observed by de Waal and others were emitted by subordinates, not dominants, as the original term implied, and because of data indicating that in some macaque species, such as Barbary macaques (see Preuschoft, 1995a), the dominant individual also signals its role in peaceful situations. Preuschoft and van Schaik (2000) identified in the data on macaques a covariation between dominance style and type of formal status signal used. Specifically, they suggested that in despotic species, only subordinates signal their roles, whereas in species with tolerant dominance styles, both the dominant and subordinate individual signal their respective roles.

We modify the status signaling concept one step further by referring to the entire class of signals as formal status signals. We do so to leave open the possibility that individuals in agonistic relationships characterized by power symmetries might exchange formal signals of equal status (Table 8.1) to indicate their accepted status in their relationship. We include brief mention of these signals here only to direct attention to their possible existence, which would make sense from a theoretical perspective if status signals generally serve to reduce uncertainty in the receiver about its state in its agonistic relationship with the sender. It is important to note, however, that signals of equal status will be particularly difficult to identify because if such signals do exist they will probably be used in peaceful greeting situations and are likely to be simultaneously bidirectional.

The form that formal status signals take differs across species. Formal status signals of subordination have been documented in several macaque species including rhesus (de Waal & Luttrell, 1985), longtailed (Angst, 1975; de Waal, 1977; Preuschoft, 1995a), and Japanese macaques (Chaffin *et al.*, 1995). In all of these species, the subordination signal is the silent-bared teeth display (SBT), when it is emitted in *peaceful* situations, outside the agonistic context (Fig. 8.1, see also Fig. 12.2). Formal dominance signals have also been tentatively documented. These include the "mock-bite" in stumptailed macaques (Demaria & Thierry, 1990), and the "rounded-mouth threat face" in

Table 8.1. *Possible relationship between dominance style, distribution of social power and conflict management*

Relationship level		Relationship-system interface		Social system level	
Dominance style and species examples	Type of status signal	Dyadic conflict resolution efficacy (resolution can be 'unfair')	Translation error (from relationship level to system level)	Distribution of social power	Political system type
Despotic Large dyadic asymmetries exist and are reinforced through severe aggression; relationships are formalized. Rhesus, longtailed, Japanese macaques.	FSS	High	If low	Uniform, such that SP increases by some constant as *Social Power Rank* (SPR) increases (Fig. 8.2a).	*Hierarchy* Resource allocation is based on SPR, which is largely determined by the number of subordinates one has. Conflict intervention reinforces this system: most intervention is against individuals that fail to act in accordance with their roles in their agonistic relationships, and is typically against the lower-ranking opponent or in favor of kin.
Tolerant Large dyadic asymmetries exist, but are reinforced through moderate to mild aggression; many relationships are formalized, some unresolved. Pigtailed, stumptailed macaques?	FSS FSD?	Medium	If low	Concentrated in a few individuals and distributed uniformly over the rest, such that SP increases by some constant as SPR increases (Fig. 8.2d).	*Informal oligarchy* Some resource allocation by SPR, powerful third-parties manage conflict by policing. Some impartial policing, most in favor of conflict participant with least social power. Other individuals intervene in pattern typical of hierarchical system.
Chimpanzees $\longleftrightarrow$	FSS FSD? FSE?	Medium	If low and controllable	Concentrated in a few individuals, other individuals have more or less the same degree of social power (Fig. 8.2c).	*Constrained* Maintained through leveling coalitions among less powerful individuals against individuals with high social power, policing by individuals with high social power and some mediation by socially influential individuals. Policing is increasingly impartial or rule-based.

Relaxed Asymmetries exist in some dyads, and these relationships are formalized, but are reinforced through aggressive displays rather than through aggression. Most relationships are unresolved. Some might be formalized egalitarian ones. *Sulawesi macaques?*	FSD? FSE?	Low	If low	Temporally stable differences in social power but these are small and subject to low amplitude oscillations (Fig. 8.2b).	*Equal outcome system* Maintained by formation of coalitions against individuals who are intolerant of subordinates; and some mediation[b] by socially influential[b] individuals. Can lead to division of labor[c] for arbiters, impartial policing to divide resources equally. If institutional roles arise, need for status signaling decreases.
Egalitarian Dyadic asymmetries are uncommon; most relationships are either unresolved or formalized egalitarian ones. Some human societies.	FSE?	Low	If low	No temporally stable differences in social power but high amplitude oscillations can occur, such that individuals temporarily have more power than others (Fig. 8.2b).	*Equal opportunity system* Maintained through meta-policing[d] of norm-breakers and retaliation. Can lead to division of labor for arbiters/impartial policing to mediate conflicts between coalitions. If institutional roles arise, need for status signaling decreases.

[a] Mediation: it has been described by de Waal (1982) in chimpanzees, who noted that some individuals, especially older females, exert influence over others in conflict situations. This influence is non-coercive in nature and presumably based on cognitive empathy, or the ability to take the perspective of others and see how to correct the situation. This ability essentially enables the mediator to be persuasive. Influence based in persuasion has been distinguished from power by Parsons (1963) (see also Weber, 1947).

[c] Division of labor: in sociology, the division of labor generally refers to individual components of a system performing different, but integrated or complementary tasks in a coordinated fashion. Durkheim (1933) emphasized that the division of labor can increase the solidarity of a society by reducing competition and increasing group-productivity.

[d] Meta-policing: this term has been used by Axelrod (1986) to refer to policing or punishment of non-policers and other norm-breakers. In systems in which there is meta-policing, all individuals are required to police and meta-police.

Abbreviations: FSS, formal signals of subordination; FSD, formal signals of dominance; FSE, formal signals of equal status.

Fig. 8.1. Exchange of silent bared-teeth (SBT) displays among pigtailed macaques in peaceful, non-agonistic contexts (Yerkes National Primate Research Center, USA). (a) An adult female forages along side the highest-ranking male; (b) she looks at him; (c) she begins to emit a SBT display while positioning her face directly in front of his; (d) she presents while emitting a more pronounced, slightly open-mouthed SBT; (e) she continues to present with a reduced intensity SBT; (f) she continues to present, but no longer emits the SBT. (Photographs by J. Flack.)

Barbary macaques (Deag, 1974). No formal signals have thus far been identified in macaque species with an apparently more relaxed dominance style, including the Tonkean macaques, according to reports by Thierry and collaborators (1989) and Preuschoft (1995b), and the liontailed macaques (Preuschoft & van Schaik, 2000). However, it is important to note that researchers have only looked for formal signals of subordination, and in some cases, formal dominance signals. To our knowledge, there has been no effort to identify formal signals of equal status in any macaque species.

Dominance style and social power

Two primary points have come out of attempts to classify macaque societies. First, it is clear that within and between societies there is variation in dominance style. Dominance style might: (1) vary across different subgroups of a population, as reported for male and female Barbary macaques (Preuschoft *et al.*, 1998); (2) vary across different communities of the same species; or (3) be species typical. Until a more comprehensive comparative data set on multiple communities of the same macaque species is available, it is impossible to say whether dominance style is species typical.

Second, there seems to be variation across societies in the distribution of social power (e.g., "dominance gradient" of Thierry, 2000), where social power is defined as the degree of implicit agreement among group members that an individual is capable of successfully using force in polyadic social situations (see also Bierstedt, 1950; Parsons, 1963). This variation is not necessarily coupled with dominance style, and, consequently, a second, higher-level concept is needed to describe it. Although many researchers have recognized the potential importance of the distribution of social power to understanding conflict management behavior, attempts to measure social power have primarily been qualitative – focusing on the so-called "steepness" of the dominance hierarchy. For example, Leinfelder and collaborators (2001) present several predictions about the distribution of grooming interactions in hamadryas baboons based on whether the rank order is "steep," "not steep," or "more or less steep." This approach, although correct in principle, has several significant weaknesses.

The most obvious of these is that the steepness of the dominance hierarchy cannot be quantitatively assessed using conventional methods, such as those developed by Appleby (1983) and elegantly extended by de Vries (1995, 1998) and collaborators (de Vries & Appleby, 2000) for calculating dominance rank. One reason for this is that the objective of the de Vries method, called the I&SI method, is to "find an ordinal rank order that is most consistent with a linear hierarchy by first minimizing the number of inconsistencies [in direction of the dominance related behavior being measured] *I* and, subsequently, minimizing the total strength of the inconsistencies *SI*, subject to the condition that *I* does not increase" (de Vries & Appleby, 2000: 239).

Measuring social power should not assume nor aim to generate a linear hierarchy in which there is one top ranking individual who dominates all others, a second ranking individual who dominates all others except the top one, and so on. This is because the distribution of social power is not necessarily coupled with rankings of dominance relationships. This is illustrated by the following hypothetical example: one individual dominates all others in its agonistic interactions, and a second individual dominates all others except the first. These two individuals would be ranked one and two in a dominance hierarchy even if there was more agreement among group members that the second individual could more successfully introduce force into social situations than the first. Measures of social power must not preclude this possibility.

Another reason why the I&SI method, and others like it, cannot be used to measure social power is that such measures produce ordinal rank orders. Measuring social power requires at least a cardinal index. Unfortunately, existing cardinal indices of dominance rank, such as Boyd and Silk's method (Boyd & Silk, 1983), and Jameson *et al.*'s (1999) Batchelder – Bershad – Simpson (BBS) method, also assume transitivity, and have other restrictive assumptions that make their application to social power questionable (see de Vries, 1998; de Vries & Appleby, 2000).

In addition to restrictive assumptions, there are at least two other problems with using ranking methods rooted in the dominance hierarchy tradition to measure social power. Behaviors, like supplant, direction of aggression, outcome of contest, and priority of access to resources, typically used to determine agonistic status and calculate dominance rank are, at best, correlates of status, rather than direct measures of it. Perhaps, however, the most significant problem is that none of the existing methods for calculating dominance rank assess the degree of implicit agreement among group members about whether an individual can successfully introduce force into social situations.

The remaining sections of this chapter are devoted to: (1) a discussion of social power and more appropriate ways to measure it; (2) discussion of the social processes leading to particular distributions of social power; and (3) developing the hypothesis that variation in the distribution of social power across societies predicts different conflict management mechanisms for those societies.

The meaning of social power

There are three classes of factors that can affect the predictability of agonistic encounters. Here, we summarize them briefly to make clear where social power fits into this scheme.

- *Contextually and temporally stable factors*: fighting ability, alliance partners (permanent or long-term supporters), and social knowledge, which are
  contextually stable and temporally invariant *proximate* factors. When linked
  through social learning to the predictability of agonistic outcomes, these factors underlie the development of dominance–subordination and egalitarian
  relationships.
- *Transient factors*: competitive motivation, coalitions (defined as opportunistic supporters), and leverage, which are temporally and contextually variable *proximate* factors. Leverage, which might be especially important to
  negotiating the outcomes of stalemates that occur when individuals in egalitarian relationships have conflicts of interests (Preuschoft *et al.*, 1998), is
  defined here as an asymmetry in a dyad in access to a resource that is possessed by one individual and desired by the other (for broader definitions that
  include socioecological factors, see Hand, 1986; Preuschoft & van Schaik,
  2000; Lewis, 2002). It is also assumed that the resource is only obtainable
  if the possessor grants access to it. If an individual cannot complete a task
  without the help of a second individual, that second individual has leverage
  over the first because it possesses something that the first needs, but which
  cannot be easily taken by force. The second individual can use its leverage
  as a bargaining chip. Transient factors can modify the outcomes of agonistic
  encounters and the expression of agonistic relationships, but they do not
  influence the formation of those relationships.
- *Socioecological factors*: such as demography, the degree of relatedness,
  and the availability of alternative living situations (i.e., ease of emigration,
  prospects of solitary survival, type of competitive regime; see Wrangham,
  1980; van Schaik, 1989; Sterck *et al.*, 1997). Socioecological factors can
  also indirectly mediate agonistic relationships in other ways. For example,
  if males in a particular species disperse and mortality among females is
  high, then it will be difficult for matrilines to form in groups, and the size of
  an individual's kinship network is unlikely to be an important factor determining its status in its agonistic relationships. These types of factors do not
  directly influence agonistic relationships, but instead influence the arena in
  which these relationships are expressed. For example, socioecological factors can influence the general degree of tolerance that dominant individuals
  show subordinates.

Only factors that are temporally and contextually invariable can lead to the
acquisition of social power. As stated, following Bierstedt (1950) and Parsons
(1963), we have defined social power as the degree of implicit agreement or
consensus among group members about whether an individual is capable of
successfully introducing force into polyadic, agonistic situations – where force
leads to the reduction or elimination of the choices of others. According to

this definition, for individual A to have social power, individuals B and C must implicitly agree that A is more capable than D of introducing force into social situations. Social power, when defined this way, is only relevant in those systems in which there are polyadic conflicts. It differs from *dyadic power*, which describes the degree to which one individual perceives another as capable of successfully using force when conflicts arise between them alone.

The concept of power (whether dyadic or social) is unlikely to be applicable in species in which individuals do not signal outside the agonistic context about their statuses in their agonistic relationships. We call such relationships informal agonistic relationships, and those relationships in which individuals do signal outside the agonistic context about their agonistic statuses, formalized relationships, following de Waal (1986a). The reason why we restrict application of the power concept to formalized relationships is that in order for an individual to exercise its power, it first needs to know that others perceive it as powerful, and thus they are unlikely to challenge its authority. As the sociologist, Bierstedt (1950: 733) wrote, power is – unlike force – "always successful . . . [and thus] symbolizes the force which *may* be applied in a social situation and supports the authority which is applied." In informal agonistic relationships, the subordinate individual communicates its willingness to yield only when conflicts arise; this means *after, or upon being threatened*, however mildly, by the dominant partner. The subordinate in informal relationships does not communicate a general willingness to yield to the preferences of the dominant partner.

In informal relationships the dominant partner therefore has to threaten force to learn whether it has power. Prior to the conflict, it only knows that subordinate has no power over it, not whether it has power over its dyad partner. Once it has learned how it is perceived, the dominant has power over the subordinate as long as the conflict lasts. As soon as the conflict ends, however, the same problem again arises because the subordinate has only communicated a willingness to yield after being threatened in a specific context. The specificity of this situation makes it difficult to generalize to others. This is not a problem in formalized relationships – those in which individuals signal about their status in peaceful situations in which there is no precipitating event or immediate functional application (such as submission or appeasement) to stimulate signal emission.

To summarize, the dyadic power concept only applies to those relationships that are formalized, and the social power concept only applies to those species in which: (1) there are polyadic conflicts; and (2) at least some individuals have formalized relationships. Later in this chapter we discuss why only some relationships need to be formalized for the social power concept to apply.

It is worth noting that the concept of power has a long and controversial history in the social sciences. For other treatments of power by ethologists, we refer the reader to Chapais (1991), Lewis (2002), and Preuschoft and van Hooff (1995b). The major difference between our approach to power and these others is that our conception, (1) applies to the larger social context that exists beyond the dyad, hence the qualification "social," and (2) is grounded in perception and acknowledgment, in the sense that A has power over B only if B perceives A to be stronger and communicates this to A outside the agonistic context. Although the stability of the perception is rooted in the validity of the perception (i.e., if A actually is a stronger fighter than B), the perception and subsequent acknowledgement count more than the validity. Thus, in our view (and in contrast to Preuschoft, 1995a) it is not appropriate to use agonistic asymmetry and power asymmetry interchangeably because these terms refer to different, albeit related, phenomena.

Measuring social power

Although most treatments of power in the ethological literature have focused on dyadic power, several researchers have discussed social power without calling it that (e.g., de Waal, 1982). That work has served as the foundation for our approach, and perhaps for the work of others who have used dominance rank implicitly as a measure of social power (for discussion of this tendency, see de Vries, 1998).

A more appropriate assay of social power than dominance rank is a measure of the degree of implicit consensus among group members about whether an individual has the capacity to introduce force into social situations. This approach has two advantages. Unlike dominance rank, the implicit consensus approach, (1) directly taps into social power, and (2) also allows for a mixture of dominance–subordination, unknown, and egalitarian relationships in groups, which, as Hand (1986) has suggested, is more realistic than assuming all relationships are dominance–subordination relationships (see Mori, 1977; de Waal, 1982; Preuschoft et al., 1998). Mori's work on Japanese macaques, for example, suggests that when groups are large, low interaction rates for many dyads keep many relationships unresolved. This intra-group variation in agonistic relationship types is likely to affect the degree of consensus about an individual's coercive potential and thus should be taken into account when quantifying the distribution of social power.

The degree of consensus among group members about whether an individual can introduce force into social situations should be quantifiable by measuring the degree of uniformity in the distribution of subordination signals that are sent and received by group members. It is important to note that this operational definition of consensus is informational not behavioral. Consensus cannot be defined behaviorally because it not a property of individuals, but of groups.

If the distribution of formal signals of subordination within the dyad tells us that the sender perceives the receiver as capable of successfully using coercion when conflicts of interest arise, the distribution of these signals across the population of senders and receivers should reflect the degree of implicit consensus among group members about whether an individual can introduce force into social situations. Thus, the distribution of subordination signals across individuals confers power in a manner analogous to the election for president in the United States: it is the cumulative actions of independent voters that confer power.

The decision to vote for a particular candidate is presumably based on that candidate's perceived capacity to run the country. The number of votes received by one candidate relative to another is thus a measure of this perception. Power is conferred to the candidate who is most widely perceived to have the appropriate skills. An important point to note is that even if a candidate does in fact have the skills to run the country, he or she will not be granted the power of the presidency unless his or her skills are recognized by the populace and acknowledged through votes received. In the case of animal societies, the social power of individual A can be conceptualized as the degree to which group members implicitly agree and acknowledge that they perceive A's coercive potential to be greater than that of B. Status signals of subordination are thus analogous to votes. J. Flack, D. Krakauer and F. de Waal (unpublished data) assessed the legitimacy of the assumption that implicit consensus is measurable using the distribution of candidate subordination signals, such as the silent-bared teeth display of pigtailed macaques, or the pant-grunt of chimpanzees.

Using data from a socially housed pigtailed macaque group, J. Flack, D. Krakauer and F. de Waal (unpublished data) evaluated this hypothesis by testing predictions about: (1) the distribution of subordination signals; and (2) the relationship between the distribution of these signals and the rate at which third parties receive requests for support from conflict participants. They found that individual i does not signal with the same frequency to all individuals dominant to it. However, they also found that there was a significant difference in how the set of subordinates signal to individual i compared to other dominant individuals, suggesting that subordinates implicitly agree that individual i differs from other individuals. This result was interpreted as support for the conclusion that although group members are not entirely in agreement about the degree to which an individual is capable of introducing force into social situations, they are in agreement about the degree to which one individual is better at this than another. This result was supported even after rank distance was controlled.

Moreover, Flack and collaborators' results indicate that individuals who frequently receive many subordination signals from many different individuals receive more requests for support from many different individuals than do individuals who receive few signals from few individuals. Thus, conflict participants disproportionately solicit intervention from individuals to whom they also send many subordination signals, suggesting that the distribution of status signals is related to the distribution of social power.

Encouraged by these results, Flack, D. Krakauer and F. de Waal (unpublished data) developed a mathematical formalization of social power, called the *Social Power Index* (SPI), which quantifies the amount of social power that each individual in a study population has depending on its distribution of subordination signals sent and received relative to the distributions of others. This formalization of social power is based partly on information theoretic principles in that it takes into account both the frequency and the "evenness" or uniformity of the distribution of subordination signals sent and received in noncontest, peaceful situations. An individual will have a high SPI value if it receives many subordination signals from many different individuals relative to others, whereas an individual will have a low SPI value if it receives relatively few signals from a relatively small number of group members. Below we review the operational definition of social power that was used to develop the formalization. The formalization is also briefly reviewed.

An operational definition of social power

The SPI was developed based on the following assumptions:

- Differences in social power do not exist among individuals who have only egalitarian agonistic relationships. Thus, in such societies, the distribution of social power is concentrated around a single value.
- The distribution of social power in most societies is likely to reflect a mixture of egalitarian, unresolved, and dominance–subordination relationships.
- Differences in social power cannot arise in a population comprised of only two individuals because social power is fundamentally about an individual's perceived capacity for coercion in polyadic situations (Bierstedt, 1950).
- Social power increases for each formalized agonistic relationship in which an individual has dominant status and decreases for each formalized agonistic relationship in which an individual has subordinate status.
- Social power is maximal when an individual has dominant status in all of its agonistic relationships and receives subordination signals with equal frequency from all other group members. Only one individual per group can have maximum social power.

- Social power is at its minimum when an individual has subordinate status in all of its agonistic relationships and gives more subordination signals to more group members than any other individual.
- Social power increases as the number of agonistic status signals received increases, and decreases as the number of subordination signals given increases.

The Social Power Index (SPI)

Briefly, the formalization takes into account two factors: (1) *information value*: the overall number of subordination signals an individual receives (r_i^T) and gives (s_i^T); and (2) *information content*: the distribution of these signals across the population, which is measured using Shannon's Information Index, $H_i(R) = - \sum_{j=1}^{N} r_{ij} \log r_{ij}$, where r is either the frequency of signals sent or received (depending on the term) from the ith individual by the jth individual. Information content is then multiplied by information value for the signaling and receiving components to produce the following index: $P_i = \alpha H_i(R) r_i^T - (1 - \alpha) H_i(S) S_i^T$, where the alpha term specifies whether the receiving (R) and signaling (S) terms are weighted equally. It is important to note that P_i is an index of social power. A positive P_i value indicates that group members implicitly agree that an individual can introduce force into a social situation, whereas a negative P_i value indicates that group members implicitly agree than an individual *cannot* introduce force into social situations.

The SPI thus has several major components: the receiving and signaling terms, and within each of those, information content and information value. J. Flack, D. Krakauer and F. de Waal (unpublished data) suggested that alpha should be set at 0.5, so that the receiving and signaling terms are weighted equally, unless data are available indicating that one of these processes is more important than the other. However, we have shown that for our group of pigtailed macaques, the most appropriate level of alpha is 1.0 (J. Flack & D. Krakauer, unpublished data). We expect that the appropriate level of alpha will vary across species.

Hypothetical distributions of social power

We have plotted several different social power functions in Fig. 8.2. In some societies social power might be distributed uniformly, such that it increases by some constant quantity across individuals. Three such distributions, each with

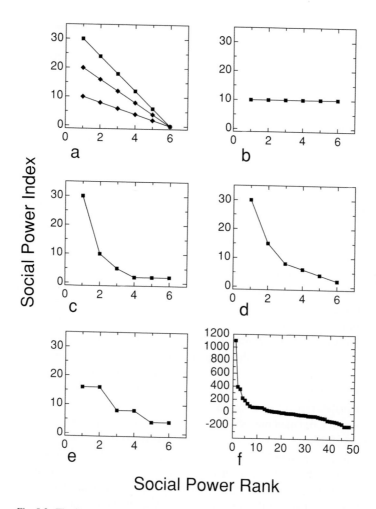

Fig. 8.2. The Social Power Index is plotted as a function of Social Power Rank for five hypothetical societies and one real one. (a) A society in which social power is distributed uniformly but hierarchically (three possible slopes are presented); (b) a society in which everyone has more-or-less the same amount of social power; (c) a society in which power is concentrated in a few individuals and distributed relatively evenly over the rest; (d) a society in which power is concentrated in a few individuals and distributed uniformly but hierarchically over the rest; (e) a society in which there is a class system; (f) the actual distribution of social power in a captive pigtailed macaque group.

a different social power by Social Power Rank slope, are shown in Fig. 8.2a. A uniform distribution of social power implicitly underlies most ideas about dominance hierarchies (de Vries, 1998), and probably best characterizes rhesus and Japanese macaque societies. As shown in Fig. 8.2b, social power might also be distributed around a single value such that it is spread more or less evenly over individuals (with the extreme case being a delta function). This distribution of social power might be found in equal outcome or equal opportunity societies (see section, Dominance style and political system type), or in societies in which many individuals have unresolved relationships, as might be the case for some of the Sulawesi macaque species. Although there are no differences in social power in such societies, some individuals might be more influential than others because of noncoercive capabilities such as social knowledge or experience. Shown in Fig. 8.2c is a society in which social power is concentrated largely in a few individuals and distributed more or less evenly over all others, as might be the case for stumptailed macaques. Figure 8.2d corresponds to a society in which social power is again concentrated in a few individuals, but distributed uniformly over the rest. This distribution of social power might be the correct characterization of societies in which some individuals are thought to be "super-dominant" (e.g., Tokuda & Jensen, 1968; Das, 2000). Using the SPI, J. Flack, D. Krakauer and F. de Waal (unpublished data) determined that the distribution of social power in their pigtailed macaque group was log-normal, such that power was disproportionately concentrated in a few individuals. Figure 8.2f, which bears considerable similarity to Figure 8.2c, shows this result. SBT matrix data reported in Preuschoft (1995a) suggest that the distribution of social power in longtailed macaques might also be of the type shown in Fig. 8.2c. In Figure 8.2e, we show a society with a class system. This type of distribution might be found in those macaque species in which different dominance styles have been reported for males and females, such as Barbary macaques (Preuschoft *et al.*, 1998), or it might only apply to human societies.

Both the shape of the underlying distribution and the slope of these graphs are likely to be important causal factors determining the types of conflict management mechanisms that are present in a social organization.

A test of the index against other potential measures of social power

If the SPI is a good measure of social power then it should predict behavioral patterns that in theory should be functionally linked to the distribution of social power in nonegalitarian societies. Moreover, the SPI should be a better predictor of these variables than other potential measures of social power. To evaluate these hypotheses, the empirical validity of the SPI was tested against other potential measures of social power, including ordinal and cardinal dominance rank indices (J. Flack, D. Krakauer & F. de Waal, unpublished data). The SPI was tested against dominance rank because dominance rank is the traditional measure of social power in studies of animal social organization. As with the SPI, Flack and collaborators used subordination signals to determine dominance rank. The difference between the SPI and dominance rank is in the way that each is calculated: the SPI uses ideas from information theory concerning the distribution of signals sent and received to measure consensus, whereas dominance rank is either determined using a statistical procedure to minimize inconsistencies in dominance related behaviors (de Vries, 1998) or using the Boyd and Silk (1983) or Jameson and collaborators (1999) BBS approach for generating a cardinal index (all of which assume transitivity).

Using data gathered from a socially-housed pigtailed macaque group, J. Flack, D. Krakauer and F. de Waal (unpublished data) tested the SPI and the cardinal and ordinal dominance indices against three intervention-related dependent variables, which in theory should be causally linked to social power: the frequency with which third parties receive requests for support, the intensity of aggression *received by* interveners (in response to their interventions), and intensity of aggression *used by* interveners. Flack and collaborators hypothesized that if an individual is widely perceived to be of greater coercive potential in polyadic agonistic situations than others, that individual should receive more requests for support from conflict participants. That individual should also not need to use much aggression to affect the behavior of others, and should rarely be challenged when intervening. Flack and collaborators found that the SPI is the most informative predictor of the measures that were tested. As expected, its fit to each of the dependent variables was better than that of either the cardinal or ordinal dominance indices: it explained at least 40% more of the variance. It was also a substantially better multivariate predictor of the set of dependent variables.

Dominance style and political system type

We have argued throughout this chapter for keeping dominance style, a relationship-level concept, separate from political system type, a social organization-level concept. In Table 8.1 we introduce five political system types: hierarchy, informal oligarchy, constrained, equal outcome, and equal opportunity. We identified these five because they span the range of possible political systems into which nonhuman animal societies, particularly macaque societies, might fall. Many human societies, and perhaps some macaque societies, however, have mixed system types, either because different rules apply to each age–sex class – as is the case in societies in which there are egalitarian relations among men but not among men and women (see Sahlins, 1958; Flanagan, 1989), or because there is a class structure in place in which each class is best described by a different system (Fig. 8.2e). A mixed system type is possible in those macaque societies in which, for example, males and females have different dominance styles, as has been proposed for Barbary macaques.

We define a hierarchical system as one in which the enforcement of status differences is the primary mechanism by which conflict is controlled. We expect that interventions occur in such systems, but are typically performed against the lower-status conflict opponent and function to opportunistically reinforce the dominance relationship between the intervener and target at low cost. Most researchers who work with macaques probably assume that the hierarchical system best characterizes the style of conflict management typical of their study species. We suspect, however, that in actuality only a few macaque species, including rhesus and Japanese macaques, have such systems.

We have borrowed the term "oligarchy" from sociology, where it means "rule by a few." Here we append the term "informal" to indicate that there is not necessarily any explicit communication among those who "rule" in this system. In Michel's (1911) sociological conception, oligarchy describes a system in which the elected come to dominate the electors in order to maintain the status quo. In such systems powerful third parties manage conflicts by policing group members. Effective policing involves conflict management and resolution but not necessarily suppression. We suspect that conflict monitoring becomes more impartial as dominance style moves from tolerant to relaxed. Policing is also sometimes called the "control role," and it might be related to the big-man/great-man concept in anthropology (Godelier, 1986; Boyd & Richerson, 1992). Policing probably facilitates the development of social bonds between individuals with unresolved relationships. Because our data indicate that social power in pigtailed society is concentrated in a few individuals, and distributed uniformly over the rest such that social power increases by some constant as social power rank (SPR) increases, it is possible that pigtails have an oligarchical system. This possibility is also supported by data indicating that the highest-ranking male intervenes using low levels of aggression (J. Flack & D. Krakauer, unpublished data), and effectively terminates conflict (Tokuda & Jensen, 1968; Oswald & Erwin, 1976; Flack & de Waal, unpublished data).

A constrained system is one in which power is again concentrated in a few individuals, but is distributed more or less evenly over the remaining group members. The remaining group members are able to form leveling coalitions and alliances that check the behavior of the powerful individuals. Leveling coalitions have been described by Boehm (1993, 1999) and de Waal (1982,

1996a) for chimpanzees (we include mention of these results for comparative purposes) and by Preuschoft and van Schaik (2000) for male Barbary macaques. We suggest here that leveling coalitions work best when the subordinates joining forces are relatively similar in social power, because this facilitates the formation of reciprocal relationships. It is not possible to say from available data whether any macaque species are best described by this system. However, this system does seem to be characteristic of chimpanzees (see de Waal, 1982; Boehm, 1999).

An equal outcome system was defined by Flanagan (1989) as one in which there are structural constraints to prevent individual differences in aptitudes from leading to large differences in resource control. This system might characterize some species of Sulawesi macaques, which have been reported to have a relatively even distribution of social power (see Thierry, 2000). An equal-opportunity system was defined by Flanagan (1989) as one in which all individuals are given equal chance to acquire resources, and any differences in resource control that result are due to individual differences in acquisition strategy. This type of system probably only applies to some human societies.

In Table 8.1, we sketch what we think is the relationship between dominance style and political system type. Although we do not know how dominance style translates into system type, three factors seems to be important:

- *The conflict resolution efficacy (CRE) at the dyadic level given the dominance style.* We suspect that the CRE determines whether (a) the agonistic relationship is sufficient as a conflict management mechanism or (b) if dyadic conflict management should shift to polyadic conflict management to maintain social stability. The CRE should be dependent on the degree of mismatch between inherent and expressed asymmetry: when there is little mismatch, such that dominants rarely tolerate subordinates, the CRE at the dyadic level should be quite high, although biased toward the interests of the dominant individual. As the degree of mismatch increases, the CRE should decrease (assuming that expressed asymmetry can never exceed inherent). The CRE should also be high when relationships are unresolved or are egalitarian (i.e., when degree of inherent asymmetry is low or unknown, or when the degree of expressed asymmetry is very low). An operational definition of CRE might be the dyadic frequency of bidirectional aggression for a given dominance style.
- *Translation error.* We use the term "translation error" to refer to errors made in formal status signaling, either because an individual has: (a) incorrectly assessed its role in its relationship based on its agonistic interaction history; or (b) because it incorrectly makes the decision to signal despite knowing its state (J. Flack & D. Krakauer, unpublished data). Status signaling error can,

under certain circumstances, generate a distribution of social power in which information about the relative coercive potential of individuals in polyadic agonistic situations has been inaccurately encoded. In such cases, individuals are perceived to have more or less social power than they should actually have given their agonistic interaction histories. Thus, when translation error is low, such that individuals make few signaling mistakes, the distribution of social power is likely to correspond correctly to the actual coercive potential of individuals in polyadic agonistic situations.

If senders have the capacity to strategically signal in that they understand the implications of signaling to one individual more frequently than to another, this would facilitate control of dominants by subordinates because subordinates would be able to partially decide in whom power should be concentrated. Because few papers on macaques report raw matrix data on the exchange of SBTs or on the social contexts (i.e., if SBTs are emitted preferentially to one individual over another when both are in proximity to the sender) in which SBTs are emitted, it is impossible to say whether this capacity is present in any macaque species. Preliminary data do suggest, however, that chimpanzees might have this capacity and do use signals to vote for the top males (de Waal, 1982; Flack & de Waal, unpublished data).

- *The distribution of social power.* In systems in which the distribution of social power is coupled to the dominance hierarchy, as is likely to be the case in hierarchical systems, the need for conflict management by third parties will probably be low. This is because the conflict resolution efficacy at the dyadic level will be high. When interventions do occur, they will either be in favor of kin or against the lower-ranking opponent. The latter type of intervention, called polarizing support by Preuschoft and van Schaik (2000), probably functions to opportunistically reinforce the dominance relationship between the third party and target at low cost. Rhesus macaque intervention data conform to this pattern (Ehardt & Bernstein, 1992).

Polyadic conflict management will probably only occur in systems in which the conflict resolution efficacy of the dyad is reduced. However, the type of polyadic conflict management mechanism that can arise should be constrained by the distribution of social power, which we hypothesize determines for potential interveners the cost-benefit ratio of engaging in conflict management. This is of course based on the assumption that the probability of receiving aggression increases the closer third parties get to ongoing conflicts, particularly if the third party approaches aggressively or targets one opponent. The cost-benefit ratio (i.e., likelihood of receiving aggression) of engaging in conflict management should be small for those individuals who have disproportionately more power than other group members. We have shown this to be the case in our study group of pigtailed macaques (J. Flack & D. Krakauer, unpublished data). In such systems, policing (impartial monitoring of conflicts by powerful third parties) should be the primary form of conflict management. On the other hand, when social power is relatively evenly distributed, the cost-benefit ratio of engaging in conflict management will be relatively large for most individuals. In this situation, impartial conflict management by third parties is unlikely to evolve unless an additional cost is imposed against those individuals who do not punish norm breakers or help resolve disputes. This kind of conflict management has been termed "meta-policing" by Axelrod (1986). Alternatively, conflict intervention style might switch from aggressive or threatening interventions to peaceful interventions that serve to appease opponents. Peaceful interventions are less likely to provoke aggressive responses from the conflict opponents.

If the distribution of social power does constrain the type of conflict management mechanism that is present in a system, the degree to which the distribution accurately reflects the coercive potential of individuals will be important. When there is a serious mismatch between the observed distribution of social power and the actual coercive potential of individuals, an inappropriate conflict management mechanism might arise. It is unlikely that inappropriate conflict management mechanisms will be sustainable over the long-term because the cost of implementing the mechanism will, over time, outweigh the benefits. This correction process will eventually reveal errors in the signaling distribution. For example, a uniform distribution of social power should not support a policing mechanism because no individual should have enough social power relative to others to make engaging in policing interventions profitable. Moreover, a policing mechanism should not be necessary in such a system, because most conflicts should be resolved using severe unidirectional aggression that reinforces the existing dominance relationships. If, however, status-signaling error leads to a log-normal distribution of social power when the distribution should actually be uniform, a policing mechanism might inappropriately arise. The policing mechanism should not be sustainable in such cases because the mismatch between the coercive capacities that individuals are perceived to have and the capacities that they actually have will probably be revealed in inefficient and potentially costly (to the intervener) interventions.

A final note. It is possible that the type of conflict management mechanism that emerges at the system level will feed back to the relationship level to effect dominance style. For example, if powerful individuals police the conflicts of group members, this might pressure dominant individuals to exert restraint when interacting with subordinates, thereby leading to a more tolerant dominance style, or even a relaxed one.

Conclusion

This framework now needs to be tested. Although we have begun testing several of its components on pigtailed macaques – namely, the ideas: (1) that social power can be measured using the distribution of subordination signals; and (2) that policing is the primary conflict management mechanism in societies in which social power is concentrated in only a few individuals – this work needs to be extended to other macaque species.

Box 8 Social space and degrees of freedom

Marina Butovskaya

Tolerance and despotism as a way of managing social relationships became an object of intensive discussions among biologists in the last decades of the twentieth century (Vehrencamp, 1983; Thierry, 1986a, 2000; de Waal, 1989a, 1996a; Butovskaya _et al._, 2000). To account for the differences between macaque societies with tolerant ("egalitarian") and rigid ("despotic") dominance styles, the degrees of freedom of individuals in their social space appear as a major intervening variable (Thierry, 1990a; Butovskaya, 1994). We can estimate degrees of freedom in terms of the size of individuals' social networks and the diversity of contacts with other group members, despite rank differences and kinship bonds. The mean number of social partners, with whom a focal individual regularly interacts at its own initiative, may be an indicator of the species-specific degree of social freedom.

  Social factors, like ecological conditions and the size of matrilines are likely to influence individuals' social choices (see Chapais, Chapter 9), but there is ground to believe that species-specific traits also influence degrees of freedom as indicated by the intervention of phylogenetic factors (Thierry, Chapter 12). The macaque world has intrinsic social limitations, in particular it is kin-centric. Females inherit their social ranks from their mothers, and it is usually assumed that there is little change in their dominance status after puberty. Social hierarchies in macaques are close to linear and quite stable (Paul & Kuester, 1987; Datta, 1992). Female kin appear to prefer the company of each other and, to this extent, relatives and friends are basically the same individuals.

  In species with despotic social relations (_mulatta_, _M. fuscata_, _M. fascicularis_), power asymmetry between dominants and subordinates is strong and formal indicators of subordination are well developed (de Waal, 1989a; Butovskaya & Kozintsev, 1996a; Preuschoft, Box 3; Flack & de Waal, this Chapter). In species exhibiting a tolerant style

Box 8 Social space and degrees of freedom 183

Fig. 8.3. In Tonkean macaques, members of different matrilines commonly stay at close proximity and groom each other (Strasbourg Primate Center, France). (Photograph by M. Butovskaya.)

(*M. tonkeana, M. nigra, M. maurus, M. arctoides*), no formal indicators of subordination are present and the proportion of friendly interactions among nonkin is quite high (Butovskaya & Kozintsev, 1996b; Aureli *et al.*, 1997; Preuschoft & van Hooff, 1997; Thierry, 2000; Thierry, Chapter 12). Lower-ranking females have a higher diversity of choices for friendship and affiliation in the latter social organization. An individual's friendly social network is broader, incorporating individuals from different matrilines (Fig. 8.3). In fact, we registered no significant kin-preferences in grooming, aggressive and affiliative support in Tonkean macaques and also to a large extent in stumptailed macaques (Butovskaya, 1993; Butovskaya & Kozintsev, 1996b; Butovskaya & Thierry, unpublished data).

Importantly, low-ranking females from tolerant societies are allowed to take and carry infants of higher-ranking counterparts (Maestripieri, Box 10). For instance, low-ranking juvenile females may stay with high-ranking infants at a distance from their mothers for half an hour and even longer. Again, low-ranking females are quite relaxed, they do not scratch more than usual, they do not attempt to leave or avoid contact when dominant females are interacting with their infants (Butovskaya & Thierry, unpublished data). Both events would hardly have been possible in despotic species like long-tailed or rhesus macaques (Butovskaya & Kozintsev, 1996a).

On the social tolerance level, the typical resting group size is smaller under despotic conditions (Butovskaya, 1993). High-ranking animals have significantly broader social networks than low-ranking ones in longtailed and rhesus macaques, while in stumptailed and Tonkean macaques this is less obvious. In tolerant species, limits in the power of high-ranking individuals and a relatively high number of interaction choices for low-ranking ones produce a situation under which the former have to develop sophisticated social manipulative strategies to establish and maintain their dominance status (see Call, Box 2). Forceful suppression of subordinates is not sufficient (Flack & de Waal, this Chapter).

In species with a rigid dominance style, low-ranking females are usually limited in initiative – whether agonistic or peaceful – while interacting with animals of a higher social status, so they keep a low profile. Rank is a main factor determining spatial proximity with other group members and initiatives in approaching them, and thus the circle of familiar partners. By contrast, in tolerant species the social situation is far less constraining for low-ranking individuals, the network of affiliative partners and initiative of special contacts depends more on the personality of individuals and their life history (Butovskaya *et al.*, 2000).

Grooming is one of the manifestations of degrees of freedom in the social space (Butovskaya *et al.*, 2000). In despotic species (*M. mulatta, M. fascicularis*) grooming is largely asymmetrical, up the hierarchy; and in accordance with the similarity principle, grooming is more frequent among partners who are near in dominance rank (de Waal & Luttrell, 1986). In tolerant species (*M. arctoides, M. tonkeana*), high-ranking individuals may groom substantially lower-ranking partners more frequently than vice versa (Butovskaya, 1993; Butovskaya & Kozintsev, 1996b), and rank similarity is not a reliable predictor of the distribution of grooming interactions. Whereas social grooming is largely kin-oriented in despotic species, we found no significant effect of kinship in stumptailed on Tonkean macaques (Butovskaya, 1993; Butovskaya & Kozintsev, 1996a,b; Butovskaya & Thierry, unpublished data). A further important difference concerns the initiative for grooming. Grooming is directed mainly up the hierarchy, but it is usually the high-ranking partner who stimulates an interaction in species with a despotic style (de Waal & Luttrell, 1986). In tolerant societies, the initiative to a large extent comes from the side of subordinates (Butovskaya & Kozintsev, 1996b; Butovskaya & Thierry, unpublished data). Thus an active choice of grooming partners is a matter of individual preference in the latter case.

Social grooming appears to be an efficient instrument for gaining allies in agonistic interactions among despotic species. Younger sisters outrank older sisters due to alliance formation and agonistic support from nonkin and close kin in rhesus and Japanese macaques (Chapais *et al.*, 1994; Datta, 1988; Chapais, Chapter 9; Chauvin & Berman, Chapter 10). Their success is largely

Box 8 Social space and degrees of freedom 185

due to a combination of their increased grooming activity in the direction of female partners and their active appeals for support from the latter in conflict situations. It remains uncertain whether grooming could be used in the same way in more tolerant species. Although grooming is definitely a powerful instrument of rank improvement in them, it is less connected with reciprocal support in aggression. Our data on a group of Tonkean macaques provides a good example (Butovskaya, unpublished data). Here, Dilie, daughter of the lowest-ranking female in the group, was able to gain a higher social position and outranked a number of females from the dominant matriline. She acted mainly by soliciting friendship with high-ranking females, not by high rate of aggression. She did not form alliances with nonkin females against her own relatives. Dilie spent a lot of time sitting in close proximity with the highest-ranking females, and was the most active groomer in the group. Small (1990c) reports similar facts for a female labeled D40 in a group of Barbary macaques. In that case the female belonged to a small matriline and so was rather limited as to the number of her potential social partners. With regard to Dilie, she belonged to a large matriline and had no limited choice of partners. Her social preferences might represent a kind of social manipulation.

9 How kinship generates dominance structures: a comparative perspective

BERNARD CHAPAIS

Introduction

Much has been written about nepotistic dominance structures in primates, but very little about their variation between groups. In this chapter I focus on the reasons why nepotistic hierarchies vary between groups and species across the *Macaca* genus, in an attempt to uncover some of the pathways through which kinship bonds generate female dominance structures. In other words, I use a comparative perspective to understand better how social structures arise.

Several studies, carried out mostly on provisioned groups of Japanese and rhesus macaques, have defined what will be called here the *classical nepotistic hierarchy* or *classical matrilineal dominance structure* (Kawai, 1958; Kawamura, 1958; Koyama, 1967; Sade, 1967; Missakian, 1972). In this system a female's rank is entirely predictable from knowledge of her kinship bonds (Chauvin & Berman, Chapter 10). A daughter eventually outranks all females that rank below her mother, and she remains subordinate to her mother. Females also outrank their older sisters (the so-called "youngest ascendancy" rule; Kawamura, 1958). Rank relations between sisters determine the relative rank of their respective daughters, hence the dominance relations between aunts and nieces and between cousins, and so on down generations. Exceptions to these rules have been reported in provisioned groups of rhesus and Japanese macaques (e.g., Missakian, 1972; Chikazawa, 1979; Mori *et al.*, 1989; Nakamichi *et al.*, 1995; Katsukake, 2000), but they tend to occur in large groups or to be related to the formation of new groups. In a classical nepotistic hierarchy, the individual attributes of females are practically irrelevant in the formation of rank relations because a female's relative rank is determined by the patterning of interventions by third parties in her conflicts with other females (references below). Alliance patterns provide the causal links between kinship and the dominance structure.

Macaque Societies: A Model for the Study of Social Organization, ed. B. Thierry, M. Singh and W. Kaumanns. Published by Cambridge University Press. © Cambridge University Press 2004.

Hence, data on rhesus and Japanese macaques provide an alliance-based model of the genesis of matrilineal dominance relations. Longtailed macaques also fit this model, both in terms of dominance rules and alliance patterns, but data on alliances are less abundant for this species (Angst, 1975; deWaal, 1977; Netto & van Hooff, 1986; van Noordwijk & van Schaik, 1999).

Not all 20 species of macaques, however, conform to this model. This should come as no surprise given the extent to which macaque species differ in several aspects of their social behavior (reviewed by Thierry 2000, Chapter 12; Preuschoft, Box 3; Maestripieri, Box 10). In particular, dominance style (*sensu* de Waal, 1989a) varies dramatically, ranging from "despotic" species with unidirectional and intense aggression between dominants and subordinates, to "egalitarian" species with bidirectional and less severe aggression and higher levels of tolerance and conciliatory interactions (Thierry, 1985a; de Waal & Luttrell, 1989; Butovskaya, 1993; Chaffin *et al.*, 1995; Petit *et al.*, 1997; Matsumura, 1998; Demaria & Thierry, 2001). The structure of dominance relations varies as well. First, the rules differ between groups of the same species. For example, in Japanese macaques the youngest ascendancy rule that characterizes captive as well as free-ranging, provisioned groups is not observed in some feral groups (Hill & Okayasu, 1995, 1996; Hill, Box 11). Dominance rules also vary between species. For instance, the youngest ascendancy rule does not hold in provisioned groups of Barbary macaques (Paul & Kuester, 1987; Prud'homme & Chapais, 1993a) and in one captive group of Tonkean macaques (Thierry, 2000).

Unfortunately, however, quantitative data on the proportions of mother–daughter dyads and sister–sister dyads that reversed rank are lacking for most macaque species, except for provisioned groups of rhesus, Japanese and Barbary macaques. Most importantly, data on the timing of reversals in relation to the age of females are also extremely scanty. For example, whether daughters outranked their mothers most often when the latter were in declining physical condition (e.g., old), or even when they were in their median adult age and well, is generally unknown. Such information is essential in order to characterize and compare the matrilineal structures of different groups, and to test various hypotheses on the causes of variation. In some cases, long-term quantitative data are available but the relevant proportions of rank reversals are not (e.g., Rhine *et al.*, 1989, for *M. arctoides*). In other cases, some information is available on mother–daughter rank relations but not on sisters (e.g., *M. arctoides*: Estrada *et al.*, 1977; Nieuwenhuijsen *et al.*, 1985; *M. maurus*: Matsumura, 1998). In still others, mother–daughter rank reversals were reported to occur frequently but the history of group formation complicates the interpretation (e.g., *M. radiata*: Silk *et al.*, 1981). For most groups and species of macaques, information on matrilineal kinship is lacking or incomplete, and quantitative data on rank relations extending over sufficiently long periods are simply nonexistent.

Keeping in mind this important limitation, the available evidence suggests that there may exist two other forms of nepotistic hierarchies, besides the classical one. In both variants, daughters initially rank just below their mother. In the first variant, tentatively exemplified by Barbary macaques, mothers remain dominant to their daughters until they reach post-reproductive age (Paul & Kuester, 1987; Prud'homme & Chapais, 1993a). In spite of this, younger sisters are unable to outrank their older sisters. Thus, the dominance order is matrilineally structured, but the level of nepotism (principally, the mother's involvement in rank relations between her daughters) is apparently lower than in the classical type. In the second variant, adult daughters are able to outrank their mothers before the latter reach post-reproductive age and even if they are in good physical condition. If the mother is outranked sufficiently early, this entails the absence of a consistent pattern of rank relations between sisters because a mother can hardly help a young daughter outrank an older daughter that ranks above herself. This second variant is tentatively exemplified by savanna baboons (Hausfater *et al.*, 1982; Combes & Altmann, 2001, see below) and also, perhaps, by bonnet (Silk *et al.*, 1981) and Tonkean macaques (Thierry, 2000). It is only incompletely matrilineal (i.e., rank is less predictable from knowledge of kinship bonds) and weakly nepotistic. Whether these two variants are truly distinct, or part of a continuum, is unclear due to the lack of quantitative data. Note that in both variants rank relations are well defined, linear and clearly nepotistic, any female initially ranking just below her mother. In contrast, in non-nepotistic dominance structures, although female rank relations are well defined and linear, daughters do not initially rank below their mother (e.g., Hanuman langurs: Hrdy & Hrdy, 1976; Borries, 1993; Koenig, 2000). For the sake of simplicity, in the rest of this chapter I will refer to the two variants as *weakly nepotistic hierarchies*.

Why do nepotistic dominance structures vary? The prevalence of the youngest ascendancy rule in provisioned groups of Japanese macaques and its absence in some wild groups may suggest that the rule is an artefact of captivity and provisioning (Hill, Box 11). But the youngest ascendancy rule also characterizes some wild groups of longtailed macaques (van Noordwijk & van Schaik, 1999) and possibly other species. Hence, provisioning is not a prerequisite for the youngest ascendancy rule, nor is the rule a correlate of provisioning, as revealed by its absence in provisioned groups of Barbary macaques. Thus, the observed variation goes beyond the opposition between provisioning and nonprovisioning.

The classical nepotistic hierarchy model provides one general hypothesis about the proximate causes of the observed variation. Given that alliance patterns generate specific dominance rules, intergroup differences in dominance rules might reflect intergroup differences in alliance patterns. Thus, one should look into patterns of alliances to understand variation in dominance structures between groups. Although detailed data on alliances in macaque species other than rhesus and Japanese are relatively scanty, three categories of explanations explicitly linking alliance patterns to dominance rules have been proposed: demographic, ecological, and phylogenetic. In this chapter, I critically examine these three explanations and discuss their inter-relationships. I then turn briefly to other possible sources of variation in dominance rules that specifically exclude the role of alliance patterns, namely differences in the intensity of aggression and differences in patterns of reproductive value. I begin by summarizing the evidence for the alliance-based model of female dominance relations issued from the detailed study of rhesus and Japanese macaques.

From alliances to nepotistic hierarchies: the classical model

Detailed studies on provisioned groups of rhesus or Japanese macaques have established that third-party interventions in conflicts play a central role in the genesis of matrilineal hierarchies. Several observational studies on rhesus and Japanese macaques (Berman, 1980; Datta, 1983a, 1988; Chapais & Gauthier, 1993) and other species of cercopithecines (Cheney, 1977; Horrocks & Hunte, 1983; Pereira, 1989) reported patterns of interventions compatible with this interpretation. Chapais (1992a; see also Pereira, 1992) summarized that evidence in the form of seven behavioral correlates of matrilineal rank acquisition and maintenance, six of which were supported by the available data. To sum up, compared to females of lower birth rank, females of higher birth rank have a larger number of allies that are more powerful (higher ranking), more active, and whose interventions induce submission more effectively. But matrilineal rank acquisition is a developmental process that begins early in life and ends around sexual maturity, at which time the females receive submission from most lower born females. Before that time, immature females have unsettled dominance relations with lower-born females from which they often receive counter-aggression (Walters, 1980; Datta, 1983a, 1988; Chapais *et al.*, 1994). Presumably, the help the immature females receive from their allies allows them to eventually outrank lower-born females. Understandably then, specific interventions in conflicts are not followed by actual rank changes, hence observational data do not provide conclusive evidence that interventions are the main factor causing rank acquisition. But experiments on Japanese macaques have provided such temporal evidence and established the direct causal connection between interventions and rank reversals. An immature female that behaved submissively toward dominant peers could outrank them in the presence of a single adult kin (e.g., her mother) just after the latter had intervened in her favor (Chapais, 1988a), indicating that interventions are sufficient to induce rank changes.

These experiments also clarified the role of specific categories of kin in matrilineal rank inheritance. Indeed, although several kin may be observed to aid a given juvenile female in natural groups, this does not mean that all allies have the same impact on that female's ability to acquire her birth rank. Experiments revealed that mothers, sisters, grandmothers and great-grandmothers consistently induced rank acquisition in experimental subgroups, but that aunts, grandaunts and cousins did not, or did so very inconsistently (Chapais et al., 1997, 2001). These results suggest that in natural groups only close kin spontaneously aid young females before they begin to act assertively with lower-born females. In doing so close kin would "assign" the females' birth ranks, thereby generating the basic rules of the system.

The fact that distant kin and nonkin did not spontaneously help submissive females in experimental subgroups suggests that they play no significant role in the assignment of matrilineal rank in natural groups. Nonetheless, distant kin and nonkin might be instrumental in the female's ability to acquire, and later maintain, her rank once it has been assigned and the female is behaving assertively with lower-born females. Several observational studies on rhesus and Japanese macaques (Chapais, 1983; Datta, 1983b; Chapais et al., 1991), other macaque species (Silk, 1982; Netto & van Hooff, 1986; Prud'homme & Chapais, 1993b) and other cercopithecine species (Walters, 1980; Cheney, 1983; Hunte & Horrocks, 1986; Pereira, 1989) support this hypothesis. Distant kin and nonkin ally in a conservative manner: when intervening in a conflict between two females, they most often support the opponent with the higher birth rank. By doing so they enforce birth ranks and contribute to stabilizing the matrilineal rank order.

But again, the existence of conservative alliances between distant kin and nonkin does not constitute conclusive evidence that these alliances are a significant factor in a female's ability to acquire and/or maintain her birth rank above potential challengers; such conservative alliances might just be an ancillary correlate of matrilineal dominance orders. However, experiments on Japanese macaques indicate that this is not the case. Any immature female that was isolated either with an older sister or with older females from a lower-ranking matriline was almost invariably outranked. However, if the same immature female was isolated with the same older lower-ranking females, but in the presence of a dominant lineage, the immature female was not outranked by her older sister (Chapais et al., 1994), nor by unrelated females (Chapais, 1988b; Chapais et al., 1991; Chapais & Gates-St-Pierre, 1997). These findings suggest

Table 9.1. *Role of alliances in the assignment, acquisition and maintenance of a female's birth rank in relation to her older sisters and to all other females that rank below her mother in a classical nepotistic hierarchy*[a]

Categories of allies	Assignment of birth rank in relation to		Acquisition of assigned rank in relation to		Maintenance of assigned rank in relation to	
	Older sisters	All other females	Older sisters	All other females	Older sisters	All other females
Close kin[b]	Yes	Yes	Yes	Yes	Yes	Yes
Distant kin[c]	No	No	Yes	Yes	Yes	Yes
Nonkin[d]	No	No	Yes	Yes	Yes	Yes

[a] See text for explanation and sources of data.
[b] Mothers, grandmothers and sisters.
[c] Aunts, grandaunts and cousins.
[d] More distant kin.

that if an immature female were to lose all her close kin she would nevertheless be able to count on distant kin and nonkin to acquire and maintain her birth rank (Table 9.1). For example, in Japanese macaques, younger sisters may outrank their older sisters even after their mother's death (Datta, 1988; Mori *et al.*, 1989). Thus, although distant kin and nonkin play no role in the assignment of matrilineal rank, they recognize a female's birth rank – presumably through the female's patterns of agonistic interactions and alliances – and they enforce it. Hence, so-called nepotistic dominance orders are far from being exclusively nepotistic.

What are the functional bases of these alliance patterns? Interventions by adult females in favor of immature kin that are hardly able to reciprocate efficiently due to their young age meet the criteria of unilateral altruism and are best explained in terms of kin selection (Chapais, 2001; Chapais *et al.*, 2001). But kin selection does not explain the existence of alliances between dominant nonkin against lower-born females. In the course of such interventions, interveners appear to protect or support the recipients and to incur a net cost in terms of energy spent and, possibly, risks taken, suggesting that they are behaving altruistically, and reciprocally so. However, when a female sides with another female against a lower-born target, the intervener herself is likely to derive an immediate net gain: the risks of retaliation are either low or nil, and the probability of inducing the target to submit is higher through a joint aggression compared to a dyadic attack (Chapais *et al.*, 1994; Prud'homme & Chapais,

1996). Thus, nonkin interventions are best regarded as opportunistic and mutually selfish. A growing body of evidence points to the mutualistic nature of alliances between unrelated females (Chapais *et al.*, 1991) and between unrelated males (Bercovitch, 1988; Noë, 1992; Widdig, 2000).

In sum, in the classical matrilineal hierarchy both kin support and nonkin support are extensive. This leads to the idea that the levels of involvement of kin and nonkin are positively correlated, strong kin support entailing strong support by nonkin. One possible explanation is that when kin support is strong, power asymmetries between females are pronounced. In these circumstances it might be to the advantage of any female (including nonkin) to opportunistically side with the most dominant of two opponents, thereby generating the support-to-the-dominant principle described above, and further increasing the degree of power asymmetry between ranks. Patterns of opportunistic and conservative support to the stronger opponent (or winner support) are common in primates (Chapais, 1995), not only between females, but between males (e.g., de Waal, 1982; Widdig, 2000), and between males and females (de Waal, 1982, 1986b; Raleigh & McGuire, 1989). According to this reasoning, low rates of kin support would entail low rates of nonkin support. On this basis, one expects weakly nepotistic hierarchies to exhibit low overall rates of female coalitions (kin and nonkin) compared to what is observed in the classical nepotistic hierarchy.

To summarize, in the classical matrilineal system, close kin (essentially the mother, the grandmother and sisters) set a female's birth rank, both in relation to the female's older sisters and to unrelated females (rank assignment). Afterwards, close kin, distant kin and nonkin all contribute to the acquisition and maintenance of the female's birth rank by performing mutualistic, conservative alliances (Table 9.1). In other words, kin altruism sets the system in motion and mutualism enforces it. In the end, these alliance patterns have generated specific and consistent dominance rules.

Alliance patterns and between-group variation in dominance structures

The role of demographic variation

The first category of factors that could generate intergroup variation in female dominance structures is demography, more specifically group composition. Datta and Beauchamp (1991) assessed the effect of demographic parameters on female rank relations by simulating two different primate populations. In the first one, females reached sexual maturity earlier, had shorter birth intervals and a lower mortality rate, this resulted in the average female producing

2.8 daughters in her lifetime (expanding population). In the other population (declining), any female produced only 0.6 daughter. Hence, in declining populations, matrilines (kin groups) were on average much smaller than in groups with high reproductive rates (see Dunbar, 1988). Considering sister rank relations, the model assumes that a younger sister will be able to outrank an older sister if the allies she needs to do so are alive, namely her mother or another sister (ranking above the older one). The simulations showed that these key allies were more likely to be alive in the expanding population than in the declining one, hence that younger sisters would outrank older sisters much more consistently in the former situation. Datta (1992) used this reasoning to explain the consistency of the youngest ascendancy rule in provisioned groups of rhesus and Japanese macaques (expanding populations) and the lower incidence of younger sisters outranking older sisters in declining populations such as savanna baboons (e.g., Hausfater *et al.*, 1982).

Datta and Beauchamp (1991) also applied their model to mother–daughter rank relations. In this case, the mother's key ally was assumed to be a dominant sister – that is, a mother with no dominant sister would be outranked by a sexually mature daughter. Again the simulations showed that mothers in expanding populations were much more likely to retain their rank above their daughters compared to mothers in declining populations. In sum, life history and demographic differences between groups would produce variation in the genealogical structure of groups, which would translate into variation in alliance networks, hence in the rules structuring female dominance relations.

The impact of group composition and of its correlate, the availability of allies, on female dominance relations has been amply demonstrated by experiments with Japanese macaques reported in the previous section. Females need specific allies both to outrank certain females (Chapais, 1988a; Chapais *et al.*, 1997, 2001) and to maintain their rank above them (Chapais, 1988b; Chapais *et al.*, 1991, 1994). For instance, with respect to sister rank relations, experiments showed that a younger sister was dominant to her older sister in the presence of their mother, became subordinate after the mother was removed and regained dominance when the mother was put back (Chapais, 1992b). These results provide strong support to the idea that consistent differences in group composition stemming, for instance, from variation in life-history patterns could result in intergroup differences in dominance rules. Other, more indirect, evidence in support of the role of demographic factors in the establishment of female dominance relations comes from exceptions to the classical rules of dominance, observed among the lowest-ranking females in large groups of provisioned Japanese macaques (Nakamichi *et al.*, 1995; Katsukake, 2000).

Hence, there is no doubt about demography and group composition determining a female's network of potential allies, and no doubt about the role of alliances in preventing rank reversals or facilitating them. However, this is not to say that demographic variation accounts for all, or even for the greater part, of the observed variation in dominance rules among groups and species. Even when their key allies are present and well, females may or may not outrank more powerful individuals, depending on the groups and the species. Dominance rules vary independently of demography and the availability of allies. I consider two examples.

First, in two provisioned populations of Barbary macaques, females were observed to inherit their mother's rank as in rhesus and Japanese macaques, but older sisters most often remained dominant to their younger sisters (Paul & Kuester, 1987; Prud'homme & Chapais, 1993a). This was not due to the absence of the mother, who was alive in most cases. Rather the contrasting dominance rules appeared to reflect different alliance patterns. Prud'homme and Chapais (1993a) found that while younger sisters were aided by close kin (including their mother) against both their older sisters and unrelated lower-born females, they were aided by dominant nonkin mostly against lower-born females, not against their older sisters (Fig. 9.1). This contrasts with data on rhesus (Datta, 1988) and Japanese macaques (Chapais et al., 1994) in which dominant nonkin females intervene in conflicts between sisters and side with the younger one. It is also likely that close kin aid younger sisters at higher rates in rhesus and Japanese macaques compared to Barbary macaques. This suggests that the contrast in sister rank relations between Barbary macaques and rhesus/Japanese macaques would stem from two aspects of intervention patterns in sister conflicts: the involvement of nonkin and the greater involvement of close kin in rhesus and Japanese macaques.

Second, Hill and Okayasu (1995) observed a natural (nonprovisioned) group of Japanese macaques in which younger sisters remained subordinate to their older sisters contrary to the rule for provisioned groups of the same species. Again, the mothers were alive and well when their younger daughters reached the age at which they would "normally" outrank older sisters. Hill and Okayasu (1995, 1996) proposed that this pattern of dominance resulted from the total lack of support of mothers in favor of their younger daughters, a pattern they attributed to the low frequencies of aggression in this group, due to its wide spatial dispersion during foraging. Thus, in both cases – provisioned Barbary macaques and wild Japanese macaques – younger sisters remained subordinate to older sisters not because they lacked the required allies, as suggested by Datta's demographic model, but because these allies were either inactive or less active.

To summarize, although life history and demographic factors determine the availability of any female's potential allies, the mere presence of potential allies such as close kin does not entail their intervening in conflicts. Sometimes they

Fig. 9.1. Looking for support in Barbary macaques (Kintzheim, France). In a contest between two adolescent females (a), one looks for her mother's support (b), who starts to groom and appease her (c), then the daughter reciprocates grooming (d). (Photographs by B. Thierry.)

do so on a consistent basis, sometimes they do not. The next question, then, regards the conditions affecting an individual's propensity to form alliances – e.g., ecological conditions.

The role of ecological variation

Van Schaik's (1989) model on the evolution of female social relationships provides the predominant theoretical framework linking the formation of alliances among females to ecological conditions. According to the model, contest competition for high-quality food patches that are monopolizable by an individual or a subgroup would induce food-related aggression and lead to the formation of stable dominance relationships between females. Then, "since rank is so important, females who give agonistic support to maturing and adult relatives, in order to help them outrank other females, raise their inclusive fitness. Thus, the hierarchy will become nepotistic . . ." (van Schaik, 1989: 202). Strong intragroup contest competition is expected to always generate well-defined patterns of nepotistic alliances and produce matrilineal dominance structures. In a more recent version of this model, Sterck and collaborators (1997) termed the corresponding class of species "resident-nepotistic." Most species of macaques are put into that class (Sterck *et al.*, 1997).

Although van Schaik's (1989) model states that alliance patterns generate nepotistic hierarchies, the model is not concerned with variation in alliance patterns between groups and species, nor with variation in the structural rules of nepotistic hierarchies. The model, however, proposes an explanation for differences in dominance styles between species. Van Schaik (1989) and Sterck and collaborators (1997) argued that when strong within-group competition for food is combined with strong between-group competition, high-ranking females are more dependent on the presence of subordinate allies to compete against other groups and therefore that they should be more tolerant and grant subordinates access to food in order to prevent them from dispersing. This situation would produce another class of species, called "resident-nepotistic-tolerant," characterized by well decided but less asymmetric dominance relations compared to the despotic resident-nepotistic species.

Few data are available to test this hypothesis but the existing evidence does not support the association between the co-occurrence of within-group competition and between-group competition, and a tolerant dominance style (Matsumura, 1998, 1999; Koenig, 2002; Cooper, Box 9). Moreover, the model predicts that both categories of species (resident–nepotistic and resident–nepotistic–tolerant) should exhibit "formal dominance relationships, stable, linear dominance hierarchies established and maintained by coalitionary support" (Sterck *et al.*, 1997: 295). In other words, despotic and egalitarian species are not expected to differ substantially in their alliance patterns and the structure of female dominance relations. This is clearly, however, not the case. Despotic species such as rhesus and Japanese macaques exemplify the classical nepotistic hierarchy whereas more egalitarian species such as Barbary and Tonkean macaques exhibit weakly nepotistic hierarchies. Thus, the basic question remains as to what ecological factors might explain the covariation between dominance styles and dominance structures, and link these characteristics to specific alliance patterns.

One possibility that has been little explored is the impact of variation in the strength (or intensity) of intragroup contest competition for food on female dominance relationships. The strength of competition is expected to vary continuously because several of its underlying factors – the size, density, depletion time, number and distance between food patches – are continuous variables. In theory, the higher the strength of competition, the steeper the slope of regression between the rank of females and their nutritional status (physical condition), and the higher the variance in female feeding success (defined as their relative food intake or energy gains) and female reproductive success (Janson & van Schaik, 1988). Rank-related differentials in access to food or physical condition among females are documented (Barton & Whiten, 1993; Saito, 1996; van Noordwijk & van Schaik, 1999; Koenig, 2000).

Sterck and collaborators (1997) argued that although the strength of competition varies continuously, the social responses to it should be rather discrete. Females are expected to either form dominance relationships or to lack them altogether. This reasoning makes evolutionary sense given that submission to a stronger opponent is advantageous regardless of the absolute amount of benefits associated with victory. Hence, females should form decided dominance relationships regardless of the intensity of competition. However, in contrast to this all-or-none reasoning, patterns of alliances may be expected to vary with the strength of competition.

One way this could happen is through the effect of the strength of competition on the profitability of kin support. Compare a situation in which contest competition is strong, therefore rank-related fitness differences are pronounced, to a situation where competition is weaker and fitness differentials are smaller. The costs (C) of aiding a kin against lower-ranking females are similar in the two situations. However, the indirect fitness benefits (B) accruing to helpers are higher when differences in rank have a stronger impact on fitness. Hence, kin support should be more profitable in the first situation compared to the second. When the benefits of aiding are higher, Hamilton's (1964) equation ($B/C > 1/r$) is satisfied for more categories of kin – i.e., for kin with lower degrees of relatedness. In the classical matrilineal hierarchy the categories of kin that play a signifiant role in assigning rank are mothers, grandmothers and sisters (see above), whose degree of relatedness with the recipient is ≥ 0.25. Thus, if $B/C > 4$ (or if $B > 4C$), these three categories of kin should provide aid. This would happen if the benefits of aiding are relatively high. However, if the benefits of aiding are smaller, it might be that only the mother should aid (if $4C \geq B > 2C$). Finally, the benefits of aiding might be even smaller so that $B \leq 2C$, in which case no kin should aid. In sum, the profitability of kin support, hence the range of kin expected to be involved in rank acquisition, should decline as the strength of competition and the benefits of dominance decline.

Figure 9.2 summarizes the consequences of variation in the profitability of kin support on female dominance relationships by considering the three hypothetical situations just described. If kin support is totally absent, rank inheritance along the maternal line should be absent as well, resulting in a dominance hierarchy ruled by the females' individual attributes only. Such linear but non-nepotistic hierarchies characterize Hanuman langurs (Hrdy & Hrdy, 1976; Borries, 1993; Koenig, 2000) and captive sooty mangabeys (Gust

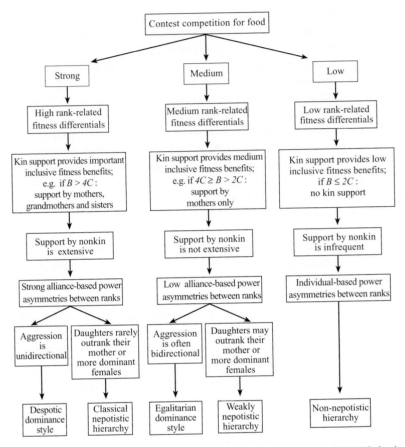

Fig. 9.2. An ecological model of female dominance relations based on variation in the strength of contest competition for food. (See text for explanation.)

& Gordon, 1994), but whether they stem from low intensity levels of contest competition remains to be ascertained.

At the other extreme, kin support could be extensive, being performed by mothers, grandmothers, sisters, and possibly other kin. In these circumstances support by nonkin should be extensive as well, as argued previously in describing the classical nepotistic hierarchy. The corresponding pattern of conservative alliances among kin and nonkin would produce strong power asymmetries between females. Two consequences would follow. First, retaliation (counter-aggression) and challenges directed up the hierarchy would be more risky, and aggression should be mostly unidirectional as a result (Thierry, 1990a). Second, females should find it more difficult to outrank their mother and other dominant females. The end result is a despotic dominance style and a strict (classical) dominance hierarchy (Fig. 9.2).

Finally, in a situation where kin support would be performed exclusively by mothers, daughters would still come to rank initially just below their mothers, producing a nepotistic hierarchy. But such low levels of kin support would entail low levels of support by nonkin. In these conditions the power asymmetry between any two females would be based mostly on the females' intrinsic (individual) power. Consequently, retaliation (counter-aggression) and challenges directed up the hierarchy would be less risky and aggression could be bidirectional (Thierry, 1990a). Females should also find it easier to outrank their mother and other dominant females. The end result is an egalitarian dominance style and a weakly nepotistic dominance order (Fig. 9.2).

The present "strength of competition model" of female dominance relationships remains to be tested, and adequate tests of it lie beyond the scope of this paper. Above all, reliable measures of the strength of competition are needed. Rates of aggression and supplants may be poor indicators of the intensity of competition in species characterized by well-established dominance relations, as are macaques. This may explain why studies testing the hypothesis that aggression rates are higher when food is more clumped have reported inconsistent results (Koenig, 2002). A better measure of the strength of competition is the regression slope between female rank and measures of feeding success, or between female rank and reproductive success. The model predicts that weak rank-related differentials in feeding success or fitness should be associated with a relaxed dominance style and a weakly nepotistic hierarchy, and that high rank-related differentials should correlate with a despotic dominance style and strict dominance rules. The characteristics of food resources should also differ between the two categories of species, with the food eaten by despotic species generating higher levels of contest competition.

Although our knowledge of the ecology of macaques is growing at a steady pace (e.g., Fa & Lindburg, 1996; Ménard, Chapter 11), the paucity of data on the relations between feeding ecology and female social relationships in macaques makes it impossible to test the above predictions. Specifically, the scarcity of regression slopes of feeding success and/or fitness measures on female rank (e.g., Paul & Kuester, 1996a; van Noordwijk & van Schaik, 1999) precludes the needed intergroup and interspecific comparisons. Other types of evidence are available but they are indirect and too fragmentary. For example, Matsumura and Okamoto (1997) found that in moor macaques kinship exerted a weak effect on proximity during feeding. Assuming that contest competition correlates positively with nepotism, they inferred that contest competition was weak in their study group. In agreement with the present hypothesis, moor macaques are characterized by a relaxed dominance style and low rates of coalitions (Matsumura, 1998). But whether the dominance structure is weakly nepotistic is unknown.

A potentially significant argument against any ecological model of female dominance relationships is the clustering of the behavioral traits of macaques according to phylogenetic proximity, rather than environmental conditions. However, as we shall see, these two categories of factors need not be incompatible.

The role of phylogenetic constraints

On the basis of morphological traits and molecular data, the 20 species of the genus *Macaca* can be classified into three main species groups corresponding to three phyletic lineages spanning 5 millions years of evolution (Fooden, 1976; Delson, 1980; Hoelzer & Melnick, 1996). Thierry and collaborators (2000; see also Matsumura, 1999) mapped the behavioral traits of macaques onto the phylogenetic tree of macaques and found a close correspondence between phylogenetic proximity and behavioral resemblances. For example, despotic species belong to the same phyletic lineage while the most egalitarian species belong to another lineage. Exceptions do occur but the fit is impressive, pointing to a substantial role for phylogenetic inertia in the distribution of dominance-related behaviors among macaque species. On this basis, Thierry (2000, Chapter 12) proposed that phylogenetic constraints had been much more important than environmental factors in shaping social relationships in macaques.

Phylogenetic constraints would have initially determined different patterns of nepotistic alliances in different lineages (Thierry, 1990a, Chapter 12). The model considers two situations and goes as follows. When kin ally frequently, a female's power is essentially dependent on her kin group, nonkin alliances playing a minor role in dominance relations. The strong kin bias generates pronounced power asymmetries between females, and the resulting hierarchy is despotic and nepotistic. "Conversely, when kin bias is less pronounced, coalitions involving nonrelatives are more common, dominance appears more dependent on individual attributes, and the individual retains some degree of freedom with regard to power networks" (Thierry, 2000: 113), in other words, adult females can more easily outrank their mother or avoid being outranked by younger sisters (Thierry, 1990a).

This "phylogenetic constraint on nepotism model" and the "strength of competition model" described in the preceding section have much in common, especially the idea that variation in the propensity of kin to form alliances (degree of nepotism) is the key factor causing intergroup differences in dominance style and dominance rules. But the two models differ in two important ways. First, the phylogenetic constraint model begs the question of the initial determinants of interspecific differences in degree of nepotism. The strength of competition model proposes that the ancestral species of different lineages would have undergone different food-related ecological adaptations, with high levels of kin support being adaptive when contest competition was more intense. Once nepotistic alliances had evolved, phylogenetic inertia would have ensured their persistence across environmental fluctuations. Thus, both ecological constraints and phylogenetic inertia would be needed to account for the observed variation in social dynamics across the *Macaca* genus. As argued by others (Sterck, 1999; Koenig, 2002), phylogenetic inertia and ecological adaptations are complementary rather than mutually exclusive.

The second difference between the two models has to do with the role of nonkin alliances. Based on empirical evidence derived mostly from rhesus, Japanese, and longtailed macaques, the strength of competition model states that when kin alliances are well developed, so are nonkin alliances, which are mutualistic and conservative. In contrast, the phylogenetic constraint model states that kin and nonkin alliances are negatively correlated, nonkin alliances being more frequent when nepotistic alliances are less frequent. This contradiction between the two models might be only apparent. In egalitarian species nonkin are freer to interact positively. Hence the proportion of friendly interactions among nonkin out of all friendly interactions (kin and nonkin) may be higher in egalitarian species compared to despotic ones (e.g., Aureli *et al.*, 1997). What is argued here, however, is that the absolute frequencies of nonkin alliances should still be higher in despotic species. In other words, even though nonkin interactions might be relatively less important in despotic species compared to egalitarian ones, both kin and nonkin alliances would be more frequent in despotic species. This remains to be ascertained.

Sources of variation other than alliance patterns

All three sources of variation previously reviewed – demographic, ecological, and phylogenetic – attribute a preponderant role to alliance patterns in determining differences in dominance style and dominance rules among groups and species of macaques. But other explanations that exclude alliances as a determining factor might contribute to the observed variation. For example, Hemelrijk (1999a, Chapter 13) has proposed a model in which artificial entities forming virtual species meet one another and establish dominance relationships following a specific set of rules. By varying only the intensity of aggression ("mild" versus "fierce" entities), the model generated several of the differences between despotic and egalitarian species, including a steeper gradient of the hierarchy in despotic species. However, the model is almost silent about alliance patterns and degree of nepotism, and provides no clear demonstration that differences in nepotism and kin support can be safely ignored.

As another possibility, Combes and Altmann (2001) reported that in savanna baboons almost all females that had a mature daughter were outranked by that daughter, whereas females with no mature daughters were less likely to be outranked by other females. On this basis they proposed that mother–daughter rank reversals did not relate to the mothers' physical decline or to coalitionary support, but that they were rather "consensual" in the sense that they involved no conflict of interest. The model invokes kin selection through relative reproductive value and the idea that, other things being equal, a mother should defer to her daughter when the mother's reproductive value becomes smaller than her daughter's reproductive value devalued by their degree of relatedness (Chapais & Schulman, 1980). Based on data on reproductive value for Amboseli baboons, mothers should experience their first rank reversal at age 15 and be more likely to experience reversals as they get older (Combes & Altmann, 2001).

According to this view, differences in the proportions of mother–daughter rank reversals between groups and species could reflect different reproductive value curves, hence life-history parameters. This hypothesis generates a promising research avenue, but a number of issues need to be considered. First, several mothers were outranked by their daughters up to 5 years before the predicted age, suggesting that the relative reproductive value of mothers and daughters is not sufficient to account for the observed reversals. Second, the model predicts that mothers should defer to their daughters, but the needed information on the dynamics of rank reversals, especially whether mothers resisted their daughters or not, is lacking. Third, it is not clear whether daughters always outranked their mothers only (a required aspect of the model), or whether they outranked other higher-ranking females as well, in which case they might have outranked their mother "transitively." Such transitive mother–daughter rank reversals are documented (e.g., in bonnet macaques: Silk et al., 1981). Fourth, the model is silent about the role of conservative alliances in the ability of mothers to maintain rank. Conservative alliances are documented not only in macaques (references above) but in baboons as well (Walters, 1980; Pereira, 1989). Presumably mothers would not make use of such alliances. This remains to be ascertained.

Conclusion

In all known groups of macaques, females begin their "lifetime ranking schedule" just below their mother. Kinship sets a female's minimal position in the rank order. But groups and species differ in whether or not females will remain subordinate to their mothers during most of their mother's lifetime, and whether or not they will outrank their older sisters. The foregoing review has shown that the basic information required to account for that variation is lacking for most species. First, quantitative data on the patterns of rank relations are needed. This includes information on the proportions of mother–daughter dyads and sister–sister dyads that reversed rank, and on the timing of reversals in relation to the females' absolute age and age difference. But also needed are data on the proportions of mother–daughter reversals in which the daughter outranked only her mother, versus higher-ranking females as well, and data on the dynamics of rank reversals (e.g., whether or not mothers resist). If daughters outrank females dominant to their mother, and their mother transitively, this might indicate that patterns of conservative alliances within and between matrilines are not well developed, allowing for a fair degree of female mobility in the rank order. If, on the other hand, daughters outrank their mother specifically, this might indicate that conservative alliances between kin and nonkin prevent further upward mobility.

Second, we need to know the extent to which dominance rules vary with ecological conditions between groups of the same species. On this basis one would be able to assess the degree of social plasticity, or reaction norm, that characterizes macaques in this respect. At present, the social plasticity of macaques regarding dominance rules is exemplified only by data on provisioned and wild groups of Japanese macaques. The youngest ascendancy rule, which is present in provisioned groups, was found to be absent in two small, wild-living groups (Hill & Okayasu, 1995), and the variation appeared to be driven not so much by demographic differences in group composition as by ecological differences affecting the intensity of feeding competition. The social plasticity hypothesis would predict that wild groups of rhesus or longtailed macaques, experiencing harsher conditions compared to provisioned groups, would also lack the youngest ascendancy rule. Reciprocally, one would predict that even though the groups of Barbary macaques studied so far lack the youngest ascendancy rule, other groups living under different ecological conditions might exhibit it. This is at present unknown.

Alternatively, some of the observed variation might reflect genetic differences between species, or even between regional populations of the same species. For example, the absence of the youngest ascendancy rule in Barbary macaques might eventually be found to characterize all populations of that species regardless of demographic and ecological conditions. The existence of genetically based behavioral differences between macaque species is supported by the correlations between phylogenetic proximity and behavioral similarities (Matsumura, 1999; Thierry *et al.*, 2000).

In sum, the observed variation in the structure of nepotistic hierarchies appears to reflect both the ecologically driven social plasticity of macaques and the evolutionary history of the genus. Both hypotheses accord a preponderant role to patterns of support by kin and by nonkin in producing dominance rules. Extensive alliance networks would generate the youngest ascendancy rule and severely limit the upward mobility of females in the hierarchy. Weaker alliance networks would allow for greater upward mobility and result in the absence of the youngest ascendancy rule. Both hypotheses are also compatible with the view that alliance networks represent adaptations to the ecological conditions affecting feeding competition. The social plasticity hypothesis views behavioral differences as reflecting adaptations to present environments, whereas the phylogenetic constraint hypothesis views the differences as reflecting adaptations to past environments.

Box 9   Inter-group relationships

Matthew A. Cooper

Interactions between groups of macaques take a variety of forms. Animals may approach and peacefully monitor other groups, affiliatively contact animals from other groups, or engage in aggression. Encounters may occur while animals are feeding (e.g., Kumar & Kurup, 1985), involve sexual interactions between nongroup members (e.g., Zhao, 1997), or be initiated by subadult and juvenile males interacting with conspecifics from other groups (Cheney, 1987). While observers can occasionally identify the context of a single encounter, what motivates animals to initiate between-group interactions, and why some interactions escalate into intense polyadic fights often remains unclear. The behavior of males and females during between-group encounters depends on a variety of functional strategies, including female resource defense, male resource defense, and male mate defense.

In mammals, female reproductive success is primarily limited by food, while male reproductive success is primarily limited by access to mates (Trivers, 1972). The female resource defense model predicts that when the distribution of food permits, female macaques will form groups to reduce predation risk (van Schaik, 1989) and cooperatively defend food resources (Wrangham, 1980). While group-living females incur the cost of within-group competition for food, they should benefit by displacing smaller groups (Wrangham, 1980; Sterck *et al.*, 1997). In support of this model some studies have found that females are active participants in aggressive between-group encounters (*M. mulatta*: Hausfater, 1972; *M. silenus*: Kumar & Kurup, 1985), but others have found that females do not play a significant role (*M. maurus*:

Box 9 Inter-group relationships 205

Matsumura, 1998; Okamoto & Matsumura, 2002), or are less involved than males (*M. fuscata*: Saito *et al.*, 1998; *M. mulatta*: Lindburg, 1971; *M. fascicularis*: Angst, 1975; *M. radiata*: Cooper *et al.*, 2004; *M. sylvanus*: Mehlman & Parkhill, 1988). While male involvement can overshadow female involvement, between-group encounters are often over food resources (Dittus, 1987; Sugiura *et al.*, 2000; Cooper *et al.*, 2004).

Models of primate socioecology assume female resource defense is an important aspect of between-group competition in macaques, and thereby influences within-group female social relationships (Sterck *et al.*, 1997; see Chapais, this Chapter). Macaques differ in the degree of tolerance displayed in their social relationships (Thierry, 2000). Macaque species that experience strong between-group competition should have more tolerant within-group social relationships than macaques that do not (van Schaik, 1989). Van Schaik reasoned that in species with strong between-group competition, high-ranking females should more readily relinquish resources to low-ranking females in exchange for their support in between-group encounters. The available data on between-group encounters in wild populations of macaques is summarized in Table 9.2. The average rate of between-group encounters is 0.64 ($SD = 0.44$) per day in tolerant macaque species, and is 0.30 ($SD = 0.016$) per day in despotic macaques (Mann–Whitney, $U = 3.0$, $n_1 = 5$, $n_2 = 4$, $P = 0.09$). The comparison includes multiple rates per species when available, but data do not permit a complete account of intra-species variation. Certainly population density and the amount of home-range overlap can greatly affect the rate of between-group encounters (e.g., Cheney, 1987; Sugiura *et al.*, 2000). This is a tentative initial comparison and, although it is in the predicted direction, the result suggests that the rate of between-group encounters does not explain the variation in macaque social tolerance.

Macaques can be more finely classified on a four-grade scale of social tolerance (Thierry, 2000; Thierry, Chapter 12), but this weakens the correlation between dominance style and the rate of between-group encounters. The percentage of aggressive between-group encounters does not appear to differ between tolerant and despotic macaques, as it can be over 50% for both groups (Table 9.2). Also, males are more aggressive than females during between-group encounters in both tolerant and despotic macaques (Table 9.2). I excluded data from provisioned groups of macaques because predictions from primate socioecology assume species differences reflect natural conditions. While the current environment for many macaques may not match their environment of evolutionary adaptedness, it seems reasonable to exclude provisioned populations. Between-group encounters occur at higher rates and are more aggressive in provisioned groups, and high levels

Table 9.2. *Patterns of between-group encounters in wild* Macaca *spp.*

Species	Study site	Domin. style	Nb groups	Nb enc.	Rate (per 12 h)	Range size (km²)	% overlap	Group domin. call	Inter-group call	Aggr. enc. (%)	Aggr. male enc. (%)	Aggr. female enc. (%)	Affiliation	Herding	Inter-group mating	Sources
M. sylvanus	Ain Kahla, Morocco	Tolerant	1	30	.56				−	50	50	0	+	−	−	1
M. sylvanus	Ghomaran, Morocco	Tolerant	6[a]	13	.31	3.1–9.0	39–100		−	23	23	0	−	+	−	2
M. silenus	Anaimalai Hills, India	Tolerant	1	31	.49	.8	43		+	55	+	+	−	+	−	3
M. maurus	Karaenta, Sulawesi	Tolerant	1	85	.42		low		+	27	12	0	−			4
M. nemestrina	West Sumatra	Despotic	3[a]	21		1.3–>1.5	79–99		+	52	40	50			+	5[b]
M. sinica	Polonnaruwa, Sri Lanka	Tolerant	2				high		+						+	6
M. radiata	Anaimalai Hills, India	Tolerant	1	102	1.40		high		+	32	32	2	+	+	+	7[c]
M. fascicularis	Ketambe, Sumatra	Despotic	3	41	.18[d]		low–high		−	+			+			8
M. fascicularis	Bali	Despotic	6				high		+	+	high	low	+	−	+	9
M. mulatta	Dehra Dun, India	Despotic	4[a]	15	.40	1–15	high		+	0	0	0	+	−	+	10[c]
M. fuscata	Yakushima, Japan	Despotic	7	151	.47[d]		high		+	46–54	high	low	−	−	+	11, 12
M. fuscata	Kinkazan, Japan	Despotic	3	63	.14[d]		low		+	13–17	high	low	−	−	−	11, 12

Abbreviations: Domin., dominance; Nb., number of; Enc., encounters; Aggr., aggressive.

Species are classified as having a despotic or tolerant dominance style, modified from Thierry (2000). The rate of encounters was calculated per day (12 h). For Aggressive encounters, Aggressive male encounters, and Aggressive female encounters, the numbers indicate the proportion of encounters in which individuals showed aggression. Plus (+) and minus (−) indicate the presence or absence of the variable, and blank spaces indicate that data are not available.

[a] Multiple groups are included in data on home range size and % overlap only.

[b] Between-group encounters occurred at baited feeding sites.

[c] Only data from forest groups are included.

[d] The value shown is the average rate of encounters for multiple groups.

Sources: (1) Deag, 1973; (2) Mehlman & Parkhill, 1988; (3) Kumar & Kurup, 1985; (4) Okamoto & Matsumura, 2002; (5) Oi, 1990; (6) Dittus, 1987; (7) Cooper et al., 2004; (8) C. van Schaik, pers. comm.; (9) Angst, 1975; (10) Lindburg, 1971; (11) Saito et al., 1998; (12) Sugiura et al. 2000.

Box 9 Inter-group relationships 207

of female participation are possible (Lindburg, 1971; Hausfater, 1972; Camperio Ciani, 1986; Zhao, 1997; see Oi, 1990).

Although the resource defense model has received support in macaques (Dittus, 1986; Mehlman & Parkhill, 1988; Saito *et al.*, 1998), the lack of an essential role for females is problematic (i.e., Okamoto & Matsumura, 2002). Male resource defense is one possible explanation. In this model, males join a group of females and defend access to food in exchange for reproductive access to females (hired guns: Wrangham & Rubenstein, 1986). Females do not necessarily have to repay males for the service of defending resources because males may directly benefit from their own actions. For example, males may inadvertently defend resources while actively defending mates (Fashing, 2001). Also, males may directly benefit from living in large groups with improved access to resources (Robinson, 1988). While male resource defense is expected to be more common in species with one-male groups (Fashing, 2001), it can occur in species with multimale groups (Harrison, 1983; Robinson, 1988).

Classical ecological theory states that the distribution of males depends on the distribution of females and that males compete for females during between-group encounters (Emlen & Oring, 1977). While male mate defense is common in species with one-male groups (van Schaik *et al.*, 1992), male mate defense can occur in species with multimale groups (e.g., Cheney, 1981; Cowlishaw, 1995; Perry, 1996). For instance, between-group aggression among males is sometimes greatest during the mating season (*M. fuscata*: Saito *et al.*, 1998; *M. radiata*: Cooper *et al.*, 2004). Also, males are sometimes aggressive to females in their own group during between-group encounters (Table 9.2; see also Zhao, 1997). One possible function of this type of male–female aggression is to prevent females from mating with non-troop males during periods of close proximity. While copulation between males and females of different groups is uncommon, it has been reported in macaques and can result in pregnancy (*M. fuscata*: Sprague: 1991b; Soltis *et al.*, 2001; *M. mulatta*: Berard *et al.*, 1994; *M. cyclopis*: Nishida, 1963; *M. fascicularis*: Angst, 1975; *M. thibetana*: Zhao, 1997). Therefore, male aggression against females from their own group may be related to herding as described for species in one-male groups, although in one-male groups herding appears to function to prevent female transfer (Byrne *et al.*, 1987; Stanford, 1991; Sicotte, 1993).

The female resource defense model does not adequately account for the nature of between-group encounters in macaques. The importance of male mate defense and male resource defense in macaques is still uncertain. In addition to the models discussed here, infanticide risk may pressure females to associate with males that provide protection and thus help explain the behavior of females and males during between-group encounters (see van Schaik, 1996). The importance of infanticide prevention for macaques remains unclear. Integrating female and male strategies will improve our understanding of the functional outcomes of between-group encounters. Intra-species variation also needs to be considered. For instance, the nature of between-group encounters depends on the distribution and abundance of food, population density, group size, age–sex distribution, and recent patterns of male immigration and group fissioning. Information from multiple populations of the same species and long-term studies of single groups will help clarify the role of between-group competition in the evolution of primate social behavior.

10 *Intergenerational transmission of behavior*

CHRISTOPHE CHAUVIN AND CAROL M. BERMAN

Introduction

The notion of a species-typical society implies a strong degree of continuity in social behavior and in social organization across generations. Although the players change with each generation, patterns of subsistence and social inter-action in a given group tend to remain constant or to vary within certain limits. Some cases of behavioral continuity have been traced to heritable behavioral tendencies that place considerable constraints on behavioral flexibility of, for example, herding behavior in hamadryas baboons (e.g., Sugawara, 1988), or the vocal repertoire of macaques (e.g., Owren *et al.*, 1992). In other cases, intergenerational continuity appears to be at least partly the result of social processes operating primarily, but not exclusively, during ontogeny. Although a variety of social influences may theoretically lead to behavioral continuity (Whiten & Ham, 1992; Avital & Jablonka, 2000), most research has focused on social transmission, in which one individual learns to perform a behavior as a result of interaction with a model of that behavior (Imanishi, 1952). This focus has grown out of a longstanding interest in the nature and origins of culture. Although biologically-oriented researchers vary widely on definitional criteria, virtually all agree that in order for a behavior to be considered cultural, there must be, at least, clear evidence that its acquisition depends on social transmis-sion. In this chapter, we examine the extent to which macaques may acquire and transmit behavior from one generation to another through social transmission.

Research in social transmission in macaques has focused on two main areas. The first has concerned the transmission of social patterns from mothers to offspring, including dominance status, maternal style and affiliative networks. The second area concerns the transmission of specific techniques or behavioral variants. For each area we ask: what behaviors may be transmitted? Who are the

Macaque Societies: A Model for the Study of Social Organization, ed. B. Thierry, M. Singh and W. Kaumanns. Published by Cambridge University Press. © Cambridge University Press 2004.

demonstrators and how does their identity influence transmission (see Coussi-Korbel & Fragaszy, 1995)? What specific learning mechanisms may be involved (see Galef, 1992)?

Definitions of terms and processes

Galef (1988) defines social transmission as instances of social learning or enhancement in which social interaction results in increased homogeneity of the behavior of the interactants that endures beyond the period of interaction between the transmitter and the recipient. This definition excludes both response facilitation (also called social facilitation or contagion) in which the presence of individual A performing an act in individual B's repertoire temporarily increases the probability of its performance by B, and matched dependent learning, in which B learns to use A's act as a discriminant stimulus (see Whiten & Ham, 1992).

Intergenerational social transmission (or more simply, intergenerational transmission) might thus be defined as instances of social transmission between members of different generations. Given our interest in the processes that lead to species-typical societies, we are most interested here in examples in which behavior is transmitted from older generations to younger generations and in which the behavior endures after the models in the older generation are no longer present or displaying the behavior. When shown by a substantial proportion of a social group or population, such examples may also be referred to as traditions (Nishida, 1987; Fragaszy & Perry, 2003).

Operationally, behavioral homogeneity may relate to specific behavioral variants (e.g., qualitatively different food processing or grooming techniques), social preferences or consistent styles of interaction (e.g., quantitative or contextual differences in interactional patterns). Thus, social transmission may involve learning a novel behavior, or it may simply involve modifying where, when, how much, or toward what (whom) an extant behavior is performed (Galef & Giraldeau, 2001). To qualify as a style or variant, different forms of behavior should be displayed consistently by different subsets of group members or by conspecific members of different social groups.

The earliest mechanism proposed to explain the intergenerational transmission of behavior in primates was identification. Imanishi (1957) derived his concept of identification from psychoanalysis, and hypothesized that some individual attributes among macaques might be socially inherited from parents because young and other group members identify the young with their mothers – that is, they "assume" the offspring possesses certain characteristics of the mother. For example, both offspring and other group members behave as if the offspring carries the mother's dominance status even when the offspring's own ability to fight is poor. Although Imanishi's specific hypotheses were never thoroughly tested and were clearly oversimplifications (e.g., Chapais, 1988a; Chapais & Gauthier, 1993), his focus on the intergenerational transmission of behavior was pioneering.

Currently, our concept of social transmission is not restricted to a specific social learning mechanism. A possible mechanism is exposure learning in which B learns a new act or preference because its association with A places it in an environment conducive to individual learning. This mechanism could encompass the effects of familiarity on social preferences, in which a mother's preferences are transmitted to dependent offspring as a result of their close association with both their mother and her preferred partners. Other possible mechanisms include: social control by conspecifics either in the form of encouragement or discouragement to behave in particular ways or with other particular conspecifics; and various forms of observational learning.

The tendency for individuals to form preferences for familiar conspecifics, frequent interactants, and even familiar inanimate objects is well documented in a variety of taxa (Hinde, 1974). In addition, there is evidence that social control by the model may lead to social transmission either when the model actively encourages the recipient to behave as it does or discourages it from doing otherwise. Social control by mothers has been implicated in the transmission of affiliative preferences (e.g., Timme, 1995; Berman & Kapsalis, 1999) and in the avoidance of particular foods or objects (e.g., Hikami *et al.*, 1990). In addition, aggressive intervention by mothers and close kin on behalf of young is thought to be an important means by which infants initially learn whom to challenge (e.g., Chapais & Gauthier, 1993) and by which older females acquire ranks in the adult female hierarchy (Chapais, 1988a).

Finally, although there is no definitive evidence for true imitation in macaques (Whiten & Ham, 1992), infants may acquire new behaviors through other forms of observational learning, including observational conditioning, response or stimulus enhancement, and/or emulation. In observational conditioning, the emotional responses of A are transmitted to B when B observes that A responds to an object, individual, or context. In response or stimulus enhancement, B's attention is drawn to an object or location by the performance of A. In emulation learning, B learns to associate an object, part of the environment or an aspect of an action (but not a particular technique) used by A with a particular goal (Tomasello & Call, 1997). Empirical examples of emulation learning are usually related to tool use rather than social behavior (e.g., Tomasello & Call, 1997), although there is as yet no theoretical reason why social behavior should be excluded. Studies of emulation learning in monkeys are just beginning to appear in the literature. Although there is evidence that some species are capable of emulation, definitive conclusions about macaques are not yet possible (e.g., Custance *et al.*, 1999; D. Custance, personal communication). Mothers are thought to be particularly salient models for observational learning (e.g., Avital & Jablonka, 2000) at least partly because of their high tolerance for offspring, although other kin and conspecifics may also serve this purpose.

Once an individual is exposed to or has its attention drawn to particular objects, conspecifics, goals or actions, it is more likely to attempt to perform the behavior or interact with the object or individual. In so doing, the recipient learns directly from the consequences of its own actions (and interactions) and is likely to achieve increased behavioral homogeneity with the model. Thus, in addition to interaction with a model, social transmission may involve both independent learning mechanisms and other social influences. In reality, it is often difficult to distinguish behavioral homogeneity arising from various social and nonsocial processes, including social learning, other social influences, independent learning, and unlearned genetically-influenced propensities, because they are likely to interact with one another in complex ways (Galef, 1995). This is particularly true of behavior observed in natural situations.

It is noteworthy that the processes guiding the maintenance of a socially transmitted behavior may differ from the processes guiding its acquisition. Galef (1995) theorizes that behavior acquired socially may be subsequently reinforced, modified, or replaced as a result of the consequences of its performance by the learner. He refers to behavior that is acquired socially but subsequently altered through independent learning as socially-biased independent learning. To be maintained in a given individual or across generations (Galef & Whiskin, 1997), the socially transmitted behavior should provide equal or greater rewards than available alternatives (see Boyd & Richerson, 1985). Other researchers maintain, however, that neither the acquisition nor the maintenance of a behavior necessarily depends on concrete rewards (e.g., Laland *et al.*, 1996; Matsuzawa *et al.*, 2001). Matsuzawa and collaborators (2001) and de Waal (2001), for example, hypothesize that motivation to perform socially-transmitted behavior may derive from a desire for social conformity with socially-bonded or otherwise salient performers.

Transmission of mothers' social styles

Close, persistent, and tolerant mother–daughter relationships, typical of female-bonded macaques, potentially provide rich contexts for the intergenerational transmission of behavior. Below we examine three hypothesized examples of maternal transmission of social style in macaques: dominance status (see Chapais, Chapter 9), maternal style, and affiliative networks. Each example is considered a candidate for maternal transmission, based on observations that: (1) individual females vary consistently in behavior; and (2) daughters show variations similar to those of their mothers.

Social inheritance of maternal dominance status

Numerous studies of cercopithecine monkeys, including macaques, document strong tendencies for females to maintain stable dominance ranks over their lifetimes (e.g., *M. fuscata*: Koyama, 1967; *M. mulatta*: Sade, 1972a; yellow baboons: Hausfater *et al.*, 1982). Described first in Japanese macaques (Kawamura, 1958; Koyama, 1967), adult daughters in many species acquire ranks within the female hierarchy (1) just below their mothers and (2) above their older sisters (e.g., *M. mulatta*: Datta, 1983a; *M. fascicularis*: de Waal, 1977), patterns that occur so consistently they are called "Kawamura's principles" (Chapais, Chapter 9). Longitudinal observations of adopted infants who eventually acquire their foster mothers' ranks, suggest that the status a female attains is not highly dependent on genetic influences, but rather on social influences during ontogeny (Itani, 1959; Bernstein, 1969).

Rank acquisition appears to follow distinct stages, thoroughly described in rhesus and Japanese macaques (e.g., Datta, 1983a; Chapais & Gauthier, 1993). In brief, youngsters first acquire their mothers' ranks vis-à-vis peers, often during the second half of the first year of life. They then begin both to direct more aggression to older lower-born females (and their offspring) that their mothers outrank than to those their mothers cannot outrank (higher-born) and to selectively join coalitions against them. Thus they develop alliances with powerful females, particularly mothers and other older female kin. As they develop into juveniles and adolescents, they increasingly target older lower-born females, including their own older sisters, with the aid of higher-ranking allies. Gradually they establish dominance over these females, and consequently, each eventually achieves a rank immediately below her mother and above her older sisters. Careful experimental studies have confirmed the critical importance of alliance support in this later stage of rank acquisition (e.g., Chapais, 1988a; Chapais, Chapter 9).

Observational studies of macaque infants suggest that they may learn to select appropriate targets for aggression in numerous ways, some of which involve social transmission (e.g., Berman, 1980; Chapais & Gauthier, 1993). Before infants can be assigned ranks among themselves, mothers and close kin actively and effectively intervene on their behalf against lower-born females and immatures. Thus high-ranking mothers actively control the outcome of their infants' early agonistic interactions with lower-ranking individuals. Hence, infants develop "dependent ranks" that correlate with those of their mothers (Kawai, 1958), but depend on her support. In addition, mothers direct aggression toward lower-born infants both in and out of their own infant's presence. High-born infants may thus learn whom they can threaten with impunity through observation of the mother, and lower-born infants to whom they must defer independently by associating the higher-born infant with its mother. Finally, low-born infants may be subject to observational conditioning when they observe their mothers avoid higher-born females and their offspring. However, at this point, the relative importance of each proposed early learning mechanism is still unclear.

A striking aspect of rank acquisition among rhesus and Japanese macaques is the apparent assertiveness that mothers show both in threatening lower-born infants of other females and in intervening on behalf of their own infants during interactions with lower-born peers. This assertiveness appears to trigger assertiveness by offspring, as indicated by their active and persistent challenging of lower-born females until they rise in rank over them (Datta, 1983a). Given that high rank among females in many species brings increased access to limited resources and ultimately increased reproductive success (e.g., Silk, 1987), it is reasonable to hypothesize that propensities toward assertiveness in such situations may be in part a product of natural selection. However, it is unclear how a system of maternal rank transmission per se could have been selected. Recent research suggests that the fidelity of transmission is likely to be affected by the availability of allies, the benefits of alliances and other factors affecting aggression between females (e.g., Chapais, Chapter 9). Where aggression tends to be mild, mothers and close kin may have less need to intervene on behalf of young relatives, and as a consequence, youngsters are more likely to attain ranks based on their own competitive abilities rather than on birth order and mother's rank. Thus, it is reasonable to propose that maternal rank transmission is the product of adaptive tendencies to strive for high rank interacting with another functional system designed to protect immatures from harassment and injury.

Intergenerational transmission of rejection style

For decades, researchers have casually described tendencies for mothers to display similar maternal styles with successive infants and for adult daughters to display maternal styles similar to their own mothers (Altmann, 1980; Goodall, 1986). These phenomena were first demonstrated quantitatively when Berman (1990a) showed that free-ranging rhesus mothers that raised three or more infants on Cayo Santiago varied consistently from one another on several measures of mother–infant interaction, including relative rejection rates. Conversely, each mother was reasonably consistent with each of her infants. Finally, this consistency was evident even when variations in group size (Berman *et al.*, 1997) and numbers of older siblings (Berman, 1992), known to influence maternal behavior, were controlled.

Berman (1990b) compared relative rejection rates of mothers with those of their adult daughters for three 6-week infant age periods (13–18, 19–24 and 25–30 weeks). Mean rejection rates for mothers were moderately but significantly correlated with those of their adult daughters in all three age periods. Although this result raised the possibility that rejection styles may be socially transmitted by mothers to daughters, it did not establish whether or not daughters acquired rejection styles specifically from mothers or from other sources to which the mother was also exposed. For example, in two out of three age periods, rejection scores differed significantly between mothers who belonged to different lineages, raising the possibility that mothers and daughters acquired their rejection styles through their membership in a common lineage. Thus, mothers and daughters may have acquired similar rejection rates through observation or interaction with common close associates (i.e., primarily lineage mates) or as a result of immediate social and environmental circumstances affecting all lineage females similarly (e.g., lineage rank). To tease apart the independent effects of mothers and lineages, the relationship between mother's and daughter's rejection scores was re-analyzed using analysis of covariance methods. The results supported the notion of transmission through the mother over transmission through the lineage: (1) daughters' scores were associated specifically with those of their mothers' scores independently of lineage effects; and (2) differences between lineages (and conversely similarities within lineages) could be explained entirely by the association between the scores of mothers and their adult daughters. No specific common attributes nor external social factors were founded to explain the similarity of mothers' and daughters' rejection scores (e.g., similarity in size of kin support networks, dominance ranks, amounts of maternal experience, sexes of offspring, or numbers of older immatures), thus lending further support to the notion of mother–daughter transmission.

To narrow possible mechanisms of transmission, Berman (1990b) correlated adult daughters' scores separately with: (1) the rejection rates the daughters experienced as infants; (2) the rates they observed their mothers apply to their younger siblings; and (3) the rates their mothers applied to their older siblings. The only significant positive correlations were for rates mothers applied to the daughter's younger siblings. These results both suggest that daughters had been influenced by observing their mothers raise their younger siblings and argue against genetic hypotheses, because they would have predicted correlations of equal strength in the three analyses.

Berman (1996) further refined this analysis by separating rejection rates daughters may have witnessed as juveniles and as adults raising their infants alongside their own mothers. The rejection styles of mothers and daughters were similar only when mothers and daughters gave birth in the same years. This surprising result leaves open several possibilities. It may be that mothers transmit rejection styles to daughters, but that the process is highly biased toward current rather than early experience. If this is the case, the similarity of rejection styles may represent an example of social learning but not social transmission, since the behavioral similarity does not necessarily endure after the model is no longer available. It is also possible that mothers and their adult daughters exert a mutual, equalizing influence over one another. Finally, mothers and daughters may display similar rejection styles not because one learns from the other but rather because they share common attributes or social circumstances not yet examined by researchers.

The scarcity of studies on the continuity of maternal style across generations makes it difficult to generalize across species and situations. What evidence is available suggests caution. First, not all studies have been able to demonstrate consistent maternal styles, even at the level of individual mother–infant pairs (Silk, 1991). Second, probable mechanisms may vary with species and/or with the particular measure of interaction. For example, Fairbanks (1989) found positive correlations between vervet monkey mothers and adult daughters in levels of contact with infants, but her data pointed to the daughter's own experience as an infant as the most likely route for transmission between the generations. Third, even where a strong argument for similarity by social learning can be made, other factors may increase or decrease the extent to which mothers and daughters actually display similar maternal styles. Indeed, the Cayo Santiago data suggested that resemblance in rejection styles between mothers and daughters do not depend on similarities in dominance ranks, sexes of infants, number of older immature, or support networks. However, when these factors happen to coincide, as they tend to do for rank and support networks (but not for infant sex or immature siblings), they are likely to lead to greater degrees of similarity between mother and daughters than when they do not.

Similarly, even where adult daughters learn aspects of maternal style from mothers, the extent to which they resemble one another might depend on the extent to which they respond similarly to changeable environmental conditions. Some measures of maternal behavior (not rejection rates) track year to year changes in group size and density (Berman et al., 1997; see also Maestripieri, 2001). In large/high density groups, infants are at more risk of harassment and injury from others, and mothers respond by spending less time at a distance from them. Since proximity-seeking is sensitive to variations in rank and aspects of the mother's support network, and since these factors tend to be similar for rhesus mothers and daughters, we might expect both to respond to changes in group size and density in similar manners each year. However, in macaque species that are less despotic than rhesus macaques, daughters are less likely to have similar ranks and support networks than their mothers (Butovskaya, Box 8; Maestripieri, Box 10; Thierry, Chapter 12). Nor are their infants as vulnerable to harassment and injury from conspecific group members (Maestripieri, 1994a,b). Thus we would not necessarily predict that they would adjust their maternal behavior similarly to changes in group size/density. Furthermore, in relaxed species, mothers and daughters spend less time together and more time interacting with a broad range of adult females. Hence, they may be less likely to learn maternal styles specifically from one another.

Although variation in maternal care patterns is likely to have important consequences for mothers' and infants' future reproductive success, evidence that the intergenerational transmission of maternal style is adaptive or that it has been specifically selected is still lacking. In some cases, it may be clearly maladaptive. For example, it has been suggested that captive macaque mothers transmit abusive care patterns that lead to infant injury or death (e.g., Maestripieri *et al.*, 1997). Cases of abuse tend to cluster within families, but as yet, the mode of transmission (genetic and/or social) is unclear. Nevertheless, the transmission of maternal style, and particularly rejection style, is likely to have important consequences on life-history variables. High levels of rejection have been associated with high infant mortality and morbidity (Altmann, 1980), whereas low levels have been associated with long interbirth intervals for mothers (e.g., Gomendio, 1989b; Berman *et al.*, 1993). The degree to which a particular level of rejection is adaptive is likely to depend on several factors related to the nutritional conditions of the mother and infant (Hauser & Fairbanks, 1988). Nevertheless, an implication of the consistency of rejection behavior from infant to infant is that the long-term consequences of particular rejection styles may accumulate over a lifetime and have profound effects on life-history variables. To the extent that daughters carry on their mothers' rejection styles it seems that the consequences will also affect grandoffspring.

Intergenerational transmission of the maternal social network

Early mirroring of social networks

Close mother–infant relationships epitomize the early development of macaques and many other cercopithecines. Macaque mothers and infants maintain nearly constant contact with one another during the first weeks of life, after which infants gradually begin to spend time away from the mother. Within this context, infants have their first experiences with other group members and begin to interact with them. When they do, their interaction patterns closely resemble those of their mothers; they associate primarily with the same individuals the mother associates with (e.g., de Waal, 1996b; Berman & Kapsalis, 1999). In some macaque species, these are primarily, but not exclusively, close female kin and their immature offspring (e.g., Berman, 1982a). In other macaques kin bias may be more moderate or absent (Rosenblum et al., 1975; Maestripieri, Box 10; Thierry, Chapter 12). This early mirroring of maternal networks is due at first to two tendencies. First, infants are exposed primarily to the mothers' close associates and their offspring because they maintain close proximity to one another. Second, in species where mothers exert a large degree of control over their infants' social interactions, mothers are typically more tolerant of their close associates and their associates' offspring than of others (e.g., Spencer-Booth, 1968). In addition, other individuals may be more likely to initiate interaction with the infants of close associates than with others (e.g., Berman, 1982a; de Waal, 1996b). In any case, it is important to note that such mirroring can be observed even before infants are able to take an active role in initiating and sustaining their own relationships. Thus, de Waal (1996b) used the term "dependent affiliation" to describe early relationships, highlighting parallels with the notion of dependent rank (Kawai, 1958).

To what extent do early maternal patterns endure as the infant develops an independent social network? Berman (1982a) followed the development of social relationships in free-ranging rhesus macaques from birth to 30 weeks of age. This age span included both the initial period of close maternal proximity plus control and a period of gradual emancipation in which the infant spent progressively more time distant from its mother and in which the mother's control relaxed considerably. Moreover, by 30 weeks, the infant's relationships acquire some degree of independence from those of the mother with the same partners; approaching peers and older immature partners appeared to be attracted specifically to the infant rather than to its mother (Berman, 1982b). The results showed a remarkable degree of stability over infant age periods. Infants' networks still appeared to resemble those of their mothers at 30 weeks of age (Berman & Kapsalis, 1999) even though many of their interactions took place far from her and were less likely to be prevented (or encouraged) by her.

The enduring similarity of mothers' and infants' social networks, brought about initially through close mother–infant interaction and control, suggests a process of maternal transmission in rhesus macaques. Furthermore, the intensity of kin bias among infants parallels that of their mothers and is related to the amounts of time the infants spend near their mothers. This is also linked with the mother's relative role in maintaining proximity to her infant, highlighting the importance of the mother in bringing about this resemblance (Berman et al., 1997; Berman & Kapsalis, 1999). Additional studies illustrate the gradual nature of the development of kin bias when infants are not under the full control of the mother (de Waal, 1996b; Berman, 2004) and suggest that the similarity between mothers' and daughters' social networks endures throughout juvenile development and into adulthood in both rhesus and Japanese macaques (de Waal, 1996b; Nakamichi, 1996).

Our understanding of the mechanisms that lead to the maternal transmission of affiliative networks is still fragmentary. It is possible that several hypothesized mechanisms play a role. They include familiarity based on differential amounts of exposure to and interaction with the mother's frequent and infrequent associates, observational learning with the mother as model as well as active and selective maternal intervention (see review in Berman, 2004). The tendency for individuals to form active preferences for familiar conspecifics has been studied intensively, primarily in the context of parental attachment (e.g., Rajecki *et al.*, 1978; Suomi, 1995). Studies of infants describe initial periods of broad social openness when attachments are formed easily and there is little fear of strangers, followed by a period when unfamiliar conspecifics are more likely to be avoided. The advent of "stranger anxiety" may serve to translate early patterns of exposure while under the direct control of the mother into longer-term social preferences (Hinde, 1974).

In a hypothesis based on observational conditioning, Altmann (1980) suggests that infant baboons may be specially prepared to learn about their mothers' relationships with other females through her differential display of fear or distress in their presence. Alternatively, mothers could serve as foci for directing infants' attention to preferred conspecifics, in a process of stimulus enhancement similar to that by which they learn the mother's preferred foods (Hikami *et al.*, 1990). Finally, mothers may influence their infants' development of social preferences by disrupting interaction with infrequent associates (Hrdy, 1976) or by actively encouraging infants to interact with close associates and their offspring (e.g., Timme, 1995).

Factors that may influence maternal transmission of social networks
Tendencies for mothers to seek proximity with infants and to regulate their early interactions with other group members can be seen as a manifestation of a functional system by which mothers protect infants from injury and harassment (Berman, 2004). Given that these tendencies also appear to lead to the transmission of maternal social networks, we may also view this transmission as an outgrowth of the same functional system (see Berman, 2004).

In situations in which mothers seek less proximity and are less protective, infants tend to associate with and receive handling from a wider range of group members than in situations in which mothers are more protective. As a result, we might expect infants with less protective mothers to develop social networks that

Fig. 10.1. Maternal protectiveness in rhesus macaques (Strasbourg Primate Center, France). (Photograph by B. Thierry.)

do not closely resemble those of their mothers in comparison to infants with highly protective mothers. This prediction might be particularly relevant for understanding differences in the way social networks arise in macaque species that differ in dominance style. For example, mild resemblance between mothers' and offsprings' social networks might be expected in relaxed or tolerant species that exhibit a high degree of social tolerance such as stumptailed and Sulawesi macaques. Several comparative studies have found higher levels of maternal protectiveness and more highly constrained social networks among infants of despotic than among infants of tolerant macaque species (Rosenblum *et al.*, 1975; Thierry, 1985b; Maestripieri, 1994a,b, Box 10; Mason *et al.*, 1993). However, no data are yet available to support or refute the prediction that infant social networks among despotic species resemble those of their mothers more than do those of tolerant species.

Berman and Kapsalis (1999) investigated the relationship between maternal protectiveness (Fig. 10.1) and resemblance of social networks within species by testing the prediction that free-ranging infant rhesus macaques who spent relatively more time near their mothers would come to resemble them more than other infants in terms of intensity of kin bias. However, the prediction was not supported. Whether or not similar predictions would be supported for other aspects of social networks, settings, or species needs testing.

Although the evidence implicating maternal transmission in the development of daughters' social networks is strong for macaques, recent evidence from free-ranging rhesus macaques suggests that other processes may also contribute (Berman & Kapsalis, 1999; Berman, 2004). First, by 25 to 30 weeks, individual infants display degrees of kin bias that are only moderately correlated with those of their mothers. Thus by this age, infants tend to modify rather than duplicate their mothers' patterns. Second, as described above, the degree to which individual infants display patterns similar to those of their mothers is not related to the amount of time they spend together. Berman and Kapsalis (1999) hypothesize that infant kin networks at this age represent maternally transmitted behavior patterns that have been subsequently modified by the consequences of their performance by the infant (socially biased independent learning, *sensu* Galef, 1995). Under this hypothesis, infants learn to prefer kin partly through the quality (as opposed to quantity) of their direct experiences with kin and nonkin. Through direct experience with kin and nonkin, infants could either reinforce or moderate patterns of affiliation acquired initially through maternal transmission.

In addition, over time, offspring take increasingly more initiative in their interactions with group members (Fairbanks & Pereira, 1993). Who they choose to pursue is likely to be influenced by individual, sex-specific and species-typical temperamental characteristics (Clarke & Boinski, 1995). Such temperamentally-based attractions may be inherited to some extent from mothers and interact with maternal care patterns, but may also be influenced by other kinds of experience during infancy. Finally, Widdig and collaborators (2001) speculate that paternal siblings may be attracted to one another based on similar inherited personality profiles, thus further modifying maternal patterns of association.

Transmission of material techniques

New behaviors or new variants of pre-existing behaviors may appear spontaneously in some individuals. These innovations usually remain idiosyncratic. In a limited number of cases, however, such behaviors appear to be transmitted to other group members, including members of different generations. The establishment of apparent local traditions has been described in several provisioned troops of Japanese macaques (e.g., Nishida, 1987; Tanaka, 1998). Three modes of social transmission have been hypothesized: (1) vertical (parent to offspring); (2) horizontal (among members of the same generation); and/or (3) oblique (from nonparents in the mother's generation to offspring) (Cavalli-Sforza *et al.*, 1982).

Transmission of food-associated techniques

Kawai (1965) described two food-processing techniques in the provisioned troop of Japanese macaques on the Koshima Islet. The first technique, sweet-potato washing, consists of holding a sweet potato in one hand, dipping and/or scrubbing it in the water with the other hand. These actions remove the sand before the potato is eaten. Kawai (1965) also reported variants of this technique that consist of using salt water instead of fresh water and of dipping the potato into the water after taking one or two bites. The newly developed variants, first seen in 1953, have been maintained over six generations and nearly 50 years (e.g., Kawai *et al.*, 1992; Hirata *et al.*, 2001). In recent years, however, few individuals display rubbing behavior, presumably because only clean potatoes are available for provisioning. The mode of transmission before 1959 was primarily horizontal, that is, from one individual to peers or playmates, often from younger to older immatures. However, two mothers also apparently learned from their immature offspring.

After 1959, sweet-potato washing was observed among a variety of age–sex classes, although it was rare among adults over the age of 11 years. Infants and juveniles appeared to learn from their mothers, but in fact vertical and horizontal (and then oblique) modes of transmission were difficult to distinguish with certainty. There were no sex differences in acquisition rates among immatures, but adult males were less likely to learn than adult females, probably because of their more peripheral social positions within the group. Several additional variants appeared during this period, including digging a little pool to collect water before sweet-potato washing (Kawai *et al.*, 1992), but particular variants were not disproportionately associated with particular lineages (Hirata *et al.*, 2001). Sweet-potato washing also appeared in other provisioned populations of Japanese macaques, but failed to spread beyond a few individuals in each group. K. Watanabe (personal communication) suggests that this was due to the manner of provisioning; potatoes were distributed closer to water sources on Koshima than elsewhere.

The second technique described in the Koshima troop was called wheat washing or placer mining. It consisted of taking a fistful of sandy wheat from the ground and placing it in water (Kawai, 1965). Because the wheat floats, it is then easy to collect it without the sand. The spreading of this technique was slower than that of sweet-potato washing and fewer individuals finally displayed it. As for sweet-potato washing, there were sex differences in acquisition rates among adults but not among immatures (Hirata *et al.*, 2001). In addition, this behavior appeared to disappear when the rate of provisioning was greatly reduced in 1972. With less provisioned food available, monkeys that washed wheat were likely to lose their bounty to higher-ranking or stealthy individuals. Interestingly, at this point a number of scrounging and counter-scrounging variants for this technique appeared and became common (Kawai *et al.*, 1992), suggesting that social factors can both discourage the maintenance of some variants and encourage the acquisition and spread of new ones.

Similar food washing techniques, accompanied by the spontaneous appearance of new variants have also been described in captive settings (Scheurer & Thierry, 1985). The patterns of diffusion in captive Japanese macaques were similar to those described on Koshima for sweet-potato washing and wheat washing: initiation by a young individual, spreading among kin, especially peers, but no acquisition by adults aged more than 11 years (Scheurer & Thierry, 1985).

Interestingly, on Koshima, acceptance of two foods that required little or no processing was spread markedly differently from those for sweet-potato washing and wheat washing. Both raw fish eating and candy eating were acquired first by older adult males, then by adult females and finally by youngsters. Rates of spread were very slow for fish eating but were rapid for candy eating (e.g., Watanabe, 1989).

Transmission of grooming techniques

Technical traditions may also concern activities not associated with food. Using video tape recording, Tanaka (1995), documented the consistency of lineage-based grooming techniques in Japanese macaques. He observed variations in techniques used to remove lice eggs from fur. Three out of five matrilines displayed single variants almost exclusively. In the fourth matriline, most members used a single variant, but some individuals used one of two minor variations. In the fifth matriline, two variants were common, but some individuals used quite different variants (Tanaka, 1995).

In a further investigation, Tanaka (1998) described the social diffusion of a newly acquired variant within one of the matrilines previously mentioned. A 22-year-old female member initiated the process. Within six months, her

20-year-old sister, two daughters (11 and 5 years old) and one 5-year-old granddaughter adopted this variant. All members of this matriline subsequently abandoned the new variant in favor of a second new variant. Tanaka actually witnessed the transition for two out of five focal individuals and described the acquisition of the second variant as a result of serendipitous movements during the performance of the previous variant. Interestingly, a third variant that involved skin pinching failed to spread or even to be sustained for long, apparently because of the painful reactions it produced in groomees. Hence, this case serves as an example in which social feedback discouraged rather than encouraged the propagation of a variant. In any case, the spreading pattern of the "successful" variants appeared to differ oddly from that described for food-washing techniques in that the initiator was an older individual and the technique spread quickly among relatives as old as 20 years. To our knowledge, innovation and acquisition of a complex technique has never been described elsewhere in individuals more than 20 years old.

Transmission of non-utilitarian techniques

Some hypothesized examples of social transmission in Japanese macaques involve apparently non-adaptive behavior. The best studied one concerns stone handling techniques (Huffman, 1984, 1996). Huffman (1984) recognized eight basic stone handling variants including gathering, picking up and hand rolling (Fig. 10.2). As in Koshima washing techniques, the initiator was a young female. Within 4 years after their first appearance, the stone handling techniques were exhibited by more than 75% of the 115 group members. However, only three were older than the initiator, and each of the three was a female only 1 or 2 years older than the initiator. Moreover, two out of the three belonged to the initiator's matriline. Although the rate of transmission was high among playmates and kin, no individuals who were adult at the time of initiation acquired the technique perhaps because of its playful nature. This horizontal spreading pattern recalls that of sweet-potato washing and wheat washing techniques, but the rate was faster. The ease and attractiveness of the behavior for youngsters, coupled with free time supplied by provisioning may have contributed to the high acquisition speed (Huffman, 1996). At the same time, the extent to which stone handling was transmitted via the vertical mode was uncertain. By 1985, all offspring of stone handling mothers and most (72%) offspring of nonhandling mothers also handled stones (Huffman, 1996), suggesting that maternal transmission may have contributed to its spread, but that youngsters were not dependent on learning from their mothers.

Fig. 10.2. Stone handling as a form of socially transmitted play in young Japanese macaques (Arashiyama, Japan). (Photograph by B. Thierry.)

In other examples, Machida (1990) reported the spontaneous standing of poles against a smooth wall and subsequent climbing up on them in a captive group of Japanese macaques. The behavior had never been observed before, despite the constant availability of poles. The initiator was a juvenile female. Two years later, two other juvenile females performed "standing and climbing a pole." As they grew older, these three individuals stopped displaying the behavior and there was no evidence of transmission to any other individuals. Similar behavior with poles and its limited spread was also observed in captive Tonkean macaques (Ducoing & Thierry, 2004). This behavior had never been observed during the 15 years preceding this first occurrence. Contrary to other examples, the initiator was a subadult male. Within 3 months, three other males acquired the technique. Again there was no sign of further spreading or of intergenerational transmission.

Another reported case of apparently transmitted behavior concerns vocalizations. Differences in the vocalizations uttered at the initiation of provisioning were found between different study sites in Japanese macaques (Green, 1975). The author suggested that such locale-specific "dialects" were displayed by both sexes but were confined to young individuals. Older individuals uttered the same general tonal pattern but never included the locale-specific modification (Green, 1975). At present, it is not known whether these vocal variations were socially acquired and/or resulted from genetic differences between populations.

Proposed mechanisms and processes

Definitive data on social mechanisms that may have been involved in the spread of techniques among provisioned natural groups are not available. Although many researchers originally accepted that the spread of techniques involved imitation or at least some form of social learning, Galef (1992) argued that they may not have. Galef suggested that the rates of spreading of sweet-potato washing and wheat washing among Koshima monkeys were too slow to implicate social learning. Moreover, rates of transmission did not accelerate over time, as one might expect as the number of potential models of the behaviors within the groups increased (Boyd & Richerson, 1985). Rather than learning new techniques from observation, Galef suggested that the monkeys independently learned to reorganize existing behaviors (e.g., food rubbing) when confronted with provisioned food near water (see also Visalberghi & Fragaszy, 1990), although stimulus enhancement may have initially drawn their attention to the potatoes and water. Finally, Galef suggested that human provisioners may have inadvertently shaped monkeys to use the new variants by directing food to monkeys whose behavior approximated sweet-potato washing or wheat washing (see also Green, 1975), a practice denied by at least some of the provisioners (de Waal, 2001).

In other studies, new techniques have spread much more quickly (e.g., Huffman, 1984; Visalberghi & Fragaszy, 1990; see also Hauser, 1988) and at accelerating rates (Lefebvre, 1995). And in any case, researchers dispute the importance of speed and acceleration as criteria for social learning (e.g., de Waal, 2001), especially when evaluated in a natural setting (Laland *et al.*, 1996). Stone handling among juvenile Japanese macaques (Huffman, 1996) offers an example not only of fast spreading but also of behaviors involving objects not specifically provided by humans (although human provisioning may have provided the monkeys with sufficient time and energy to pursue an apparently nonproductive activity). Beyond this, the specific mechanism responsible for the spread of stone handling is still unclear, but the example suggests that tangible rewards are not always necessary.

When authors speculate on specific social learning mechanisms, if any, that may be involved in the spreading of techniques among free-ranging macaques, most cite a form of stimulus enhancement, at least partly because definitive evidence is lacking that macaques are able to learn by emulation or imitation (Whiten & Ham, 1992; Tomasello & Call, 1997; D. Custance, personal communication). An exception is Tanaka (1998) who described lice-egg removal techniques as consisting of hierarchically organized subroutines. He speculated that the monkeys learned the specific structure of subroutines through observation, and did not merely have their attention drawn to relations between hand and hair through stimulus enhancement. Thus Tanaka suggests a form of imitation learning. Clearly the questions of emulation and imitation in macaques need further work, especially in view of recent findings of emulation and/or imitation in tufted capuchin monkeys (Custance *et al.*, 1999), and other animals (e.g., Galef, 1998). Finally, Matsuzawa and collaborators (2001) and de Waal (2001), citing examples of social transmission with no tangible reward, propose socioemotional (as opposed to cognitive) mechanisms based on social bonding, identification and a desire to act like the model to explain nut cracking in chimpanzees and stone handling in Japanese macaques.

Experimental evidence for social transmission

Although observational studies, particularly those in free-ranging populations, have been useful in describing spreading patterns in naturally-organized groups, experimental studies on captive groups are potentially better able to confirm the role of social influences in behavioral continuity and to address questions of mechanism. The number and rigor of such experimental studies has blossomed in recent years (e.g., Fragaszy & Perry, 2003), although only a few studies have focused on macaques. Below we describe some of them.

Quantitative similarity

In an experiment dealing with the transmission of behavioral style, de Waal and Johanowicz (1993) asked whether juvenile rhesus macaques might increase their rates of reconciliation as a result of co-housing with juvenile stumptailed macaques. Stumptailed macaques normally display substantially higher reconciliation rates than rhesus. After 5 months of co-housing, the rhesus macaques were returned to their original groups consisting of only conspecifics. After co-housing, the rhesus juveniles showed marked increases in their reconciliation rates compared with both their rates before co-housing and rates of a control group of rhesus juveniles. Since the groups were observed for only 6 weeks after the period of co-housing, it is unclear how long increased reconciliation rates would have been maintained. Hence we cannot be sure whether this apparent example of horizontal social transmission had the potential to be carried on over generations. Interestingly, the rhesus and stumptailed macaques used different specific behaviors during reconciliation, suggesting that propensities to reconcile may have transmitted, but not specific behaviors used to reconcile.

Qualitative similarity

Studies of fear responses to snakes have demonstrated the potential for social transmission through observational conditioning. When confronted with various kinds of snakes (real, toy or model), naïve laboratory-reared rhesus macaques typically show no strong emotional reactions and do not appear to either avoid or prefer them to neutral objects. In contrast, wild-reared individuals display an almost complete avoidance of snakes (Mineka *et al.*, 1984). To investigate the basis of this difference, these authors compared the responses of naïve laboratory-reared offspring rhesus macaques before and after the youngsters observed older wild-born monkeys confront a snake. After a single trial, the subjects themselves reacted with intense fear to snakes. In nature, models are likely to be parents or other older group members, and in this sense, the observational conditioning of snake fear appears to represent a *sensu stricto* case of intergenerational transmission of behavior. In reality, genetically influenced factors also play a role in that observational conditioning of fear specifically for snakes appears to be more easily achieved than for other objects. To show this, Cook and Mineka (1990) used videotape to compare the responses of naïve laboratory-reared monkeys to images of monkeys behaving fearfully to snakes and without fear to flowers, and vice versa. Interestingly, the naïve offspring, afraid of neither snakes nor flowers before the tests, acquired a fear of snakes, but not of flowers, suggesting that they were inherently prepared to respond with fear to such experiences with snakes but not with flowers. The extent to which observational conditioning in macaques occurs for other risky objects, individuals or situations is unknown. Nor is it clear whether macaques actively look for the emotional responses of mothers or others upon encountering novel objects, individuals or situations, as do human infants (Feinman, 1982).

Examining social transmission of feeding techniques in Tonkean macaques, Drapier and Thierry (2002) provided five mothers with four kinds of fruit they had never encountered. The technical variants used by each mother were recorded as their offspring were allowed to observe only their own mother's consumption of the fruits. In subsequent trials, isolated offspring were provided with the same fruit types, but they used different techniques to process them than did their mothers. In a second experiment, the authors recorded different variants used during the consumption of familiar fruits in the two maternal lineages of the group (Drapier & Thierry, 2002). They found some significant differences in variant usage between the two maternal lineages. The technical variants appeared arbitrary in that no variant was more efficient than another. The authors suggested that the differences between the two sets of results may have been due to differences in the number of opportunities the youngsters had to observe the techniques used by others. The probability of social transmission may also have been affected by the constraints of the experimental design. In this case, the youngsters had very limited opportunities to interact freely with both the demonstrators and the new foods. Social transmission of feeding techniques may not necessarily involve one-trial learning like the observational conditioning of snake fear. Rather it may require several exposures and opportunities for refinement via trial and error to be integrated in the individual repertoire.

In an attempt to throw light on the mechanisms that may have led to the propagation of food-washing techniques at Koshima, Visalberghi and Fragaszy (1990) documented the acquisition of a similar technique in captive longtailed macaques. They gave four naïve hand-reared juveniles access to a water-filled basin with toys floating on the surface. After four hours of habituation to the apparatus, they provided the monkeys with sandy fruits, a novelty for them. Three of the four subjects acquired a food-washing technique within three 30-min sessions. The fourth subject was frequently threatened when attempting to gain access to the basin, but nonetheless played with the food in the water and retrieved food from the water as early as the first session. The authors concluded the three individuals learned to wash fruit through independent learning mechanisms, but that response facilitation might have aided their acquisition of the new behaviors.

Conclusions

While each example we have reviewed offers some evidence for the influence of social learning on behavioral continuity across generations, in all cases the evidence is incomplete. In many cases, we are able to describe patterns of propagation in the field (e.g., sweet-potato washing), but we have little or no understanding of specific learning mechanisms that may be involved. In others, we have identified a specific learning mechanism (e.g., observational conditioning of snake fear), but lack a clear idea about the range of objects or situations to which it may apply in a natural situation. And in most examples we have only a fragmentary appreciation of the various social and environmental factors that affect the fidelity and duration of transmission and the likely ultimate effects of transmission on the evolution of the species. Clearly much more research needs to be done, particularly with ontogenetic and experimental approaches.

In recent years, theoretical modelers have explored the possibility of complex evolutionary interaction between socially and genetically transmitted information in the form of dual-inheritance models (e.g., Boyd & Richerson, 1985; Laland *et al.*, 1996). In general these models have suggested that a mixture of intergenerational social transmission and individual learning is likely to have advantages over both purely genetic modes of transmission and purely individual learning under a wide range of ecological circumstances, especially when environmental change occurs at intermediate rates relative to the frequency of transmission. As such, we might expect the widespread occurrence of intergenerational transmission among macaques.

In our examples of maternal transmission of social style, however, there is little evidence that transmission per se has been specifically selected. Rather they appear to be outcomes of selection for other behavioral propensities and motivations (e.g., to strive for high rank, to protect infants from harassment and injury, to tolerate close associates) within particular social and demographic contexts. In this sense, the maternal transmission of social behavior may be considered the end product of the interaction of a number of flexible behavioral propensities channeled through the constraints and opportunities presented by a particular social organization.

Nevertheless, where transmission has reproductive consequences for individuals, a potential for natural selection to enhance or moderate intergenerational transmission is present, producing genetically biased social transmission systems (e.g., Richerson & Boyd, 1989). The specificity and efficiency with which naïve monkeys learn to fear snakes through observational conditioning suggests that such a process may have occurred at least once. However, so far no such mechanisms have been linked to other examples of intergenerational transmission in macaques.

Finally, one aspect of transmission that has received increased attention recently is niche construction – i.e., the potential for the introduction and transmission of technical variants to open up new adaptive niches for exploration and exploitation (Avital & Jablonka, 2000; Laland *et al.*, 2000; Perry, 2003). For example, Hirata and collaborators (2001) suggest that the introduction of provisioned foods near and in the water led not only to sweet-potato washing and wheat washing, but also ultimately to the exploitation of new food sources (e.g., fish) and new ways to play and escape the heat (e.g., sea bathing, swimming, and diving). However, at present, the extent to which socially transmitted behavior in nonhuman primates leads to niche construction independently of human encouragement, and the extent to which it creates new selective pressures for the next generation is still unknown (Thierry, 1994b; Russon, 2003).

Box 10 Mother–infant behavior and socialization   231

Box 10   Maternal behavior, infant handling, and socialization

Dario Maestripieri

Macaque females typically produce a single offspring after a period of gestation of about 5.5 months and care for their infant with little or no help from other individuals. Feeding, transport and protection represent the main forms of maternal care prior to weaning, which usually occurs within one year. Mothers continue to provide grooming and agonistic support to their offspring throughout their lives, or in the case of males, until they emigrate from the group. Adult males typically avoid young infants but often affiliate with male juveniles and adolescents. In Barbary macaques and to a lesser extent in some closely related species (e.g., stumptailed, Assamese, bonnet, toque and Tibetan macaques), adult males hold and carry unrelated young infants as a buffer against aggression from other males (see Fig. 15.1) (Maestripieri, 1998). The taxonomic distribution of agonistic buffering with infants suggests that this behavioral phenomenon emerged quite early in the evolutionary history of the tribe Papionini and subsequently disappeared in some of most recently evolved macaque species (Maestripieri, 1998).

Sex differences in interest in infants emerge early on and persist throughout the lifetime, with females being far more involved in infant handling than males (e.g., *M. arctoides*: Bruce *et al.*, 1988; *M. fuscata*: Glick *et al.*, 1986a,b; *M. mulatta*: Lindburg, 1971; *M. nemestrina*: Wheeler, 1986; *M. radiata*: Silk, 1999; *M. silenus*: Kumar & Kurup, 1981; *M. sinica*: Baker-Dittus, 1985; *M. sylvanus*: Small, 1990b). Most female interactions with other females' infants are brief and do not carry any obvious benefits to infants or their mothers (Maestripieri, 1994c). Infant handling by female juveniles may be a way to acquire mothering skills that will subsequently improve offspring survival (Lancaster, 1971). Infant handling by older and experienced females, however, requires a different explanation. One possibility is that infant handling is a by-product of female responsiveness to infant stimuli (Paul & Kuester, 1996b; Silk, 1999). Another possibility is that infant handling reflects a form of reproductive competition among females (Silk, 1980; Maestripieri, 1994c).

Different explanations for infant handling need not be mutually exclusive and may apply to different types of interactions or age classes of individuals. For example, the reproductive competition hypothesis may only apply to a subset of infant handling interactions in which reproductively experienced females harass or kidnap other females' infants (Maestripieri, 1999).

Fig. 10.3. In Tonkean macaques, high levels of maternal tolerance allow juvenile females to handle infants and carry them for extended periods of time (Strasbourg Primate Center, France). (Photograph by B. Thierry.)

The quality of female infant handling in macaques tends to covary across species in relation to maternal protectiveness or permissiveness as well as to the quality of adult female relationships (Maestripieri, 1994c; Thierry, 2000; Thierry, Chapter 12). Infant harassment and kidnapping have been best documented in rhesus macaques and, in this species, mothers are generally described as very protective and intolerant (e.g., Hinde & Spencer-Booth, 1967). Rhesus macaques are also often described as the most despotic and nepotistic macaque species, in that their social interactions are strongly affected by dominance rank and kin support (de Waal, 1989b). Japanese and longtailed macaques are closely related to rhesus macaques and share with them some similarities in adult social relationships, the quality of infant handling, and the degree of maternal protectiveness (Hiraiwa, 1981; Eaton et al., 1985; Thierry, 1985b).

Rhesus macaques are similar to pigtailed macaques in terms of the risks associated with infant handling and these two species also exhibit similarities in maternal protectiveness and intolerance (Rowell et al., 1964; Wheeler, 1986; Maestripieri, 1994a,b). Pigtailed macaques, in turn, appear to resemble a closely related species, the liontailed macaque, for patterns of infant

Box 10 Mother–infant behavior and socialization   233

handling and maternal protectiveness (e.g., Kumar & Kurup, 1981), although direct comparisons between these two species have not been made. The Sulawesi macaques (e.g., *M. nigra* and *M. tonkeana*), which are also closely related to pigtailed and liontailed macaques, exhibit relatively benign infant handling and high maternal tolerance (Fig. 10.3) (Thierry, 1985b). Their adult female relationships also tend to be affiliative and egalitarian (Thierry *et al.*, 1994). Bonnet, stumptailed and Barbary macaques are another cluster of closely related species in which infant handling is mostly benign, mothers are permissive, and female social relationships are generally affiliative and egalitarian (Rosenblum & Kaufman, 1967; Blurton-Jones & Trollope, 1968; Caine & Mitchell, 1980; Small, 1990b; Maestripieri, 1994a,b; but see Paul, 1999 for Barbary macaques, and Simonds, 1965, and Silk, 1980, for bonnet macaques).

Taken together, these observations suggest that the evolution of infant handling and maternal styles in macaques can be viewed, in part, as an adaptation to evolutionary changes in adult social relationships and social organization, and in part as the result of phylogenetic inertia (e.g., Thierry *et al.*, 2000). Although differences among species are often striking, there is also a great deal of variation between groups and populations of the same species. The relation between social environment, quality of infant handling, and mothering style, however, seems to hold also at the intraspecific level. For example, rhesus macaques living under conditions of high social density exhibit higher risk of infant kidnapping and harassment and higher maternal protectiveness than rhesus macaques living in less crowded environments (Maestripieri, 2001).

Both infant handling and parenting styles are likely to have multiple determinants and be subject to multiple selective pressures and constraints. Variation in other dimensions of parenting style such as frequency and timing of maternal rejection seems to be better accounted for by ecological and reproductive factors than by social factors. For example, in seasonally breeding species, maternal rejections begin earlier and are more frequent than in nonseasonal species (e.g., Worlein *et al.*, 1988; Maestripieri, 1994a). This is because among seasonal breeders mothers are under pressure to wean their infants early to be able to conceive again during the mating season (Simpson *et al.*, 1981). Differences in maternal rejection also occur between seasonal breeders such as rhesus and Japanese macaques, in part as a result of differences in climate (and hence infant thermoregulatory needs) or in seasonal availability of food (Hiraiwa, 1981).

Intra- and interspecific differences in interactions between infants and other individuals, including their mothers, are likely to have important consequences for social development. Unfortunately, aside from the information provided by social deprivation studies, little is known about the influence of early social experiences on behavioral development in macaques. Macaque infants typically spend their first years of life in close interaction with their mothers, other matrilineal kin, and peers. Sex differences in grooming, play, and in the extent to which infant behavior is biased toward kin, infants, or adult males emerge predictably during development and across different environments (e.g., Eaton *et al.*, 1985; Glick *et al.*, 1986a,b). Such early sex differences are a prelude to the different life histories of males and females, with males preparing for emigration from the natal group and females preparing for integration into their matrilineal social network. It is likely that the developmental trajectories taken by macaque infants are, in part, genetically determined and, in part, the result of social and learning processes. Although macaque mothers initially play an active role in encouraging infant independence (Maestripieri, 1995, 1996a), infants are not actively encouraged or instructed by adults to acquire skills that will prepare them for their adult social roles. Instead, the development of macaque behavior is probably shaped by individual learning processes that occur in the context of interactions between the infants, their mothers and peers, and other group members (e.g., Berman & Kapsalis, 1999; Chauvin & Berman, this Chapter). Comparative studies of macaque social development are rare (e.g., Thierry, 1985b) and the evolutionary relationship between social organization and infant socialization remains poorly understood.

Part IV *External and internal constraints*

Introduction

Are there no limits to the forms of social organization that animals can produce to cope with environmental constraints, or do internal constraints define the space of possibilities open to them, limiting social organization to finite sets of forms? By forming groups, animals may protect themselves from predators, defend resources, improve breeding and get information about the environment (Alexander, 1974; Dunbar, 1988). Social groups are basically cooperative units, but living in close proximity also entails disadvantages for the individuals. It enhances the competition for resources and gives rise to multiple bonds. We may explain the differences in group size observed among primates by the balance between the costs and benefits of group-living (van Schaik, 1983; Terborgh & Janson, 1986). However, it is more problematic to account for the patterns of a social organization only by the external factors (Chapter 11). Historical contingencies and phylogeny channel a species' adaptation. Though macaques show a fair amount of intraspecific variation, their style of social relationships appears quite stable within each species. Comparative studies indicate that macaque societies are covariant sets of characters, which travel together through the evolutionary process (Chapter 12). Such patterns betray the existence of systemic constraints, internal to social organization. Computer simulations show that complex structures may arise as by-products of the behavior of artificial agents implemented with simple decision rules. This generates sets of cascading effects, which mirror those found in real macaques (Chapter 13). Such results have important implications with regard to the evolution of macaques and their societies. As Raff (1996: 428) puts it, if internal factors constrain evolution this is not a minor issue, it means that the variation presented to selection is not random.

235

11 Do ecological factors explain variation in social organization?

NELLY MÉNARD

Introduction

Predation risk is considered as one of the main factors leading to gregarious-ness in primate females (van Schaik, 1989). However, living in a permanent group possibly induces direct or indirect feeding competition among females. The negative effect of within-group competition can be compensated by ben-efits of obtaining more food through between-group competition (Wrangham, 1980; van Schaik, 1989). The degree and type of within- or between-group competition result in different styles of female–female social relationships. According to the "ecological model," initially developed by Wrangham (1980) and then expanded by van Schaik (1989), the abundance and distribution of food resources may determine the type of competition. When food resources are patchily distributed, foraging contest competition should occur (van Schaik & van Noordwijk, 1988). On the other hand, when food resources are evenly distributed, or patches are wide enough to allow many animals to feed together or items are too small to be shared, within-group feeding competition should be low.

Among primate species, several styles of female relationships have been described varying from a "dispersal–egalitarian" system when competition is weak, where females show tolerant relationships with other females and dis-perse, to a "resident–nepotistic" social organization characterized by a high level of within-group contest competition and a low level of between-group competition (Van Schaik, 1989; Sterck et al., 1997). In the latter case, females support kin during conflicts, are philopatric, and a clear dominance kind of rela-tionship is expected. Most of the macaque species are considered as resident–nepotistic, characterized by a high level of within-group competition and a low level of between-group competition. But the crested macaque corresponds to

Macaque Societies: A Model for the Study of Social Organization, ed. B. Thierry, M. Singh and W. Kaumanns. Published by Cambridge University Press. © Cambridge University Press 2004.

the resident–nepotistic-tolerant category (van Schaik, 1989; Sterck *et al.*, 1997) as do Tonkean (Thierry, 1985a) and stumptailed macaques (de Waal & Luttrell, 1989). In these latter species, dominants possibly need the cooperation from subordinates due to a high level of between-group competition. They should thus become more tolerant (van Schaik, 1989).

More recently, Thierry (2000; Chapter 12) proposed a gradient within the genus Macaca distinguishing four categories of species on the basis of their style of social relationships within groups, from highly hierarchical and nepotistic species (equivalent to the previous resident–nepotistic species, such as rhesus and Japanese macaques), to tolerant species (equivalent to the previous resident–nepotistic-tolerant species such as Sulawesi macaques). Most conclusions about the style of relationships are drawn from studies conducted on captive groups and focusing mainly on female–female relationships.

Another model based on mating strategies took into account the potential role of males in between-group competition (Caldecott, 1986a). Indeed, within and between-group contest competition may exist for access to fertile females. On the other hand, the abundance and distribution of resources may indirectly affect the quality of inter-male social relationships and male grouping tendencies through female sexual behavior. Available reproductive resources (i.e., fertile females) depend on group size, adult sex ratio, synchronization of estrus, female sexual behavior, and receptivity.

The styles of social relationships in macaques, as documented, were considered as specific features that emerged and persisted through phylogenetic constraints (Thierry *et al.*, 2000; Thierry, Chapter 12). However, a question remains about the origin of the observed inter-specific variations of social organization. Could these divergences be partly explained by differences in ecological conditions? Indeed, the broad geographical distribution of the genus reflects the great variety of ecological conditions in which the different macaque species or populations live (evergreen and deciduous forests, swamp forest, rocky mountains, urbanized sites, habitats located in tropical or temperate climates). The noticeable remnants of distribution of an earlier colonization is what probably limits our knowledge of the entire adaptive potential of macaque species, and that of the initial conditions in which existing social organization emerged.

The aim of this chapter is to investigate to what extent the inter-specific variability of the type of social relationships in macaques could be explained by ecological factors. One could expect that the level of within-group contest competition would be high when the food is distributed in small patches and could be monopolized by high-ranking individuals. On the other hand, if patches are wide enough to contain entire groups, allowing groups to defend resources against other groups, between-group competition should be high, encouraging tolerance within groups. A comparison will be made of the abundance and distribution of food resources between habitats colonized by different macaque species or populations. Information on these parameters allows approximating the degree of defensibility of food resources. Dietary compositions and time spent in different activities may also provide valuable, although indirect, information on food availability. We will compare the results on these parameters from different studies in order to document their variations at specific, population, and group levels. We will pay special attention to highlighting, as much as possible, the existence of intra-specific flexibility of the ecological responses of animals to variations in their environmental conditions. According to the ecological model (van Schaik, 1983), we could expect more within-group competition in species or populations living in habitats characterized by patchily distributed resources, or species feeding on fruit (assuming that they are clumped resources) than in species or populations living in habitats where resources are evenly distributed, thus non-defensible, or feeding mainly on leaves (considered as uniformly widely distributed resources). In the first case we could also expect a higher rate of locomotion than in the second one, due to travel between patches of fruit. According to these hypotheses, the most intolerant macaque species should be mainly frugivorous, living in environments with small patches of food. The most tolerant species should be mainly folivorous, living in habitats where resources are widely distributed and non-defensible. Styles of social relationships within and between sexes result from both kinds of resources: food and sexual partners. Thus, we will also examine group size and composition in macaque species as an indicator of abundance and distribution of sexual partners, notably females, within groups.

Geographical distribution and ecological conditions

Is each species confined to a particular habitat?

In his review of the distribution of macaque species, Fooden (1982a) considered that five species (*M. silenus, M. nemestrina, M. assamensis, M. thibetana* and *M. arctoides*) inhabit primary broadleaf evergreen forests while the others mostly exploit swamps or secondary, deciduous, coniferous, or degraded forests. These habitat preferences of macaques appear to be independent of phylogeny. When they are sympatric, species generally use different habitats or occupy separated ecological niches. For example, pigtailed macaques live in evergreen forests whereas longtailed macaques live in deciduous forests in Sumatra, Malaysia, and Borneo. Equivalent separation is found between liontailed and bonnet macaques in India, and between Assamese and rhesus macaques along the flanks of the Himalayas. When species live in the same habitat, competition probably has induced the use of different parts of the forest. Stump-tailed macaques forage on the ground while Assamese macaques forage mostly in the canopy (Fooden, 1982b). This global schema, however, suffers some exceptions, which might provide valuable information on the actual adaptive abilities of monkeys. Are they actually exceptions or do they reflect, because of the lack of studies in certain parts of the distribution areas of species, our limited knowledge of habitat use by monkeys? For example, rhesus macaques have never been observed in broadleaf evergreen forests that are the preferential habitat of Assamese macaques (Fooden, 1982b). By contrast, it seems that Assamese macaques could adapt to mixed forests, the habitats colonized by rhesus macaques. In fact, groups of these two species have been observed in proximity to each other in the same habitat type (letter of McNeely, 1973, in Fooden, 1982b: 26), and it could be that this occurs more frequently than has been thought. The Assamese macaque could be more flexible in its ecological requirements than has been supposed. More recently, the Tibetan macaque was found to use subtropical evergreen broadleaf forest, mixed deciduous broadleaf forest and evergreen coniferous forest, exploiting the first for 50% of the observation time. Hence, as judiciously underlined by Zhao (1996), "to describe the habitat of Tibetan macaques as subtropical evergreen broadleaf type . . . is questionable." The same categorization may be also in doubt for other macaque species when studies are conducted in more or less provisioned populations or groups or when the studies do not cover the entire annual cycle, thus not providing representative results of the ecological features of species.

Same habitat categories may include different ecological realities

Most of the habitats where macaques occur have been modified to varying degrees by human activities. Consequently, without a clear description of habitats and associated disturbances caused by humans, it is difficult to appreciate the ecological factors that may or may not allow monkeys to colonize a given habitat type. A single category of forest could include a variety of habitats produced by different degrees of disturbance by humans. In Barbary macaques for example, evergreen cedar–oak forests represent different ecological conditions for monkeys depending on whether they are located in Algeria or in Morocco (Ménard & Vallet, 1988; Ménard & Qarro, 1999). Algerian oak forests support populations of Barbary macaques, whereas human exploitation of oak trees in Morocco has rendered the habitat unsuitable for monkeys for either feeding or refuging. A study in Morocco alone might wrongly conclude that Barbary macaques prefer pure cedar forests to oak forests. As abundance and distribution of resources are difficult to assess, most studies give only a qualitative description of habitats. Attempts to quantify resources in wild, undisturbed habitats remain relatively rare. They exist for pigtailed (Caldecott, 1986b), Barbary (Ménard & Vallet, 1988), rhesus (Richard *et al.*, 1989), Japanese (Maruhashi *et al.*, 1998) and longtailed macaques (Wich *et al.*, 2002).

Quantified resource availabilities

Japanese macaque
Available resources for Japanese macaques were quantified with the same methods in two contrasted habitats, one in a warm-temperate forest (in Yakushima Island, Japan) and the other in a cool-temperate forest (in Kinkazan Island, Japan; Maruhashi *et al.*, 1998). Tree crowns were measured along transects and regularly visited to quantify their production. Tree density was found 13 times higher at Yakushima than at Kinkazan. Yakushima forest was composed of several types of vegetation: primary forest, mixed primary and secondary forest. One species (*Myrica rubra*) had the highest density among the potential fruit-food trees (i.e., mature trees that have equal to or larger crown area than mean crown area of the food trees) with 11 fruiting trees per hectare while it composed up to 73% of the diet depending on the month. *Rhus succedanea* and *Lithocarpus edulis* had the highest density among the seed-food trees (4–6 trees per hectare) while they composed up to 42% and 13% of the diet respectively. By contrast, *Ficus superba* had a density of 0.3 per ha but its fruit composed up to 41.3% of the diet. Thus monkeys actively searched for the fruit of this latter species. Resources from trees were generally clumped and constituted the staple food for monkeys leading them to travel long day-range distances (Agetsuma, 1995a; Agetsuma & Nakagawa, 1998). Most of the plants had seasonal fruit or flower production. At Kinkazan, unfortunately, only the tree layer was sampled although the mean annual diet of monkeys was composed of 15% herbs (up to 50% per month) and of 30% underground food from the species *Oplismenus undulatifolius* (Agetsuma & Nakagawa, 1998). Consequently, it is difficult to compare the distribution of resources, and even their global abundance, in the two sites. In addition, any such comparison should include to what extent monkeys found clumped resources in the course of the year. Nevertheless, one can reasonably assume that resources at Kinkazan, as they included more herbaceous items, were more evenly distributed than at Yakushima.

Rhesus macaque
Most studies on rhesus macaques were conducted in disturbed environments where monkeys had easy access to human supplied food, due either to direct provisioning or to availability of crops in agricultural areas. One study was

carried out in a "natural" habitat at Dunga Gali (Pakistan) that was composed of five types of forest showing various degrees of degradation (Goldstein & Richard, 1989). Animals in this area showed a clear preference for disturbed habitats where 48% of their feeding was recorded even if these sites represented only 24% of the total available vegetation. Due to human exploitation, the area was a mosaic of patches of various vegetation types, which probably increased the diversity of available resources. Most of these resources were small sets of widely dispersed ground herbs that can hardly be monopolized.

Barbary macaque

A comparative study of available resources was conducted in two different habitats occupied by Barbary macaques, in a deciduous oak forest (Akfadou, Algeria) and in a cedar–oak forest (Djurdjura, Algeria; Ménard & Vallet, 1988) (Fig. 11.1). In the Akfadou forest, the tree layer constituted 93% of the forest surface. It was composed of two oak species (*Quercus faginea* and *Q. afares*) rather evenly distributed. Food resources at the shrub and herbaceous layers are also evenly distributed. In addition, food items were too small-sized to be considered defensible. Clearings could constitute patches of herbaceous resources but they were large enough to contain one entire group. Some resources, such as flowers of the shrub species *Calycotome spinosa* that were eaten over a short period in May, or mushrooms, that were actively searched for in autumn, were distributed in defensible patches. In addition, the existence of small springs gave opportunities for monopolizing access to water and could induce between-group competition during summer. In Djurdjura forest, trees covered 74% of the surface area, dominated by two evergreen species (*Cedrus atlantica* and *Quercus ilex*) while grassland areas represented 24% and shrub areas 5%. The habitat was composed of a mosaic of different vegetation types. Shrub species were patchily distributed but these patches contained several individual shrubs and they could generally contain one group of modal size. Production of food items was highly seasonal. They were generally too abundant or too small-sized to be efficiently monopolized by one individual. Only large roots, bulbs, mushrooms, or water in hollow trees could induce within-group contest competition.

Pigtailed macaque

Vegetation was sampled in both a logged and an unlogged tropical rain forest (Peninsular Malaysia) colonized by pigtailed macaques, along transects where trees were measured and phenology was monitored monthly. In such forests, where Dipterocarpaceae dominated, individuals of any given species occurred singly and were widely spread (Caldecott, 1986b). In the unlogged forest, resources were more abundant than in the logged due to a higher proportion of fruiting trees, a higher production rate, and larger trees, possibly indicating

Fig. 11.1. Habitats of macaques. (a) Barbary macaques foraging in a cedar–green oak forest (Djurdjura National Park, Algeria; photograph by D. Vallet.) (b) Crested macaques progressing in lowland rainforest (Tangkoko Nature Reserve, Sulawesi; photograph by B. Thierry.)

the presence of a larger individual food source. In both habitats, there was a flower and fruit shortage from October to February.

Longtailed macaque

A study of the abundance and distribution of food trees was conducted in the tropical rain forest colonized by longtailed macaques at Ketambe (Gunung

Leuser National Park, Sumatra, Indonesia), based directly on measures of the phenology. This was one of few studies measuring the size of food patches. The sizes of actually fruiting trees and of trees in which macaques forage were larger than the sizes of all other possible fruiting trees. Moreover, macaques foraged in trees larger than the mean size of all fruiting trees, which indicated that they avoided foraging in small fruiting trees. These results suggested that they could reduce competition by choosing trees the size of which may allow the individuals of one entire group to forage together (Wich *et al.*, 2002). On the basis of observations of monkey feeding behaviors instead of tree phenology, the number of fruiting trees would be underestimated and the mean size of the fruiting trees would be overestimated.

Categorization of species according to their resource distribution

Any attempt to make a quantitative comparison of macaque species in terms of the abundance and distribution of their food resources is confounded by the fact that different studies do not give comparable categories of measures. When resources are known to be patchily distributed, one may easily fail to find all of the required information: size of patches, distances between patches, and temporal variations in resource clumping depending on the seasonality of production. It would be necessary to know how many monkeys can eat together at the same patch and for how long a time. Nevertheless, results of the studies presented above do allow attempts to categorize the species mentioned. Pigtailed and longtailed macaques depend the most on clumped resources (but we do not know to what extent over the year) whereas rhesus and Barbary macaques get food from evenly distributed resources (if cases of provisioned groups are excluded). Depending on the habitat, the Japanese macaque finds clumped or rather evenly distributed resources.

Diet

There have been few analyses of macaque diets that are both quantitative and comprehensive. Most studies have presented largely qualitative descriptions, or have been conducted over a short period, often less than one annual cycle, or were carried out on at least partly provisioned groups. Consequently, the general situation is that we have a limited knowledge of mean annual diets, of flexibility of diets depending on seasons or years, and of intra-specific variations depending on habitat types. Thus, studies on only some macaque species provide reliable quantitative analyses of diets for comparative analysis (Table 11.1).

Table 11.1. *Diet of wild* Macaca *species*

Species (study duration)	Fruits	Seeds	Leaves and buds	Roots	Flowers	Herbs	Invertebrates	Sap and resin	Fungi	Lichens	Others	Sources
M. sylvanus (Akfadou, Feb '83–Mar '85)	0.8 (0.4–8.4)	32.2 (9.2–84.1)	8.8 (0.5–48.3)	6.9 (0–33.5)	3.5 (0–36.2)	18.5 (0–49.3)	10.5 (0–83.5)		4.1 (0–23.7)	14.2 (0–43.1)	0.5	1
M. sylvanus (Djurdjura, Feb '83–Mar '85)	4.3 (0–19.3)	26.7 (0–61.5)	13 (0–31.4)	7.7 (0–29.5)	1.6 (0–26.6)	35.1 (0–75.8)	5.6 (0–66.6)		1.5 (0–6.5)	1.9 (0–9.3)	2.6	1
M. nemestrina (Jan '79–May '81)	74.2		11.1		1.1		12.2				1.4	2
M. fascicularis (Jan–Dec '85)	66.7		17.2		8.9		4.1				3.1	3
M. fascicularis[a] (Kalimantan, Oct '74–Jun '76)	87 (58.2–100)		1.6 (1.8–7.7)		3.3 (4.3–25)	2.1 (1.8–17.6)	4 (1.8–25)		0.7 (1.1–1.9)		1.3	4
M. mulatta (autumn '78–summer '79)	8.5 (0–32)		84.4 (60.3–99.4)	2.2 (0–23.7)	3.7 (0–16.9)			1.1 (0–8.4)			0.1	5

(cont.)

Table 11.1. (*cont.*)

Species (study duration)	Fruits	Seeds	Leaves and buds	Roots	Flowers	Herbs	Invertebrates	Sap and resin	Fungi	Lichens	Others	Sources
M. fuscata[a] (Yakushima, Jan '90–May '92)	28.6 (5–60)	28.2 (3–60)	22 (0–65)		4.9 (0–35)		8.9 (3–25)				7.4	6
M. fuscata[a] (Kinkazan, Nov '84–Aug '92)	10 (0–30)	44 (0–80)	14 (0–55)		3 (0–40)	15 (0–55)	2 (0–10)	5 (0–15)			7	7
M. fuscata[a] (Yakushima, Dec '87–May '89)	30.2 (3–75)	13.2 (0–40)	35 (2–82)		5.5 (0–28)		10.3 (1–32)		4.6 (0–18)		1.2	8
M. cyclopis[a] (Oct '91–Sep '92)	53.8 (0–92)		29.1 (18–100)		7.2 (0–73)		9.8 (0–25)					9

Values indicate the mean percentage of feeding time spent eating different food items. They are calculated from a 12-month period except for studies 4, 6 and 7. Values in brackets are monthly variations.

[a] Monthly variations were drawn from figures.

Sources: (1) Ménard, 1985; Ménard & Vallet, 1986, 1996; (2) Caldecott, 1986b; (3) Yeager, 1996; (4) Wheatley, 1980 (averaged from data for 18 months); (5) Goldstein & Richard, 1989; (6) Agetsuma, 1995b (mean diet calculated from 10 months of data); (7) Agetsuma & Nakagawa, 1998 (averaged from percentages calculated for 20 half-month periods covering 11 months); (8) Hill, 1997; (9) Su & Lee, 2001.

Mainly herbivorous species

Rhesus macaque

This is one of the macaque species presenting the greatest ability to adapt to habitats linked to human activities. In agricultural or urbanized areas, natural vegetation constituted from 7% to 53% of diets when people fed monkeys and when monkeys raided agricultural crops (Malik & Southwick, 1988; Siddiqi & Southwick, 1988). In the natural habitat at Dunga Gali, rhesus macaques appeared mainly folivorous as leaves composed 84% of its mean annual diet (Table 11.1). Diet diversity was relatively low, with monkeys feeding mainly on 11 plant species while 24 others were only occasionally consumed. In addition, they were selective feeders, with 36% of the available 59 trees and shrub species included in their diet, while 14% of the herbaceous plants were eaten. There were great seasonal variations, in particular in fruit eating. Grass and leaves of *Trifolium* sp. amount to up to 35% of the annual diet and remained the staple food all the year round. In winter, the sap of conifers and roots of grasses and of *Oenothera* constituted up to 8% and 24% of the diet respectively. Rhesus macaques showed a preference for disturbed areas where they took 50% of their diet (17% in winter and 77% in spring), attracted by the *Trifolium* and *Viburnum* grasses (Goldstein & Richard, 1989). As confirmed on the basis of the rhesus diet in a natural habitat, nothing indicated that food items could be clumped resources. Most of them were small-sized, generally evenly distributed and could not be monopolized by one individual. In addition, displacements of individuals away from food sources by conspecifics were very infrequent (Goldstein & Richard, 1989).

Barbary and Tibetan macaques

Barbary macaques appeared mainly either granivorous or folivorous depending on the habitat (Table 11.1). They showed a highly flexible diet with marked seasonal and inter-annual variations. The proportion of herbaceous plants consumed was relatively high. Animal prey consisted mainly of caterpillars that constituted a monthly staple food depending on the year. When they were absent, monkeys supplemented their diet with the flowers of oaks or shrubs. Relatively few items could induce contest competition. These are mushrooms or roots of *Smyrnium perfoliatum* that composed only 1.5–4.1% and 0–0.4% of the mean annual diet respectively, depending on the habitat (Ménard, 1985; Ménard & Vallet, 1986, 1988). In addition, during dry periods, drinking in hollow trees occasioned conflict situations within groups but related aggressive interactions occurred in a very limited period of the year (N. Ménard, unpublished data). Thus, in both habitats, opportunities of direct food competition remained relatively limited. Contest competition within groups was not observed over any of the major foods in any month.

While our knowledge of its diet remains partial, the Tibetan macaque seems to be mainly a leaf eater, displaying feeding strategies comparable to that of the Barbary macaque (Zhao, 1996).

Frugivorous–granivorous species

Japanese macaque

Clearly, the diet of Japanese macaques cannot be assigned to a simple category as it is very flexible depending on the forest, the season and the year (Hill, 1997; Table 11.1). Groups in warm-temperate forest (at Yakushima) were more frugivorous (essentially *Ficus* fruits) than groups in cool-temperate forest (at Kinkazan), which were more granivorous (Agetsuma & Nakagawa, 1998). However, great seasonal variations seemed to be the rule in both types of forest (Iwamoto, 1982; Agetsuma, 1995b; Hill, 1997). Staple foods were leaves, seeds, fruit and flowers or fruit and insects depending on the month. Monkeys consumed a great diversity of plant foods (up to 147 from 93 species) although they were also selective feeders since only three to 21 food items accounted for 90% of the food items eaten depending on the season (Hill, 1997). They ate seeds of Fagaceae, the most abundant trees of the forest at Yakushima. The composition of the diet at Kinkazan suggested that, most of the year, resources were not defensible since they were small-sized items, produced seasonally in large quantities and evenly distributed.

Mainly frugivorous species

Taiwanese, pigtailed, and longtailed macaques appeared as the most frugivorous among macaque species whose diet was known (Table 11.1). Fruits corresponded to 54–87% of their feeding time. Little information was given on seasonal and inter-habitat variations. Nevertheless, longtailed macaques ate fruit from January to May, and from June to December they switched their diet to a mixed one composed of leaves, flowers, and fruit (Wheatley, 1980; Yeager, 1996). Although the pigtailed macaque was classified as mainly frugivorous, one can assume that during periods of fruit and flower shortage (Caldecott, 1986b), monkeys included mainly leaves or invertebrates in their diet. The Taiwanese macaque is the least frugivorous of these three macaque species since it is almost completely folivorous during 5 months every year (Su & Lee, 2001). In addition, Su and Lee (2001) indicated that this species had a mainly folivorous mean annual diet at other sites.

Liontailed, Assamese, and bonnet macaques were considered as frugivorous species although insects could compose a noticeable part of the diet (up to 30%), depending on the season, in addition to mushrooms and flowers (Fooden, 1981, 1982b; Kurup & Kumar, 1993; Menon & Poirier, 1996).

Tentative classification of species

Pigtailed and longtailed macaques can be considered as the most frugivorous among the species documented in Table 11.1 whereas Japanese, rhesus, and Barbary macaques can be considered as nonspecialized eaters. The diets of the latter species show considerable variation depending on habitat and season. Our knowledge of macaque diets remains relatively limited, mainly due to difficulties of observing them in tropical rain forests; the most documented species, in terms of mean annual estimates, monthly variations, and inter-population variations, being Japanese and Barbary macaques. The degree of flexibility of the species categorized as frugivorous is not known but it could be assumed that, as they feed on fruit trees sparsely distributed within the forest, within-group contest competition could exist at a higher level than in leaf and seed-eaters.

Time budget

Comparing time budgets between different studies is generally a difficult task because of different definitions of the same behavioral categories. The "foraging" category sometimes includes moving or not, the "feeding" category includes foraging or not and the "social" category either includes resting in contact with a group member or not. In addition, results may be misinterpreted if the range of variations in day lengths over the year is not taken into account while comparing studies at different latitudes. Thus, percentages may not be entirely accurate concerning time budgets and comparisons have to be considered cautiously. Table 11.2 presents only data referring to activity categories comparable between studies.

Species with the most moving time

The comparison between species shows that the two most frugivorous (*M. fascicularis*: Wheatley, 1980; *M. nemestrina*: Caldecott, 1986b) spent the most time moving (45–61% of time) and the least time feeding and foraging. However, data on longtailed macaques were only collected during 3 days of observations on a

Table 11.2. *Percentage of time spent on each activity for different Macaca species*

Species (study duration)	Habitat	Feeding	Foraging	Resting	Locomotion	Social	Other	Sources
M. sylvanus (Feb '83–Mar '85)	cedar–oak forest, Djurdjura	25.4 (3.3 h: 2–5 h)	6.2 (0.8 h: 0.2–1.6 h)	36.9 (4.8 h: 2.2–8.2 h)	20 (2.6 h: 1.7–3.6 h)	11.5 (1.5 h: 1–2.9 h)		1
M. sylvanus (Feb '83–Mar '85)	deciduous oak forest, Akfadou	23.8 (3.1 h: 1.6–5.5 h)	3.9 (0.5: 0–1.6 h)	40 (5.5 h: 3–8.7 h)	22.3 (2.9 h: 1.6–3.5 h)	10 (1.3 h: 1–2 h)		1
M. silenus (Sep '90–Aug '91)[a]	disturbed forest fragment	17.9 (10–27)	23.7 (15–33)	16 (9–26)	34 (26–45)	(+ other) 8.4 (2–18)		2
M. silenus[a]	undisturbed protected forest	27.8 (22–41)	26.7 (18–36)	27 (22–45)	15 (10–24)	2.4 (2–8)	1.1 (0–4)	3
M. nemestrina (Jan '79–May '81)	tropical rain forest	16	16	19	61	4		4
M. radiata (1 year)[a]	urban area		31	11	2	34	4	5
M. radiata (1 year)[a]	rural area		24	22	8	43	3	5
M. radiata (1 year)[a]	forest		37	21	7	30	5	5
M. fascicularis (only three days)			13	42 (+ social)	45			6

M. mulatta (autumn '78–summer '79)[a]	deciduous forest	45	34	11	10		7
M. mulatta (Jul '80–Aug '83)	rural area	20.8 (2.5 h: 2.2–2.8 h)	30 (3.6 h: 2.8–4.4 h)	18.6 (2.2 h: 1.8–2.6 h)	11.5 (1.4 h: 1.2–1.5 h)	19.1	8
M. mulatta (Jun '74–Aug '78)	montane forest in Nepal	27	28	25	20		9
M. fuscata (Feb–Mar '72)[b]		26.7	40.9	17.3	14.8	0.3	10
M. fuscata (Aug '89–Apr '92)[a]	warm-temperate forest, Yakushima	30.8 (3.8 h: 2–6 h)	22.1 (2.8 h: 1.7–4.5 h)	22.6 (2.8 h: 2–4 h)	20.7 (2.5 h: 1.3–3.2 h)	3.7	11
M. fuscata (Nov '84–Aug '92)[a]	cool-temperate forest, Kinkazan	53.9 (6.2 h: 4–8.5 h)	17.6 (2.2 h: 1.2–4.2)	16.8 (2.1 h: 1–4.1 h)	11.5 (1.3 h: 0.9–1.9 h)	0.3	11
M. fuscata (Dec '87–May '89)[a]	warm-temperate forest, Yakushima	20–48 (2.8–5.5 h)	?	10–22 (0.1–3.1 h)	?		12

Values in brackets give the variation range in percentage or in hours (when available, the mean annual daily time in hours is added before the range).
[a] Values drawn from figures; [b] the category 'huddling together' was included in the social category.
Sources: (1) Ménard & Vallet, 1997; (2) Menon & Poirier, 1996; (3) Kurup & Kumar, 1993; (4) Caldecott, 1986b; (5) Singh & Vinathe, 1990; (6) Wheatley, 1980; (7) Goldstein & Richard, 1989; (8) Malik & Southwick, 1988; (9) Teas et al., 1980; (10) Wada & Tokida, 1981; (11) Agetsuma & Nakagawa, 1998; (12) Hill, 1997.

focal adult male and may not be representative. The activity budget of pigtailed macaques was constructed on the basis of group activity assuming that individuals were stationary for one-third of the time when groups were recorded in slow travel. In order to take into account individual's local movements while the group was recorded as static, one-quarter of this time was assigned to movement (Caldecott, 1986b). The activity budget of pigtailed macaques is nevertheless taken into account assuming that this method yielded results comparable with those of other studies.

Species with the most feeding time

In all other minimally frugivorous species, the mean annual feeding and foraging time was 34% of the total time on average (varying from 21% to 55%) while moving accounted for 18% of the time on average (7–34% depending on the species and the habitat). The highest feeding and foraging values (42–55%) were found in liontailed (Kurup & Kumar, 1993; Menon & Poirier, 1996) and Japanese macaques (Agetsuma & Nakagawa, 1998) in cool-temperate forest while the lowest (20–37%) was found in rhesus macaques (Teas *et al.*, 1980; Malick & Southwick, 1988), Japanese macaques in warm-temperate habitat (Agetsuma & Nakagawa, 1998), bonnet macaques (Singh & Vinathe, 1990) and Barbary macaques (Ménard & Vallet, 1997). In fact, we noted that liontailed and Barbary macaques, for which foraging and feeding activities can be distinguished, did not differ greatly in their feeding times (between 20% and 28%) while the first species spent more time foraging than the second one (24–27% vs. 4–6%). It is possible that the actual feeding time of other species, such as the Japanese (at Kinkazan) and the bonnet macaques, does not exceed 30%, but this cannot be confirmed since not all studies provided distinct foraging and feeding time values. In the bonnet macaque, the presence of humans probably modified considerably the activities of monkeys who spent the most time in social activities (30–43%) while reducing significantly their moving time (7–20%) when compared with the other macaque species (Singh & Vinathe, 1990).

Seasonal variations

In all species where information was available, great seasonal variations of time budgets were observed. Those variations showed comparable ranges across species for feeding and foraging as well as for moving. Monkeys spent a daily feeding and foraging time varying between a minimum of 2–4 h to a maximum of 3–8 h depending on the month, the most common range being 2–6 h. The highest values were found in Japanese macaques at Kinkazan (4–8.5 h) but we do not know what the actual part of the feeding time is. The most common monthly variations in moving time range around 2–4 h. Whatever the species, animals devoted the least amount of time to social activities.

The main factor in variation is change in resource availability. Japanese macaques spent less time moving but more time foraging when more leaves were included in their diet, as in autumn and winter compared to the summer period. By contrast, they moved more when they fed on animal matter as well as when they had to move between fruit and seed sources in late spring and early summer. Moving time could then increase by more than 100% (Agetsuma, 1995a; Hill, 1997).

In liontailed macaques, the time spent moving and foraging increased when fruit production was low while the resting time decreased (Kurup & Kumar, 1993; Menon & Poirier, 1996).

When resources were maximal in spring, the foraging effort of Barbary macaques decreased while they maximized their feeding time (from 2 h to 5 h) and spent more time in social interactions (from 1 h to 3 h). At the same time, moving time decreased from 3.5 h to 1.6 h. During the summer period of food shortage, foraging and moving time doubled although monkeys did not succeed in maintaining a high feeding time (Ménard & Vallet, 1997).

Inter-habitat variations

Rhesus macaques seem the most adaptable among nonhuman primates. They show great variation in their time budget depending on habitat type. When they were fed by people, these monkeys spent two times less time feeding and foraging while their moving time increased and time spent in social behavior was not very different (Malik & Southwick, 1988) to that of rhesus living in deciduous forest (Goldstein & Richard, 1989). However, results must be taken cautiously because not all the categories were well defined, notably the category "others."

In Japanese macaques, the time spent on feeding varied depending on habitat from 27% in a warm-temperate forest to 54% in a cool-temperate forest (Agetsuma & Nakagawa, 1998). In the first habitat type, feeding time reached 37–49% in autumn and winter (Iwamoto, 1982). The time spent moving increased when animals included more fruit in their diet at Yakushima, whereas at Kinkazan, animals spent more time feeding to compensate for poorer quality food. At the same time, resting and social time decreased. By contrast, in rich habitats monkeys increased their grooming time. Consequently, we can suppose that in a poor environment, a decrease in grooming time probably changes the within-group relationships (Agetsuma & Nakagawa, 1998).

The comparison of the time budget of bonnet macaque groups living in different conditions showed that in urban areas, where there was high disturbance by people, monkeys spent more time in locomotion and less time in social activities than in rural areas or in forest. Females spent more time in social activities in rural areas than in forest or urban areas. The increase in locomotion in urban areas was probably linked to less availability of resources than in forest or rural areas (Singh & Vinathe, 1990). The social time was very high compared with other macaque species but we do not know whether or not sitting in passive contact was included in the social behavior category.

In degraded habitat, liontailed macaques spent less time resting in disturbed conditions than in protected forests while monkeys were compelled to move constantly to ensure optimal foraging (Kurup & Kumar, 1993; Menon & Poirier, 1996). Despite a comparable foraging time, monkeys in disturbed areas spent less time feeding.

The mean annual feeding time for Barbary macaques was similar at both study sites (Ménard & Vallet, 1997). Monkeys spent approximately 37–40% of their time resting and around 24% feeding.

In brief, it seems that the most frugivorous species spend more time on average per year in locomotion and less time in feeding behaviors than the other macaque species. However, it would be interesting to know the range of monthly variations in their time budgets. Indeed, whatever the other studied species, their time budget appears highly flexible depending on the season or the habitat and we do not observe higher variations of time budgets between species than within species.

Group size and composition

Group size

The mean size of social groups is quite comparable across the different macaque species (Table 11.3). In natural undisturbed habitats, the maximal size above which groups tended to split into smaller groups was found to be around 80–90 individuals for rhesus (Lindburg, 1971) and Barbary macaques (Ménard & Vallet, 1993b), and around 50 or 70 individuals for Japanese macaques (Yamagiwa & Hill, 1998). The maximal group size of toque macaques was observed to be around 35 individuals before splitting (Dittus, 1988). Group size reached more than 100 individuals mostly in provisioned conditions or rural areas where monkeys could raid crops. There were great intra-specific variations

Table 11.3. *Group size and composition of* Macaca *species*

Species	Habitat type	Adult sex ratio M:F	Ratio immature	Group size	Number of groups	Sources
M. sylvanus	cedar oak forest	1:0.7–1.8	0.41–0.59	13–88	3	1
	deciduous oak forest	1:0.8–1.9	0.42–0.58	33–54	1	1
M. sylvanus	cedar oak forest	1:0.7 (1:0.8–1.4)	0.24–0.30	30–80	4	2
M. silenus	degraded forest	1:6	0.67	43	1	3
M. silenus	protected forest	1:3–5	0.50–0.63	12–31	4	4
M. nigra				30 (10–60)	13	5
M. tonkeana				20 (10–40)		5
M. maurus		1:2–3		22–40	1	6
M. ochreata				20 (10–40)		5
M. nemestrina	tropical forest	1:8 (1:5–9)		24 (15–55)	13	7
M. nemestrina	tropical forest	1:2.7 (1:2–3.3)[a]		49–81	3	8
M. sinica	semideciduous forest	1:1.7–2.5[a]		24.8 (8–43)	18	9
M. radiata	forest	1:1.3	0.50	14.7	7	10
	urban area	1:1.7	0.50	17.3	11	10
	rural area	1:1.4	0.49	25.4	18	10
M. assamensis		1:2.3		12–50	20	11
M. thibetana	evergreen forest	1:3 (1:2.5–6.5)	0.60	38.3 (28–65)	6	12
M. arctoides	evergreen forest	1:5.7 (N = 5)		2–60	17	13
M. fascicularis	tropical rain forest	1:2.1 (1:1–4)	0.60	27.4 (6–42)	14	14, 15
M. fascicularis	tropical rain forest without felids			12.5 (10–15)	10	15

(*cont.*)

Table 11.3. (cont.)

Species	Habitat type	Adult sex ratio M:F	Ratio immature	Group size	Number of groups	Sources
M. mulatta	forest or urban areas in tropical or arid climate	1:1.4–2.8		21–30	8	16
M. mulatta	small forest patch in agricultural area rural area			70–140		17
M. mulatta	moist forest	1:3–4	0.50–0.54	32 (9–98)	14	18
M. mulatta	mixed deciduous and coniferous forest	1:3.4	0.50	49.8	5	19
M. mulatta	rural area	1:0.7	0.52	61	1	20
M. mulatta	moist Himalayan temperate forest	1:3.5		23–25	1	21
M. mulatta	cool temperate oak forest, more or less	1:2.1–4.6 (N = 2)		82 (50–120)	26	22
	provisioned	1:1		24	1	23
M. fuscata	cool-temperate forest	1:1.4 (1:0.7–3.3)		35 (8–86)	34	24
M. fuscata	warm-temperate forest	1:2.3 (1:0.8–10)		75 (17–161)	18	24
	warm-temperate forest	1:1.3 (1:1–2)		27 (13–47)	17	24
M. cyclopis	mature and secondary evergreen forest, coniferous and grassland areas	1:0.6–3		46.8 (9–86)	15	25

Values in brackets give the variation range when available.

[a] M: F is (adult and subadult mature males):(mature females).

Sources: (1) Ménard & Vallet, 1993b, 1996; (2) Ménard, 2002; (3) Menon & Poirier, 1996; (4) Kurup & Kumar, 1993; (5) MacKinnon, 1983, quoted in Caldecott, 1986a; (6) Okamoto & Matsumura, 2002; (7) Caldecott, 1986a; (8) Oi, 1996; (9) Dittus, 1975; (10) Singh & Vinathe, 1990; (11) Fooden, 1982b; (12) Zhao & Deng, 1988b; (13) Fooden, 1990; (14) van Noordwijk & van Schaik, 1985; (15) van Noordwijk & van Schaik, 1985; (16) Seth & Seth 1985; (17) Siddiqi & Southwick, 1988; (18) Lindburg, 1971; (19) Southwick et al., 1965; (20) Malik & Southwick, 1988; (21) Goldstein & Richard, 1989; (22) Qu et al., 1993; (23) Wada & Tokida, 1981; (24) Yamagiwa & Hill, 1998; (25) Kawamura et al., 1991.

in group size depending on the habitat occupied or the history of groups. For example van Noordwijk and van Schaik (1985) showed that groups of longtailed macaques tended to live in smaller groups when predators were absent. Similarly Yamagiwa and Hill (1998) found in Japanese macaques that group size was significantly greater in heavy snowfall areas than in light snowfall areas and that it was greater in habitats characterized by scarce, low-quality food and a low-density population.

Adult sex ratio

The adult sex ratio was in favor of females in most species (Table 11.3). According to the available data, three species may be distinguished by their low adult sex ratio (M:F) with more than four adult females per male on average: pigtailed (Caldecott, 1986b; Oi, 1996), liontailed (Kurup & Kumar, 1993; Menon & Poirier, 1996) and stumptailed macaques (Fooden, 1990). In contrast, Barbary macaques presented the most balanced sex ratio with one female per male on average, the remaining species showing intermediate patterns (Ménard & Vallet, 1993a, 1996; Ménard, 2002).

To conclude, these comparisons should be considered cautiously as, in most cases, the available data were drawn from few groups observed at only one study site. Thus, they were perhaps not representative of the species under study. Indeed, there were non-negligible intra-specific variations in sex ratio. In the rhesus macaque, for example, the adult sex ratio varied from 0.7 to 4.6 females per male depending on the group or the study site. Yamagiwa and Hill (1998) showed a negative correlation between group size and the number of males per female within groups and a relation between sex ratio and habitat types. To appreciate more accurately the divergence between species, it should be useful to compare the modal distribution of adult sex ratios between species, thus taking into account sex ratio variations among groups for each species.

Existent ecological factors alone do not explain variation in social organization

Macaque societies cannot be classified according to the ecological model

Macaque species were found to be characterized by specific patterns of their social organization, notably dominance styles between females, on the basis of studies on captive groups (Thierry, 1990a, Chapter 12). Available data on the characteristics of habitats and modes of exploitation by monkeys does not indicate species-specific features of known ecological conditions that appear to be correlated with their characteristic dominance styles. Indeed, following the ecological model (van Schaik, 1983), if we assume that the abundance and patchiness of resources would, while determining their degree of defensibility, also influence the degree of competition and dominance styles of species, then we would expect the Barbary macaque and, to a lesser extent, the Japanese macaque, to be rather tolerant species with relaxed relationships (Table 11.4). By contrast, pigtailed and longtailed macaques should appear as despotic species. The existence of high between-group competition has been proposed to explain more relaxed relationships within groups. However, low between-group competition involving defense of resources by females was observed in "tolerant" species such as moor (N. Matsumura, 1999) and Barbary macaques (N. Ménard, unpublished data). By contrast, Saito and collaborators (1998) observed between-group competition involving females in wild groups of Japanese macaques, a despotic species, although between-group competition varied according to the habitat. Unfortunately, this study did not provide the relative frequency of peaceful inter-group encounters compared with agonistic ones. In all these species, aggressive inter-group encounters commonly resulted in a defense of females by males rather than defense of resources by females (see Cooper, Box 9). Thus, predictions based upon ecological reasoning are generally at odds with Thierry's classification (1990a, 2000), which found rhesus and Japanese macaques to be the most despotic–nepotistic species in terms of symmetry of aggression and high effect of kinship on social interactions (Table 11.4).

Why such a discrepancy between ecological and behavioral data?

If present ecological characteristics do not explain specific social organization, we might consider at least three possible explanations: (1) specific traits may have been selected in the past under different ecological conditions, which may have persisted despite the disappearance of these initial ecological conditions because they did not entail any significant disadvantages; (2) different social organizations may be equally efficient in a given environment; and (3) we cannot exclude the possibility that studies of a few groups under captive conditions were unable to detect the degree of flexibility of social organization in any given species, thus giving only a partial view of the full social potentialities of that species.

Table 11.4. *Ecological factors, group composition and category of social organization of some Macaca species, (1) predicted following ecological reasoning, (2) as observed and assumed to result from phylogenetic constraints*

Species	Resource distribution	Diet	Defensibility	Sex ratio M:F	Social organization (1)	Social organization (2)
M. sylvanus	even	not specialized	weak	balanced	relaxed	moderately relaxed
M. silenus	?	?	?	low	?	moderately relaxed
M. nemestrina	patchy	frugivorous	yes	low intermediate	despotic	moderately despotic
M. arctoides	?	?	?	low	?	moderately relaxed
M. fascicularis	patchy	frugivorous	yes	intermediate	despotic	moderately despotic
M. mulatta	even	not specialized	weak	intermediate	relaxed	despotic
M. fuscata	patchy or even	frugivorous or not specialized	yes or weak	intermediate	moderately relaxed	despotic

Ecological conditions and flexibility of social relationships?

Following the first and second assumptions, one could expect that species-specific styles of social relationships should emerge whatever the environmental conditions. In fact, macaques show a great flexibility in their behaviors in terms of modalities of resource exploitation, time spent in different activities, and modalities of grouping (group size and composition), depending on ecological conditions (see Hill, Box 11). Flexibility is evident between populations as well as for any given individual. Thus, one could expect that social disposition must also be highly flexible to allow for variations in the degree of competition, which would depend on the abundance and degree of defensibility of resources and mating partners. Indeed, different groups of wild Japanese macaques showed variations in their degree and style of competition depending on resource distribution. When the size of patches decreased, the degree of within-group contest competition increased. Although actual attack remained rare, the number of submissive behaviors became more frequent (Saito, 1996; Yamagiwa & Hill, 1998). Similarly, Hill (1999) showed that the frequency and intensity of aggressive behaviors increased within Japanese macaque groups when resources were clumped – in provisioned conditions. The distribution of resources also influenced the quality of social relationships between females, between males and between males and females (see Hill, Box 11). Contrary to provisioned groups, the pattern of dominance acquisition by "youngest ascendancy" was not observed in wild non-provisioned groups because of low frequency of aggressive behaviors and low kin support (Hill, Box 11). Dominant–subordinate relationships were much less frequently reinforced; proximity between males was more common; subordinate males showed less submissive behaviors and females did not maintain proximity with dominant males as protectors (Hill, 1999). Even in species such as pigtailed or longtailed macaques, whose resources in the wild are considered patchily distributed, investment of monkeys in establishing patterns of social relationships based on a strict dominance style is expected to vary depending on the duration of resource patchiness over the year (which is not documented).

Availability of females and social relationships

Availability of resources probably also influences male–female relationships, and male–male relationships in turn. Caldecott (1986a) suggested, for example, that when resources are patchily distributed and patches are small-sized, there would be an advantage for females to limit the number of males within groups in order to reduce food competition. This could explain the low sex ratio observed in pigtailed macaque groups living in Asian forests of low quality. The male grouping tendency is likely to be influenced by the quality of inter-male relations that probably varies according to the males' abilities to monopolize females. This may depend for example on the synchrony of female estrus. In highly seasonal species, synchronous females may experience more female competition for mates in groups with skewed sex ratios than in groups with balanced sex ratios. Thus, the composition of groups could result from a trade-off of female strategies between competition for food and competition for mates. Therefore, the low seasonality of reproduction and low synchrony of female estrus in pig-tailed macaques (Oi, 1996) suggest that a skewed sex ratio does not imply a high sexual competition among females. By contrast, in Barbary macaques, characterized by highly seasonal reproduction, competition for mates between females was very high in a group composed of only one adult male and five adult females. During female estrus, the only male did not succeed in monopolizing the females from foreign males. Consequently, after one year of tenure by this male, the studied group reached a balanced sex ratio through male immigration (Ménard & Vallet, 1993b, unpublished data).

Is social organization so specific?

All the above findings, although based on few studies, provide arguments to reconsider the third assumption. In fact, even in captive conditions, divergences were shown to exist in the social organization of different groups of the same species. In a stumptailed macaque captive group, reconciliation and grooming time were more frequent than in a rhesus macaque group, which resulted in a more relaxed dominance style (de Waal & Luttrell, 1989). Nevertheless, the two rhesus macaque groups that were studied differed in their patterns of social relationships. One of the rhesus groups showed grooming relations comparable to the other group whereas it showed the same patterns of social tolerance as the stumptailed macaque group. More recently, Butovskaya and collaborators (1996) found that two captive groups of longtailed macaques of similar composition, observed under identical environmental conditions, displayed different dominance styles.

Conclusion

Much information is still needed to allow efficient comparison of social organization between macaque species. Such comparative studies require the knowledge of the range of flexibility of each studied species. Social organization being the product of individual social dispositions, it seems obvious that even under similar captive conditions (standardized in terms of group size, composition, and environment), different responses between groups of different species could either be due to actual differences between species or result from the effect of differences between groups (and not species) because of particular individual temperaments. Results from laboratory studies also have to be verified in the light of field studies. An effort needs to be made to study the intensity and quality of competition in different conditions in the wild and should be conducted using standardized methods and sufficiently large group sampling to allow reliable comparisons between populations within species and between macaque species.

Box 11   Intraspecific variation: implications for interspecific comparisons

David A. Hill

Several studies of macaques have revealed consistent interspecific differences in the quality of their social relationships, and particularly in the nature of dominance (e.g., Thierry, 1985a; de Waal & Luttrell, 1989). Much less attention has been paid to the potential for variation in social behavior within species, particularly with reference to macaques in their natural habitats. If intraspecific variation is substantial, then it may bring the validity of species-specific "profiles" of social behavior into question.

Environmental variation

The geographical distribution of some macaque species is extensive, incorporating a great variety of habitats and environmental conditions (Ménard, this Chapter). The most extreme example is the rhesus macaque, whose distribution extends from the mountains of Afghanistan in the northwest, to the coast of China in the east, and the plains of central India to the south (Fooden, 1980; Southwick *et al.*, 1996). Within this vast area, the species can be found in semi-arid regions and moist forest, and from tropical to cool temperate climes. The longtailed and pigtailed macaques also have extensive distributions incorporating a variety of habitat types, although mostly variants of tropical forest. Other macaque species with less extensive distributions may also be found in diverse habitats. For example, the Barbary macaque in Algeria, is found in both evergreen and deciduous forests, as well as in less forested montane habitats (Ménard & Vallet, 1996), and the Taiwanese macaque is found in lowland areas, as well as the highland forests of Taiwan (Lee & Lin, 1991).

Box 11 Intraspecific variation 263

In addition to natural variation in environmental conditions, many macaque species have experienced extensive habitat loss and alteration as a result of human activity. Some species, such as rhesus and bonnet macaques are able to adapt to the modified habitat, to the extent that they live commensally with humans throughout much of their range (Richard *et al.*, 1989). Others, such as the liontailed macaque, have a more limited propensity to adapt (Singh *et al.*, 2001), and their range continues to shrink as human influence grows (Easa *et al.*, 1997).

Environmental influences on demography and time budgets

Habitat quality can influence aspects of social behavior and social organization in a variety of ways. For example, for several species of macaque, regional variation in demographic parameters, such as group size and composition, has been attributed to environmental variation (e.g., Takasaki, 1981; Singh & Vinathe, 1990; Lee & Lin, 1991; Southwick *et al.*, 1996). Group size and composition will determine the number and nature of potential partners in an individual's group, which in turn will govern the potential for complexity of the social network. There are strong indications that maximum group size is limited by food availability. When macaques are regularly provided with food by human observers, groups generally grow, often very rapidly, and to much larger sizes than are normally observed in non-provisioned groups (Asquith, 1989). Larger groups also tend to have more adult females per adult male than smaller ones (Yamagiwa & Hill, 1998), although the mechanisms underlying this pattern are not clear.

The distribution of food resources within the environment also influences the area that a group needs to range over (Takasaki, 1981) and the amount of time individuals spend foraging. Japanese macaques in the cool temperate deciduous forests in the north of Honshu spend much longer foraging than those in the evergreen forests of the warm temperate south (Agetsuma & Nakagawa, 1998). As time spent moving and foraging tends to be inversely correlated with time spent resting and grooming (Singh & Vinathe, 1990; Hill, 1997; Agetsuma & Nakagawa, 1998), macaques in relatively poor habitats will have less time available for social behavior than conspecifics in richer habitats.

Environmental influences on macaque social behavior

Intraspecific variation is not limited to patterns of grouping, ranging behavior, and time budgets. There is also evidence that environmental conditions can have a profound influence on the quality of social relationships within a species. Most of the evidence comes from comparisons of provisioned and natural populations of Japanese macaques.

Under natural conditions, Japanese macaques are usually dispersed when foraging. Direct competition over food resources is observed, but infrequently, because very clumped resources are few. Nevertheless, higher levels of aggression are associated with feeding than with most other activities (Mori, 1979b; Furuichi, 1983; Hill & Okayasu, 1995). A key characteristic of the provisioned situation is that the major food resource is highly concentrated in both space and time (Fig. 11.2). This routinely brings individuals into close proximity during feeding and heightens competition over food, resulting in increased frequency of severe aggression (Mori, 1979b; Hill & Okayasu, 1996). Even when efforts are made to scatter food over an area, aggression is much more frequent than when feeding on natural resources (Mori, 1979b).

Japanese macaques are generally classified as being among the most nepotistic of the macaques, with a high degree of asymmetry in aggression (Thierry, Chapter 12). Mothers frequently support their daughters in aggression with others and, around adolescence, daughters inherit their mother's dominance rank (Kawamura, 1958). This means that, in most cases, females subordinate to the mother are also subordinate to her nonjuvenile daughters. Where aggression occurs between sisters, the mother usually supports the younger daughter who, as a consequence, comes to outrank all of her older sisters.

These characteristics of female dominance relations have been repeatedly observed in various provisioned populations of Japanese and rhesus macaques. Comparable data are not available for most natural populations because few have been studied for long enough to establish matrilineal kinship. However, in one small group of Japanese macaques for which long-term kinship data were available, in all four pairs of sisters the older clearly outranked the younger (Hill & Okayasu, 1995, 1996). Datta and Beauchamp (1991) presented a model which shows how variation in demographic parameters, such as interbirth interval could account for differences in patterns of acquisition of female dominance rank (Chapais, Chapter 9). However, the model relies on the probability of dominant female supporters being alive at the time when the younger daughter reaches adolescence. This cannot explain the four exceptions noted above, as in each case the mother was still alive and dominant to both daughters when the younger reached adulthood. The alternative explanation offered by Hill and Okayasu (1995) was based on the fact that aggression in the study group was infrequent, and that agonistic aiding was very rare indeed. They concluded that mothers did

Box 11 Intraspecific variation 265

Fig. 11.2. Provisioning a free-ranging group of Japanese macaques. The population of Arashiyama, near Kyoto (Japan), has become a tourist attraction. (Photograph by B. Thierry.)

not preferentially support their younger offspring, because aggression was rarely severe, so the young females were rarely in danger.

This isolated result clearly needs confirmation from further studies. Nevertheless, it shows that social mechanisms that are routinely observed in captive and provisioned groups, are not necessarily active under natural conditions. Furthermore, there is evidence from comparison of studies of natural and provisioned groups, that environmental factors can have a profound effect on the nature of social relationships between males and females, and on those between males (Hill, 1999). As the distribution of food resources, and consequently levels of competition over food, may vary enormously within the range of a macaque species, there is clearly potential for intraspecific variation in the quality of social behavior and relationships under natural conditions. Unfortunately, even for the most-studied species, testing this proposal from existing data would not be feasible. Studies of macaque species living under various natural conditions, which incorporate detailed data on both social behavior and the distribution of food resources in the environment, simply do not exist.

Implications for interspecific comparisons

A common approach to the study of interspecific variation in macaque social behavior has been to contrast "dominance styles" (de Waal & Luttrell, 1989). Subsets of species may be assigned to groups according to characteristics such as degree of nepotism, asymmetry in agonistic interactions, and patterns of conflict resolution (Aureli *et al.*, 1997; Thierry, 2000; Thierry, Chapter 12). Any classification of this kind involves establishing profiles that encapsulate species-typical patterns of behavior. For example, dominance relations among rhesus macaques have been described as nepotistic and asymmetrical, whereas those among Tonkean macaques tend to be more egalitarian (Thierry, 1985a). However, in most cases the data on which these behavioral profiles are based have come from studies of captive or provisioned groups, and so represent quite extreme environmental conditions in terms of concentration of food resources. In order to understand the evolutionary significance of species differences in dominance styles, we need first to establish the extent to which they are apparent in natural populations, and how much they vary within each species, under the range of natural conditions in which it occurs.

12 *Social epigenesis*

BERNARD THIERRY

Introduction

To understand how evolutionary forces shape the societies of primates, we have to trace the patterns of their social organization back to the behavior of individuals and beyond, up to the determinants of their phenotypes and genotypes. We are far from reaching such goals as yet. Whereas a complete theory of evolution should include developmental laws (Lewontin, 1974; Futuyma, 1979), we do not really know how the genetic machinery produces organisms. In the last decade, the growing field of evolutionary developmental biology has started to actively fill this gap by focusing on the internal processes that govern modes of change (Raff, 1996). But this stream has little permeated the study of animal social behavior. We continue to study the proximate and ultimate causes of behaviors separately. This leaves little room for epigenesis, the very place where the inherited and acquired characters of organisms channel future evolutionary pathways.

Epigenesis is the set of processes that bring the individual phenotype into being. Waddington (1957) introduced this concept to distinguish between the direct biochemical action of genes and their indirect effects. Indirect effects are the interactions among the constituent parts of the organism at any level of organization (see Gottlieb, 1992). "Genetics proposes, epigenetics disposes" (Medawar & Medawar, 1983), one may expect that the role of epigenetic processes becomes all the more important as the distance from the level of expression of the genome increases. It is thus at the level of the social phenotype that we should expect the most powerful epigenetic effects. On the scale of a society, social epigenesis is the set of processes responsible for the construction of the social phenotype. A full understanding of social organization requires taking into account the mechanisms at play in the emergence of both the individual and the social phenotypes. Some might say that such a definition covers everything, and that it would be simpler to talk in terms of proximate causes and

Macaque Societies: A Model for the Study of Social Organization, ed. B. Thierry, M. Singh and W. Kaumanns. Published by Cambridge University Press. © Cambridge University Press 2004.

developmental mechanisms. This, however, would miss the point. By giving emphasis to epigenesis, we mean more than the study of proximate causes. We aim to investigate how the nongenetic machinery and its self-organizing processes constrain the action of selective pressures and the course of evolution (Thierry, 1997).

Our knowledge of the wide interspecific variation that features the social relationships of macaques provides a unique tool to assess the role of epigenetic constraints in the evolutionary changes undergone by social organizations. I will first specify the different constraints responsible for the linkage of social characters at the epigenetic level. I will then review the evidence supporting the covariation of the components of macaque social styles. In the closing section, I will consider the evolutionary implications of the interdependence of characters.

Epigenetic constraints

An epigenetic constraint is a limitation on the variability of the phenotype caused by the structure and dynamics of the epigenetic system (see Maynard Smith *et al.*, 1985; Schlichting & Pigliocci, 1998). It may be called an internal or a structural constraint as well. Different constraints on the social phenotype may originate from the genomic, organismic and social levels.

Genomic level

The coupling of characters may arise from the architecture and regulation processes of the genome. In linkage disequilibrium, the proximity of two gene loci on a chromosome reduces their rates of recombination, leading certain allelic effects to be correlated (Price & Langen, 1992). Another source of coupling involves the regulation processes of the genome. Since several genes share the same regulatory circuits, this makes it harder for characters to shift independently (Weiss, 2002). The functioning of the genetic system produces further linkages. A single gene locus can affect a variety of phenotypic characters. This pleiotropic effect of genes produces multiple correlations between characters. As an example, several behaviors may be influenced by a single hormone or neuromediator (Price & Langen, 1992; Ketterson & Nolan, 1992). In rhesus macaques, a genetic component has been demonstrated for levels of serotonin (Trefilov *et al.*, 2000; Champoux *et al.*, 2002). Reduced serotonin activity is associated with high levels of impulsivity and aggression, a short duration of social contact, and an earlier dispersal of males (Kaplan *et al.*, 1995; Mehlman *et al.*, 1995; Higley & Linnoila 1997; Capitanio, Chapter 2). The pleiotropic effect of the alleles responsible for decreased levels of serotonin may play a major role in such correlated variations.

Organismic level

Phenotypic linkage

The organism is built through a number of trade-offs and structural constraints, which produce intercorrelations between morphological, physiological and life-history characters (Gould, 1977; Stearns, 1992). Since Darwin, the scientists have wondered about the origin of the laws of allometry, which remains poorly understood (Gould & Lewontin, 1979; Western & Ssemakula, 1982; Harvey *et al.*, 1987; Raff, 1996: 428). Whereas trade-offs are adaptive devices, other compromises may occur through development and result from structural constraints. Phenotypic linkage is not limited to physical characters, it also extends to psychological, emotional, cognitive and communicative abilities. Could there be no structural relationships between behavior and life-history characters? And would it be possible that the degree of tolerance in inter-individual contacts are independent from aggression and reaction systems? Conversely, certain characters may be incompatible (see Lott, 1991). To state that the various components of psychobiological systems are not separate is rather to state the trivial. Rhesus macaques that inhabit cities are more aggressive, less fearful, and affiliate less frequently with unfamiliar conspecifics than their counterparts living in a forest environment (Singh, 1966, 1969). Rather than looking for a different origin for each of these differences, it is more parsimonious to consider that the distinct experiences undergone by urban and forest monkeys affect interdependent psychobiological components. This produces contrasted temperaments, which then trigger different sets of behavioral outputs (see Mendoza & Mason, 1989; Capitanio, Chapter 2).

Social pleiotropy

A single individual character can affect several features of the social organization. Since several individuals may share the same character, their iterative action generates multiple effects at the population level. Social pleiotropy may account for a number of behavioral features (Thierry, 1997). In Japanese macaques, the mounts and consortships reported among females result from sexual motives usually oriented toward opposite-sex partners (Vasey, 1995, Box 7). In several macaque species, immature individuals interfere in matings by approaching and contacting the copulating partners. There is no obvious function for such a behavior. Interference may be a secondary output of a device contributing to protect social relationships, no matter how we would label it, "jealousy" or "anxiety" system. One psychobiological system can be involved in different behaviors (Capitanio, Chapter 2). Anxiety may be activated by the management of the risks incurred by various social interactions such as social conflicts, reconciliations, and mothers' infant retrievings (Maestripieri, 1993, 1994b; Aureli & Schino, Chapter 3). We may expect that two individuals differing by their baseline anxiety will show systematic variations in their behavioral responses in these three types of interactions.

Social level

Behavioral constraints

Individuals influence each other through social interactions. Their behaviors may complement or reinforce one another, or on the contrary oppose or even exclude one another. Because of such constraints, some patterns are made possible whereas others cannot arise (Thierry, 1997). For instance, tactical considerations indicate that competitors' readiness to struggle for a resource depends on the risk they incur (Parker, 1974; Popp & DeVore, 1979; van Rhijn & Vodegel, 1980; Enquist & Leimar, 1983). If the risk of being wounded in a conflict is elevated, the better tactic for the weaker individual is to submit or flee rather than to counter-attack the opponent. Conversely, when the dominance gradient is low, the threatened individual can easily retaliate, forcing his adversary to avoid potentially dangerous attacks. Thus, in animals which are able to use graded threats, a high intensity in aggression should be associated with strong asymmetry of contests at the interaction level, and a steep dominance gradient and marked submission at the relationship level (see Preuschoft & van Schaik, 2000; Flack & de Waal, Chapter 8). This means that in stable social conditions a high proportion of retaliations in conflicts is not compatible with frequent biting and wounding because that would correspond to poor tactics. Following the same rationale, any association between intense aggression and a high degree of mother permissiveness appears unlikely. A female living in an intolerant social environment has to protect her infant from potential dangers. She must limit her infant's social contacts and so behave as a restrictrive mother (McKenna, 1979; Thierry, 2000). Coupling between agonistic and maternal patterns is the outcome.

Social inheritance

Individuals inherit not only genes but also their environment (Hailman, 1982; Odling-Smee, 1988). The development of immatures occurs in a social milieu provided by relatives and other partners. Therefore, the phenotypic characteristics of individuals are influenced by the behavior of conspecifics, producing a correlation between generations. The dominance status and the networks of relationships can be passed on from mothers to daughters, and some behaviors carried on through intergenerational transmission, perpetuating a set of developmental pathways (Kawamura, 1958; Missakian, 1972; Berman, 1982a, 1990; de Waal, 1996b; Chapais, 1992a, Chapter 9; Chauvin & Berman, Chapter 10). In experiments where young rhesus macaques were co-housed with young stumptailed macaques, the conciliatory tendencies of the former were multiplied threefold, stabilizing at levels comparable to those of the latter (de Waal & Johanowicz, 1993). The characteristics of the social environment may shape the temperament of offspring and it may also affect their life-history traits. Demographic dynamics are a major feature of development, they may give rise to delayed life-history effects (Beckerman *et al.*, 2002). In rhesus macaques, a high dominance status provided by maternal protection accelerates reproductive maturation both in males and females (Silk, 1987; Bercovitch, 1993). Offspring are influenced both by the acquired and inherited characters of adults. The genome intervenes twice in development, directly through the individual's own genes, and indirectly through the genes of the individual's conspecifics (Altmann & Altmann, 1979; Wolf *et al.*, 1998). Some component of the social environment is therefore heritable. This may tighten the variation range of individual and social characters through successive generations.

Covariation of characters

In view of the numerous constraints liable to link the characters of a social organization, the *covariation hypothesis* states that any significant variation in a single character induces a set of correlated changes in the social organization. Thus, we may predict that the comparison of social relationships between macaque species will evidence a consistent series of interspecific variations.

Aggression and appeasement

The study of agonistic interactions among macaques has shown marked contrasts between species. In rhesus and Japanese macaques, most conflicts are unidirectional, high-intensity aggression is common, and reconciliations are infrequent: the conciliatory tendencies measured among unrelated individuals rate between 4% and 12% (Thierry, 1985a, 1990b, 2000; de Waal & Luttrell,

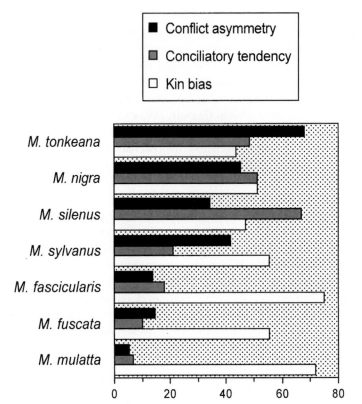

Fig. 12.1. Levels of conflict symmetry, reconciliation and kin bias across 7 captive populations of macaque. Conciliatory tendencies are calculated according to Veenema *et al.* (1994), only dyadic conflicts occurring between pairs of unrelated females (older than 3.5 years), and between pairs of females and unrelated juveniles (between 1.5 and 3.5 years of age) were entered into the analysis. Kin bias is a ratio representing the degree of preference for matrilineal relatives; it is calculated at the group level as the conciliatory tendency between relatives divided by the sum of the conciliatory tendencies between relatives and nonrelatives multiplied by 100. The level of conflict symmetry is the percentage of conflicts in which the aggressed individual protests or counter-attacks; it is calculated by pooling all conflicts having occurred between mature females (i.e., at least 4 years of age) in a given population. (Data from Thierry (1986a, 2000); Abegg *et al.* (1996); Petit *et al.* (1997); Demaria & Thierry (2001); and C. Marengo (unpub. data on Barbary macaques, Kintzheim, France).)

1989; Aureli *et al.*, 1993; Butovskaya, 1993; Chaffin *et al.*, 1995; Petit *et al.*, 1997; Schino *et al.*, 1998; Kutsukake & Castles, 2001). These results depart from those found in Sulawesi macaques (crested, Tonkean and moor macaques) in which a majority of conflicts are bidirectional, that is, most aggressive acts induce protests or counter-attack. Aggression is generally of low intensity, and measuring rates of reconciliation yields especially high values: conciliatory tendencies score around 50% among unrelated partners (Bernstein *et al.*, 1983b; Thierry, 1985a, 2000; Matsumura, 1991, 1996; Petit *et al.*, 1997; Thierry *et al.*, 1997; Demaria & Thierry, 2001). Behavioral constraints may account for this covariation of aggression patterns and reconciliation rates at a functional level (Fig. 12.1). On one side, asymmetric conflicts and increased risk of injury may inhibit the occurrence of affiliative contacts between opponents, while approximate symmetry and uncertainty about outcomes create room for negotiation (Silk, 1997). On the other side, conciliatory behaviors may reduce the probability of conflict escalation by facilitating information exchange between adversaries (de Waal, 1986a; Thierry, 1986a; Aureli *et al.*, 2002).

Other macaques are located intermediately between the previous species. Longtailed and pigtailed macaques are more similar to rhesus and Japanese macaques (Angst, 1975; Judge, 1991; Thierry, 1986a, 2000; Castles *et al.*, 1996; Aureli *et al.*, 1997), whereas stumptailed, Barbary, and liontailed macaques tend toward the Sulawesi macaques (de Waal & Luttrell, 1989; Butovskaya, 1993; Abegg *et al.*, 1996; Aureli *et al.*, 1997; Abegg, 1998; C. Marengo, unpublished data). I have proposed to arrange macaque species along a 4-grade scale based on patterns of aggression and reconciliation (Table 12.1) (Thierry, 2000). Species from grades 3 and 4 display high rates of specific affiliative contacts like clasps and embraces (Thierry, 1984; de Waal, 1989b; de Waal & Luttrell, 1989; Matsumura, 1991; Abegg, 1998). They are also characterized by the development of special behaviors that may reduce social tension and facilitate social contact. In stumptailed macaques, a ritualized soft biting performed by higher-ranking individuals toward subordinates may bring about positive interactions and the end of conflicts (de Waal, 1989b; Demaria & Thierry, 1990). In Barbary macaques, adult males use infants as buffers to facilitate approach and affiliation (Deag, 1980; Paul *et al.*, 1996). In Sulawesi and liontailed macaques, individuals regularly intervene in and occasionally terminate conflicts by appeasing one of the opponents (Petit & Thierry, 1994; Abegg, 1998). Such behavior patterns are basically absent in species from grade 1 (rhesus and Japanese macaques). The use of infants in triadic interactions can occur in longtailed macaques albeit at a low rate (de Waal *et al.*, 1976; B. Thierry, personal observation). Clasping interactions also occur at a low rate in grade 2 (longtailed macaques: Thierry, 1985c; pigtailed macaques: Castles *et al.*, 1996; Maestripieri, 1996b).

Table 12.1. *Classification of* Macaca *social styles according to 14 behavioral characters. Interindividual tolerance increases from the left to the right*

Grade 1	Grade 2	Grade 3	Grade 4
M. mulatta 1,2,3,4,5,6,7,8,9,10, 11,12,13,14	*M. fascicularis* 1,2,3,4,5,6,7,8,9, 12,13,14	*M. sylvanus* 2,4,7,8,10,12,13	*M. tonkeana* 1,2,3,4,5,6,7,8,9, 10,11
M. fuscata 1,2,3,4,5,6,7,8,9, 10,11,12,13	*M. nemestrina* 1,2,3,4,5,7,8,9, 10,11,13,14	*M. silenus* 1,2,3,4,6,8,14	*M. maurus* 2,4,5,6,8,11
M. cyclopis 4,8		*M. arctoides* 1,2,3,4,5,6,7,8,9, 10,11,14	*M. nigra* 1,2,3,4,5,6,7, 8,11
		M. radiata 4,5,6,7,9,10,11, 12,13,14	*M. nigrescens* 4
		M. assamensis 4,7,8	*M. hecki* 4
		M. thibetana 4,7,8,13	*M. ochreata* 4
		M. sinica 4,6,12,13	*M. brunnescens* 4

Figures indicate the characters for which data are available in the species: (1) degree of asymmetry in conflicts; (2) conciliatory tendency; (3) intensity of aggression; (4) meaning of bared-teeth displays; (5) dominance gradient; (6) rate of specific affiliative contacts; (7) development of behaviors reducing social tension; (8) rate of immatures' interference in mounts; (9) degree of mother's permissiveness; (10) amount of alloparental care; (11) degree of kin preference among females; (12) patterns of female rank inheritance; (13) patterns of male dispersal; (14) temperament.
From Thierry, 2000.

Dominance–subordination relationships

Comparisons between macaques further showed that dominance relationships covary with patterns of agonistic interactions. The dominance gradient is the steepest in species of grade 1, in which social life is governed by strict rules (Kawamura, 1958; Sade, 1972a; Kurland, 1977; de Waal, 1989b). Power asymmetry determines who may interact with whom. It affects how an individual chooses partners for proximity, affiliation, or play, and whether the distribution of choices is skewed in favor of higher-ranking individuals. Subordinate individuals may be inhibited from approaching or contacting higher-ranking individuals because of the possibility and subsequent cost of attack. This picture differs from that observed in Sulawesi macaques where the dominance gradient is substantially less steep. In these species, status differences do not hinder contacts between group members and have little effect on grooming distribution (Thierry *et al.*, 1990, 1994; Matsumura, 1991; Petit *et al.*, 1992). A dominance relationship represents the formalization of a balance of power (de Waal, 1986a). Hence, inequality is functionally related to the level of asymmetry observed in agonistic interactions. Computerized models confirm that dominance gradient and aggression intensity may be linked functionally (Hemelrijk, 1999a, Chapter 13).

Both the dominant and the subordinate actively maintain their social relationship (Rowell, 1974; de Waal, 1986a). Formal signals of subordination vary consistently along the 4-grade scale (Thierry, 2000; Preuschoft, Box 3; Flack & de Waal, Chapter 8). In macaques from grades 1 and 2, subordinates retract the lips and expose the teeth to express submission (Figs. 8.1 & 12.2a) (rhesus, Japanese, longtailed and pigtailed macaques: Angst, 1975; de Waal, 1977; de Waal & Luttrell, 1985; Chaffin *et al.*, 1995; Maestripieri, 1996b; Preuschoft, 1995a, Box 3). By using this silent bared-teeth display outside the context of conflict, individuals formally acknowledge their lower status relative to higher-ranking conspecifics. In grade 3, the same facial expression often expresses subordination (e.g., stumptailed macaques: Bertrand, 1969); in Barbary macaques, however, it sometimes occurs in affiliative contexts (Fig. 12.2b), and in liontailed macaques its meaning is positive in a significant proportion of interactions (Preuschoft, 1995a; Box 3). In species from grade 4, silent bared-teeth displays no longer communicate social status, instead they signal the sender's peaceful intentions and serve to initiate affiliative interactions (Fig. 12.2c) (crested, Tonkean and moor macaques: Dixson, 1977; Thierry *et al.*, 1989; Petit & Thierry, 1992; Preuschoft, 1995b; Box 3). The meaning of silent-bared teeth displays is related to the dominance gradient typical of a given social organization. It should be added that displays being under strong genetic determinism, their production is quite fixed in a given species (Preuschoft, Box 3).

Kinship networks

Strong kinship networks mark the social relationships of rhesus and Japanese macaques. Females have a high preference for relatives and strict rules of inheritance determine the acquisition of dominance rank within matrilines (Kawamura, 1958; Sade, 1972a; Chapais, Chapter 9; Chauvin & Berman, Chapter 10). Females achieve, by adulthood, a dominance rank just below their mothers and rarely outrank them. Because females choose to support their youngest relatives in contests, dominance ranks are ordered inversely to

Fig. 12.2. The different meanings of the silent bared-teeth display. (a) A female rhesus macaque acknowledges her lower dominance status by addressing this facial expression towards a higher-ranking individual (Strasbourg Primate Center, France; photograph by B. Thierry). (b) In Barbary macaques, the message varies according to the context, the display may have a submissive function or convey a friendly meaning according to the context, here it conveys an affinitive message between females (Kintzheim, France; photograph by B. Thierry). (c) An adult male crested macaque addresses a bared-teeth display to a juvenile to signal his peaceful intentions, the message has no link with dominance relationships in this species (Wildlife Preservation Trust, Jersey; photograph by O. Petit.)

age within matrilines, and younger daughters dominate their elder sisters. The influence of kinship on competition and proximity among partners remains significant in other grades. However, the degree to which females prefer relatives in affiliative contact, social grooming, and support in conflicts is less pronounced in the third than in the first two grades (de Waal & Luttrell, 1989; Aureli *et al.*, 1997; Butovskaya, 1993; Box 8). Kin-bias is still weaker in Sulawesi macaques (Fig. 12.1), it is not possible to recognize kinship networks using only patterns of spatial distribution in these species (Thierry *et al.*, 1990, 1994; Matsumura & Okamoto, 1997; Matsumura, 2001). We lack quantitative data about female rank acquisition in most macaque species (Chapais, Chapter 9). We know that in longtailed macaques (grade 2), rank acquisition follows the rules described in grade 1, although rank reversals between mothers and daughters are not uncommon (Angst, 1975). In Barbary macaques (grade 3), daughters often outrank older mothers, and females are usually subordinate to their older sisters (Paul & Kuester, 1987; Prud'homme & Chapais, 1993a). In Tonkean macaques (grade 4), rank reversal of mothers and daughters is not rare and the rule of youngest ascendancy is irrelevant (B. Thierry, unpublished data).

Alliances provide the causal link between kinship and dominance structures (Thierry, 1990a; Chapais, Chapter 9). We may account for the linkage between the degree of nepotism and dominance gradient by the frequent occurrence of coalitions within groups of macaques. When most alliances involve relatives, the dominance status of individuals depends primarily on the power of the kin subgroup to which they belong. This increases rank differences between non-relatives and further develops kin-alliances, generating group structures based on strong hierarchies. Conversely, when kin-bias is less pronounced, coalitions involving nonrelatives are rarer, dominance appears more a question of individual attributes and the individual retains some degree of freedom with regard to power networks (Thierry, 1990a). Dominance relationships remain balanced among group members and close ties exist even between nonrelatives. Comparisons between macaque species support the view that dominance gradient and nepotism level among macaque females are connected by positive feedback (Fig. 12.1) (Aureli *et al.*, 1997; Thierry *et al.*, 1997; Demaria & Thierry, 2001; Butovskaya, Box 8). The same rationale may explain the dominance of adult males over all adult females in the last grades 3 and 4, owing to their superior physical strength, whereas it is far from the rule in the first grades 1 and 2 where females can recruit strong coalitions (Thierry, 1990a).

It is worth adding that Hemelrijk (1999a; Chapter 13) has developed a self-organization model in which the strength of dominance and kinship relations is accounted for by aggression intensity only, without resorting to the existence of alliances between relatives.

Mothering behaviors and socialization

Except for the highest-ranking females, mothers in the first two grades are protective of infants, they frequently retrieve their infants, restricting their interactions mostly to relatives. The amount of care provided by females other than the
mother is limited. By contrast, mothers belonging to species from grades 3 and
4 are permissive, and allomaternal care is common, that is, many females in the
group may handle and carry infants from an early age (Kaufman & Rosenblum,
1969; Kurland, 1977; Hiraiwa, 1981; Thierry, 1985b; Small, 1990b; Mason *et
al.*, 1993; Maestripieri, 1994a,b; Box 10). These facts may be explained by two
arguments: (1) females are interested in all infants in any species; but (2) only
those permitted by mothers may have access to them.

From the first argument, maternal and allomaternal care is a pleiotropic effect
of a basic female tendency: a strong attraction to their own infant. This is the
character that is selected (Quiatt, 1979). It is vital in species where females
give birth to a single offspring undergoing an extended period of development.
Infants have distinctive morphological features. It would be difficult to understand why females who pay considerable attention to their own offspring would
show no interest in other females' infants. For that to be possible, one would
have to postulate that selective attachment processes occur during brief sensitive
periods, which is incompatible with what we know about macaques' cognitive
abilities. Their learning performances are not contextually dependent, they generalize their knowledge from one situation to another. The consequence of this
is the existence of a phenotypic linkage between high cognitive abilities and a
general attraction of females to any immature bearing the distinctive signs of
infancy.

Following the second argument, allomothering can be expressed only when
behavioral constraints are favorable, that is, when dominance asymmetry is
weak enough for the mother to safely allow mates to take away her infant. The
patterns of maternal and allomaternal care may thus be explained by the level of
protection needed by infants in a given social milieu (McKenna, 1979; Thierry,
1985b). Whereas alternative functional explanations have been proposed to
account for the occurrence of infant handling by nonmothers (Maestripieri,
1994c; Box 10), the interspecific variations observed among macaques cannot
be satisfactorily explained without resorting to epigenetic mechanisms.

Behavioral constraints may affect the expression of other behaviors as well.
In rhesus and Japanese macaques (grade 1), immatures at play cannot endure
as high a proportion of close physical contact during wrestling as do their
crested and Tonkean macaque counterparts (grade 4) (O. Petit, C. Demaria &
B. Thierry, unpublished data). The rate of immatures' interference in mounts
also varies according to the tolerance level of the social environment. Young
individuals may enter into proximity of mating pairs only if they are allowed

to do so (Thierry, 1986b). Interference by immatures is frequent in species from grades 3 and 4 (Dixson, 1977; Niemeyer & Chamove, 1983; Thierry, 1986b; Kumar, 1987; Kuester & Paul, 1989; Matsumura, 1995). It is rare in longtailed macaques (de Benedictis, 1973; Gore, 1986), and virtually absent in pigtailed, rhesus and Japanese macaques, in which adult males would not permit immatures' interference in their mating.

Temperament and dispersal patterns

Consistent differences have been found between macaque species in response to stress and novelty when measured by arousal, alarm and exploration behaviors, corticosteroid levels, and heart rates. Rhesus and longtailed macaques are less explorative with regard to their physical environment when compared with liontailed or Tonkean macaques (Thierry *et al.*, 1994; Clarke & Boinski, 1995). As a general rule, species from grades 3 and 4 are less easily aroused than are species from grades 1 and 2 (de Waal, 1989b; Clarke & Boinski, 1995). We may expect that macaque species differ in a number of psychobiological variables. For instance, lower serotoninergic activity is associated with higher aggression intensity in rhesus compared with pigtailed macaques (Westergaard *et al.*, 1999). Interspecific variations in temperament have the potential to explain many interspecific differences in social behavior by genetic effect, phenotypic linkage and social pleiotropy. It should be stressed, however, that our knowledge of between-species differences in the temperament of macaques is quite incomplete and remains preliminary at best (see Capitanio, Chapter 2).

Males break their social ties when emigrating from their natal troop (Bercovitch & Harvey, Chapter 4; Gachot & Ménard, Chapter 6). Whereas male rhesus and Japanese macaques typically leave when they are three to five years of age, first dispersal occurs some years later in other grades. Consistent differences are observed between species in the number of migrating males of all ages and the duration of solitary phases, which seem especially long in the first two grades (Mehlman, 1986; Pusey & Packer, 1987; Oi, 1990; but see Paul & Kuester, 1985). Interestingly, low levels of inter-male tolerance characterize grades 1 and 2 (Caldecott, 1986a; Hill, 1994). We may understand the variation in male dispersal patterns between grades as a side effect of interspecific differences in inter-male tolerance and temperament, these being at least partly dependent on genetic determinants (see Trefilov *et al.*, 2000; Champoux *et al.*, 2002).

Degree of character entrenchment, caveats and predictions

Direct interspecific comparisons of social characters have been made only for a limited sample of species, groups and characters, and intraspecific variability calls for caution (Thierry, 2000; Hill, Box 11). Rating species along a discrete and bipolar scale is an idealization. Although each species is assigned to one grade, a more accurate picture would represent the various study populations of each species using a cluster of points centered on one modal location, and would allow for overlaps with other clusters centered on neighboring locations. In addition, all characters are not expected to covary to the same extent (Thierry, 2000). To account for this, Wimsatt and Schank (1988) have introduced the notion of "generative entrenchment," which they defined as some measure of the number of characters affected by changes in one character. Although quantitative measurements are not yet available, we must be ready to consider that characters may be more or less tightly bound to each other.

The 4-grade scale may be used as a periodic table (Table 12.1), which allows us to generate falsifiable hypotheses. For instance, aggression and reconciliation in the bonnet macaque have not been quantitatively compared with those of other species. But their patterns of temperament, dominance, nepotism, socialization, and male dispersal consistently place the species in grade 3 (Thierry, 2000). One can therefore predict that rates of reconciliation should be relatively high in bonnet macaques. Likewise, a few characters may be used to predict others in the least known species. For example, based on the combination of teeth-chattering, infant use by males and immatures' interference in mating, Tibetan macaques probably belong to grade 3 (Table 12.1). Intraspecific variation sometimes makes for difficulty in ranking species from intermediate grades. Assamese macaques display patterns similar to those of Tibetan macaques, thus placing them in grade 3. A recent study, however, has found relatively low rates of counter-aggression and reconciliation in a group of this species (Cooper & Bernstein, 2002), which would indicate that Assamese macaques might be set into grade 2. The data were collected in a free-ranging group living in a mostly urban environment, unlike most other groups of other species that were studied in captive settings. Therefore, it is difficult to know to what extent data are comparable. I will adopt a conservative stance by classifying the species as grade 3, but keeping a record that it could turn out to be grade 2.

The balance between characters may be influenced by multiple intraspecific variations in individuals' characters and behavioral tactics (Thierry, 2000). Differences in the means and goals of both sexes, for example, may tune their social relations at different levels (see Capitanio, Chapter 2). We may expect that adult males and females differ in their agonistic patterns (e.g., Preuschoft et al., 1998; Cooper & Bernstein, 2002), but this does not imply that their behaviors are uncoupled. In a given species, males and females share a large amount of genetic information carried by autosomes and parallel conditions of development, which makes them dependent. The extent of variations should remain within the range allowed by epigenetic constraints. In Barbary macaques, competition tests point to a rather egalitarian balance among males whereas the seizure of food is unequally distributed among females, this perhaps being related to between-sex differences in fighting ability and risks of injury (Preuschoft *et al.*, 1998). These differences, however, are not sufficient to qualify the relationships between females as despotic while those between males would be egalitarian (see Paul, 1999). Measuring counter-aggression and reconciliatory rates among unrelated adult individuals has yielded figures consistent with the position of Barbary macaques on the 4-grade scale (percent of bidirectional conflicts: 41.8% between females and 54.3% between males; corrected conciliatory tendencies: 19.0% and 21.8%; $n = 40$ females and 24 males; C. Marengo, unpublished data). In crested macaques, the behaviors of females assign the species to grade 4. It was claimed that crested males strongly contrast with their female counterparts because of the existence of a linear rank order among them, which would demonstrate they are not egalitarian (Reed *et al.*, 1997). But the ordinality of dominance brings no information about the gradient of dominance (Flack & de Waal, Chapter 8), and the occurrence of stable hierarchies among adult males is a common finding in macaques of any grade. The same study also showed that crested macaque males display regular greeting contacts toward each other (Reed *et al.*, 1997), which is in fact a characteristic pattern of the most tolerant species (see above). The use of terms like "despotic" and "egalitarian" does not help to distinguish shades. Originally proposed as asymptotic horizons in a mathematical model (Vehrencamp, 1983), these terms tend to draw social relationships in black and white. In the real world there does not exist a single group-living primate that may qualify as truly egalitarian.

Some characters may be relatively independent. This is true in particular for the mating patterns of macaques, which do not fit the 4-grade scale. Species differ markedly in the duration of the associations occurring between males and females approaching ovulation. In some species, mate guarding by the male may last several days (e.g., Tonkean, liontailed, and pigtailed macaques). In others, the male follows and mates with the female for some hours (e.g., rhesus and Japanese macaques) or about a quarter of an hour (e.g., Barbary macaques) (Caldecott, 1986a). In multimale, multifemale groups, seasonality is the key-factor influencing the operational sex ratio (i.e., the ratio of sexually active males to fertile females), and thus the mating tactics available to males. Macaques living in temperate regions experience strongly seasonal mating. When several females enter estrus simultaneously, one male cannot monopolize them all. It follows that dominance rank and paternity rate are only slightly correlated. In contrast, reproduction takes place year round in tropical species. There is generally no more than one female in estrus at a time in a group, so the dominant male is in a position to control her during the fertile period. Consequently, the probability of a male's paternity is closely correlated with his dominance rank (Oi, 1996; Thierry *et al.*, 1996; Paul, Box 6; Soltis, Chapter 7). This entails the following paradox: in nonseasonal species with limited dominance asymmetry (Tonkean and stumptailed macaques: grades 3 and 4), dominance rank has more influence on the reproductive success of males than in species in which hierarchical differences are marked but where females' fertility is synchronous (rhesus and Japanese macaques: grade 1). The outcome of male reproductive competition cannot be predicted from the social style typical of the species.

Evolutionary implications

Evolution is commonly defined as random mutation followed by selection, what leaves no room for the epigenetic processes that produce phenotypes (Ho & Saunders, 1979). Mutations do not produce random variations in the organism, "The internal organization of the existing system must significantly affect the evolutionary outcome because it can constrain the range of possible phenotypes available for sorting by selection" (Raff, 1996: 324). The idea that changes in social organizations should also be considered at a systemic level (Mason, Box 13) has long been resisted on the basis that the producers of societies are the individuals, who themselves pursue their own goals. If the variation of social organizations is nonrandom, as the covariation hypothesis states, parametric modeling should apply to the social level as well, "If a species succeeds in adapting by introducing a structural change compatible with its preexisting structure, it thereby alters a component of its social, behavioral, or morphological organization. The result may be disharmony within the organization, so that other components must be altered to accommodate the main change and to reestablish a functioning entity. Such secondary adjustment also must be within the capacity of the heritage" (Kummer, 1971: 91). We should expect that a social organization reflects some balance between internal and external constraints – i.e., between epigenetic and selective factors.

Phylogenetic correlates

Styles form systems (Lévi-Strauss, 1955: 203). The empirical finding that macaque social styles represent covariant sets of characters indicates that these styles belong to a single family of forms. Put another way, the social organization of species range within a *sociospace* defined by the interconnections between characters. By limiting the changes possible to the social organization, they act as constraints that channel evolutionary processes and allow only a subset of organizations to arise (Thierry, 1990a, 2000). Thus, it should not come as a surprise that the three phyletic lineages of macaques have a different distribution on the 4-grade scale, and that variations in the social style of macaques correlate with their phylogeny (Fig. 12.3) (Thierry *et al.*, 1994, 2000; Matsumura, 1999). By tracing each of the characters on the phylogenetic tree of macaques, it is additionally possible to recognize their most ancient states and so reconstruct the typical ancestral organization of macaques. The resulting set of characters closely matches grade 3 on the scale, which may be tentatively considered as the ancestral state (Thierry *et al.*, 2000). The fact that Barbary and liontailed macaques are located in grade 3 reinforces the previous finding since these two species come closest to the root of the phylogenetic trees established from morphological and molecular data. The other species of the first lineage have diverged, moving either to grade 4 (Sulawesi macaques) or grade 2 (pigtailed macaques). Members of the second lineage have remained on grade 3 – with the possible exception of the Assamese macaques, which may have diverged to grade 2. The third lineage evolved toward grades 1 and 2 (Fig. 12.3). The location of every lineage is mostly restricted to two grades in the scale.

The good match between macaque phylogeny and the 4-grade scale indicates that the core of the species-specific systems of interconnections underwent limited changes during several hundred thousand years. We may conclude that macaques display a fair degree of "phylogenetic inertia" in their social patterns. This statement, however, represents no more than a descriptive account of the slowness of evolutionary changes. Our conclusion would not gain any explanatory power if we add that this reflects the action of "phylogenetic constraints." The latter term covers an ill-defined concept as it only repeats that we have found phylogenetic correlations (Plavcan, 2001). To account for these correlations, we have to resort to nothing other than the genetic and epigenetic processes underlying the renewed development of phenotypes with each generation.

Selective processes

A social organization may be considered as an "adaptive complex" (see Mayr, 1963). They are the resultant of the various individual characters, which are submitted to "particulate" or "system" selection (Mason, Box 13). Regardless of the exact strategies pursued by individuals, we may formulate three different hypotheses about the selective processes involved in the emergence of macaque social styles.

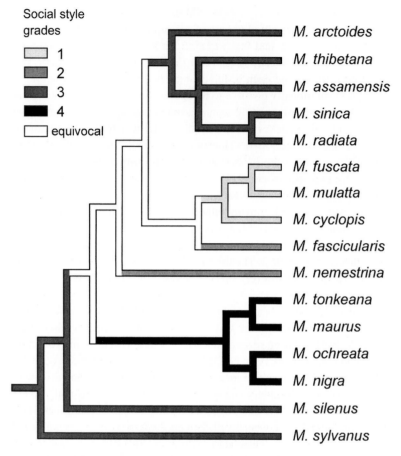

Social style grades

□ 1
■ 2
■ 3
■ 4
□ equivocal

M. arctoides
M. thibetana
M. assamensis
M. sinica
M. radiata
M. fuscata
M. mulatta
M. cyclopis
M. fascicularis
M. nemestrina
M. tonkeana
M. maurus
M. ochreata
M. nigra
M. silenus
M. sylvanus

Fig. 12.3. The distribution of social styles on a 4-grade scale is mapped onto the phylogenetic tree of macaques (Purvis' tree, 1995). The social style is significantly associated with phylogeny (calculated consistency index: 0.75, which is larger than the expected consistency index: 0.60) (Thierry *et al.*, 2000). (Note: To account for the fact that Assamese macaques could be located in grade 2 or even in grade 1, the analysis was re-run for each of these assumptions. This yielded a consistency index of 0.60 for both ($P = 0.05$), the ancestral macaque still having a score of 3 (A. Iwaniuk & S. Pellis, pers. comm). This demonstrates that the fit between phylogeny and social style is quite robust. See also Matsumura, 1999.)

Trade-off hypothesis

There is a trade-off when a benefit realized through a change in one character is linked to a cost paid out through a change in another (Stearns, 1992). In such a perspective, a social organization is a set of coadapted characters and social styles result from trade-offs between the selective advantages of different characters. The similar clustering of characters found in different species results from a convergent adaptation induced by a similar clustering of environmental pressures. Resorting exclusively to trade-offs corresponds to the null hypothesis regarding the action of epigenetic constraints: characters are held structurally independent of each other (see Antonovics & van Tienderen, 1991). To thus account for the social styles of macaques, we must provide an adaptive function for every character – i.e., one for the rates of alloparental care, another for the occurrence or non-occurrence of immatures' interference in mating, still another for the different rules of rank inheritance, etc. Such a position cannot be maintained for long. Adaptation is an onerous concept (Williams, 1966), and trade-off hypotheses require multiple adaptive explanations (Gould & Lewontin, 1979). In addition, it predicts that the components of the social style can be decoupled, which remains to be proved.

Pacemaker hypothesis
It may be hypothesized that external factors can act on some key character(s), which would then work as pacemaker(s) for the whole adaptive complex. Other characters would be shaped by correlated response. If we favor this pacemaker hypothesis, the main task is to identify the pacemaker character(s). Until now, the preferred candidates have been those related to dominance relationships and agonistic interactions: their asymmetry, intensity and nepotistic character (Sterck *et al.*, 1997; de Waal & Luttrell, 1989; Aureli *et al.*, 1997; Hemelrijk, 1999a; Matsumura, 1999). It has become fashionable to subsume species-specific social relations under the label "dominance style" (de Waal & Luttrell, 1989; Flack & de Waal, Chapter 8). This stress is understandable in view of the importance of competition in the evolution of species. But social relationships cannot be reduced to food competition and social dominance. If we look at progeny output in rhesus and Japanese macaques, two of the most dominance-oriented species on earth, we realize that the proportion of offspring sired by higher-ranking males is lower in these species than in more tolerant aseasonal breeders (see above). Moreover, the lifetime reproductive success of females primarily depends on their longevity, and social rank has no significant impact on it (in provisioned groups at least: Fedigan *et al.*, 1986; Bercovitch & Berard, 1993; see also van Noordwijk & van Schaik, 1999, for limited effects of dominance in wild longtailed macaques). We could point to further potential pacemakers by asserting that macaques differ in their "personality style," their "conciliatory style," their "dispersal style," their "maternal style" or their "nepotistic style" (see Thierry, 1990a; Capitanio, Chapter 2; Chapais, Chapter 9; Chauvin & Berman, Chapter 10). It is safer to use "social style" as a more neutral term – germane to "the style of group life" of Kummer (1971: 150) – to label a set a social characters.

Burden hypothesis

The fact that constraints internal to organizations channel evolutionary changes in some directions does not mean that external determinism should be reduced to a few causes. It could be that there is no pacemaker and that a variety of characters count. Social styles are packages that may be quite stable evolutionarily and hinder individual adaptiveness. Given the number of trade-offs and epigenetic constraints that keep a social organization within a certain range, the burden hypothesis states that any shift in individual strategies will entail a series of advantages and disadvantages. Whereas some characters have positive effects, others have neutral or even negative effects. In a given environment, the sum of effects on the individual would differ according to the social style. When comparing the respective consequences of a less tolerant and a more tolerant style on the individual, we may speculate that: (1) a tough temperament allows better resistance to stressful conditions, whereas a more tractable one is physiologically less costly; (2) a high rate of male dispersal favors gene flow, but a low rate diminishes mortality in bachelors; (3) clear-cut contests decrease the number of potential conflicts and shorten their duration, but elaborated negotiation skills favor the resolution of conflicts and diminish the occurrence of wounds; (4) a higher degree of nepotism allows individuals to get reliable allies, while a lower degree may augment the number of potential allies; (5) higher maternal restrictiveness guards an infant against short-term dangers, but allowing alloparental care increases the number of potential protectors; (6) a lower level of tolerance corresponds to an appropriate cautiousness when facing the unknown, but a higher level enhances social contacts and information transmission between group-mates. Each class of individuals is liable to be differentially affected by the tuning of characters. Even if we could recognize the consequences of every character, we would be unable to calculate the attached costs and benefits of a given social style. Note that natural selection is not really more skilled than us, it simply works another way, relying on the fitness of individuals over a large number of generations. We should do something similar, by comparing the lifetime reproductive success of individuals from different groups and species in a given environment. Needless to say, the comparative data necessary to test the hypothesis will not be available in the foreseeable future.

It may be noted that the pacemaker and the burden hypotheses only differ by the number of characters relevant to the selective process. Nonetheless, they contrast with regard to their consequences on the speed of the process. The cascade effects induced by pacemakers could produce rapid shifts in social styles. On the contrary, the genetic and epigenetic burden may slow down the selective process, which would explain the relative evolutionary conservatism observed in the social style of many macaque species.

External determinants

The powerful action of internal attractors indicated by phylogenetic inertia does not mean that external factors play no role in the determinism of social relationships. Several attempts have been made to correlate the contrasting social styles of macaques with the main ecological features of their habitats, predation risk and food distribution (Caldecott 1986a; de Waal & Luttrell, 1989; van Schaik, 1989; Sterck *et al.*, 1997). These socioecological models fail, however, to account for the interspecific variations observed in macaque social relationships (Petit *et al.*, 1997; Matsumura, 1999; Thierry, 2000; Cooper, Box 9; Ménard, Chapter 11). Furthermore, the ecological preferences of macaques appear to be independent of their phylogeny (Fooden, 1982a; Richard *et al.*, 1989).

If social styles are indeed species-specific (but see Ménard, Chapter 11; Hill, Box 11), several reasons may be proposed to account for the failure of ecological hypotheses. Some relevant ecological factors may still go unidentified. It could also be that social styles were selected under past ecological conditions, and that phylogenetic inertia has maintained them in spite of subsequent ecological changes (see Sterck *et al.*, 1997; Chapais, Chapter 9; Ménard, Chapter 11). This argument of anachronism is not satisfactory, however, in that by forbidding any possibility of falsification, it makes the hypothesis of ecological determinism a dogma.

A close examination of the paleoecological circumstances of the deployment of the genus *Macaca* provides further insights into the influence of environment. The colonization of Asia by the three successive waves of macaques was long approached from the standpoint of competitive exclusion between species (see Abegg, Box 12). A basic assumption was that populations from younger lineages would have replaced those from older lineages because they were better adapted to more recent conditions of environment (Fooden, 1976; Delson, 1980). Only geographical barriers would have prevented a later lineage from invading some refuges where remnant populations of an earlier lineage still survived. This scenario was consistent with the expectation that macaque social styles should fit with ecological conditions. It appears, however, that climatic cycles may have had more influence on the evolution of macaques than interspecific competition. Quaternary periods of aridity associated with periodic glaciations induced repeated shifts in the distribution of primates (e.g., Eudey, 1980; Brandon-Jones, 1996; Jablonski, 1998). Pleistocene periods of cool drought eliminated species from their distribution ranges and split up areas where relict populations were able to survive. A reappraisal of the circumstances of macaque dispersal in Southeast Asia indicates in particular that the disappearance of the first lineage from large areas was caused by glacially induced deforestation (Abegg & Thierry, 2002). Some populations took refuge in relict moist habitats. This may explain the presence of liontailed, Mentawai, and Sulawesi macaques in remote areas today (Abegg, Box 12). Subsequent re-expansion of the forests during wetter and warmer periods would have favored the dispersal of species able to take opportunistic advantage of a wealth of habitats devoid of primates.

The adjective "opportunistic" may be understood in two ways. In one sense, opportunism is chance-taking (Abegg, Box 12). Some macaque populations may have encountered strong geographical barriers whereas others were lucky enough to find colonization routes. The first were held in check in marginal zones. The second were able to colonize broad areas by land and sea rafting, which might have given rise to bottlenecks and founder effects – i.e., chance effects (Abegg & Thierry, 2002; Abegg, Box 12). This interpretation would support the null hypothesis with regard to the influence of ecology on social styles. Different styles can be ecologically equivalent, or else regular habitat shifts may have precluded the action of strong directional selection.

In the other sense, opportunism is a quality displayed by the individuals themselves. Those displaying higher dispersal abilities would have been favored by the conditions prevailing at the time of reforestation. We may notice that the most successful colonizers in recent geological times are the members of the last wave (*fascicularis* lineage: longtailed and rhesus macaques in particular), plus the pigtailed macaque (first wave: *silenus–sylvanus* lineage), which invaded an important part of continental and insular Southeast Asia (Fig. 1.1). These species all belong to grades 1 and 2 on the 4-grade scale. A quick colonization process requires high rates of dispersal. This gives an advantage to individuals whose temperament drives them to emigrate earlier and frequently. This character could be the pacemaker having pulled species toward the less tolerant side of the scale. The individuals sorted by the selection process would not have been those in (past or present) stable ecological conditions but those better adapted to the conditions of the change, that is, the availability of habitats open to the fastest invaders. Admittedly, this is mere speculation. We have no proof that higher male dispersal is sufficient to enhance colonization processes. After all, females must also be present for reproduction to occur, and females' choices are influential in group settlements (N. Ménard, personal communication). By offering this further hypothesis, I wish to underline the wide range of possibilities left. We do not know which selective processes or ecological factors are liable to move social styles in one direction or another. We have to keep an open mind concerning the influence of external determinants upon the social organization of macaques.

Conclusion

Our knowledge of the covariation of social characters is basically correlational. A great deal of work remains to be done to uncover its mechanisms. The coupling between agonistic patterns and maternal protectiveness, for instance, may be generated at several levels. Heightened aggression by individuals and low rates of infant rejection by the mother are associated with reduced serotonin activity (Lindell *et al.*, 2000). The latter character has both a genetic and a nongenetic component (Higley & Linnoila, 1997). It may be associated with higher anxiety at the temperament level, which increases the rate of infant retrieving by the mother. Maternal behavior may also be influenced through social inheritance, since there may be social learning of maternal style. Behavioral constraints may additionally intervene, the mother having to be protective when the infant faces a relatively intolerant social milieu. Our experiments and models should be devised so as to investigate not only the functional value of social characters but also their mutual bonds and side effects (see Hemelrijk, Chapter 13).

The systemic way of thinking is sometimes criticized as circular. It is indeed. How could it be otherwise? It is a well-acknowledged fact that a social organization is made up of multiple feedback loops (Hinde, 1976b; Altmann & Altmann, 1979). It is impossible to account for the diversity and limits to diversity that we see in primate sociodemographic forms by resorting to trade-offs exclusively (see Strier, 1994; Janson, 2000). Because of the multiple constraints and trade-offs that constitute them, the social organization responds in a nonlinear way. There is not one optimal solution in a given environment, but rather multiple selective peaks corresponding to stable system states. If we want to understand how sorting processes affect social organizations, we need to know the rules of transformation governing their changes in a given sociospace, and how the number of dimensions of the sociospace comes to be modified.

Whereas the present chapter has focused on the interdependence of characters, this does not mean that all characters should be similarly entrenched in any social organization. In mostly solitary species (e.g., orang-utan), individuals are directly exposed to ecological constraints, and the influence of social epigenesis on them is minimal. In group-living animals, the action of external factors upon individuals is screened by the group dynamics. We may predict that the strength of social epigenesis depends on the *degree of interactiveness* displayed by a given social organization. The degree of interactiveness can be computed from the number of group members, the rate of social interactions, and the development of coalitionary behaviors. We may expect a weaker influence of social epigenesis in the small and loosely structured groups found in some animals (e.g., lemurs: Roeder *et al.*, 2002), compared with the strongly integrated groups formed in species like macaques and baboons. By enhancing the emergence of self-organization processes (see Hemelrijk, Chapter 13), a high degree of interactiveness should act as a further feedback, tightening the epigenetic bonds of social organizations.

Box 12   The role of contingency in evolution

Christophe Abegg

The evolution of macaques is generally considered from the point of view of competition between species (Fooden, 1976, 1980). The gradual replacement of ancestral populations through competition is held to be the main mechanism responsible for the disappearance of the first wave of macaques in Asia and its replacement by more recent colonizers. Numerous events, however, have occurred through geological time, creating ample opportunity for the occurrence of historical contingencies – i.e., the occurrence of chance events in the course of history. In times of adverse climatic changes, such as those associated with global fall of temperatures, contingent events are more likely. For instance, populations may be drastically disjoined and reduced due to habitat fragmentation; the existence of refuges, the emergence of landbridges, and the possibility of sea rafting may then strongly affect the fate of animal populations (Haffer, 1969; Simpson, 1983). Brandon-Jones (1996) convincingly argued that habitat changes from rainforest to savanna resulting from climatic fluctuations could have been the main cause behind the diversification of Colobinae species in Asia. In light of the latest biogeographical data available (e.g., Morley, 2000), Abegg and Thierry (2002) have proposed a new model of the evolution and dispersal of macaques in insular Southeast Asia, attributing a key role to chance. Here, we will examine what could have been, in the last millions years of macaque evolution: (1) the refuges and relict areas at the time of population decreases; and (2) the role of barriers (leading either to isolation or recolonization) and competitive exclusion in periods of expansion into islands.

   The fragmented distribution of the first macaque radiation, the *silenus* lineage, which dispersed in Asia after its arrival more than 3 million years ago, has been interpreted as primarily resulting from competitive exclusion from the two Asian radiations, the *sinica–arctoides* and *fascicularis* lineages, that later emerged (Fooden, 1982a). Until now, reconstruction of the deployment of macaque species on Sulawesi and Mentawai islands as well as the long-tailed macaque on several other oceanic islands has been based on postulated

Box 12 Contingency and evolution 291

sea level changes (see Abegg & Thierry, 2002), which already stressed the role of contingency. However, as pointed out by some authors (Eudey, 1980; Brandon-Jones, 1996, 1998), past climatic changes also induced marked ecological consequences, which affected the distribution of primates. Climatic cycles induced repeated contractions in the distribution of each primate taxon that survived in available habitats, sometimes followed by expansions (Jablonski *et al.*, 2000). Such climatic consequences on primate distribution have long been underestimated, due to insufficient paleoclimatic data or lack of analysis. Their role in explaining present-day macaque distribution (e.g. Fooden, 1975; Eudey, 1980) should be re-evaluated (Abegg & Thierry, 2002).

The liontailed macaque, which is probably the closest descendant of the ancestor of all Asian macaques (Fooden, 1975; Delson, 1980; Purvis, 1995), could well have been early restricted to just about its present South Indian range (Fig. 12.4c) when an early maximal Plio-Pleistocene glaciation reduced their forested habitat (Fig. 12.4a,b). Other refuges likely to keep a wetter climate during glacials would be certain regions (like the Khasi Hills, the Annamatic Cordillera or the Dawna Range) of the Indochinese peninsula, some low altitude plains (on Sumatra or Borneo islands) of the Sunda shelf as well as the Mentawai and Sulawesi oceanic islands at its periphery (Eudey, 1980; Brandon-Jones, 1996; Morley, 2000). In the Indochinese peninsula and Sundaland, however, whether macaques completely disappeared or used one of the above-mentioned areas as a refuge during an intense glaciation remains an open question (Fig. 12.4b). Within Southeast Asia, the formerly continuous distribution of the ancestor of all Asian macaques fragmented into distinct populations that were able to survive and speciate. This leads to a dispersal scenario for the *silenus* lineage from India to Indonesian oceanic islands (i.e., South India, Eastern India and the Indochinese peninsula, on continental islands like Sumatra and Borneo as well as Mentawai and Sulawesi oceanic islands) which is simpler than previous ones (Fooden, 1975). After broad dispersal of the *silenus* lineage's progenitor in Southeast Asia (Fig. 12.4a), several populations were isolated as a result of climatic change: in South India, presumably somewhere between eastern India and the Greater Sunda islands, and on Sulawesi and Mentawai (Fig. 12.4b). Subsequently, apart from the progenitors of the Indochinese and Sundaic pigtailed macaques, each population speciated while maintaining its ancestral characteristics in the peripheral refuges. The peripheral macaque populations on Sulawesi and Mentawai islands would thus represent ancient offshoots that separated during an early glaciation in what remained relict areas while they had all disappeared in intermediate areas. Their ancestors were more likely restricted to their island refuges as a result of climatic

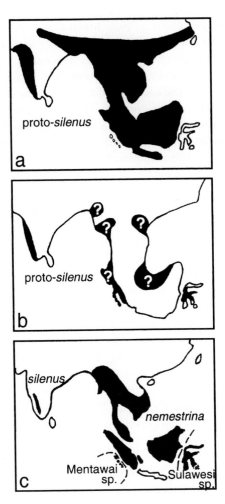

Fig. 12.4. Proposed steps for the dispersal of the *silenus* lineage in Asia. (From Abegg & Thierry, 2002.) The distribution range of macaques is indicated in black. (a) Distribution during the moderate climatic fluctuations that followed a Pliocene arrival in Asia; (b) distribution at a time of maximal glaciation during the early Plio-Pleistocene; (c) present distribution.

changes rather than being superseded by macaque competitors from a new lineage.

Pigtailed macaques are unique in the *silenus* lineage for their wide distribution throughout insular and peninsular Southeast Asia. Such a distribution indicates that they have been successful in terrestrially recolonizing a significant portion (Fig. 12.4c) of the wide range that was formerly inhabited by the ancestor of the *silenus* species group (Fig. 12.4a). The fact that these

Box 12 Contingency and evolution 293

lands were empty of competitors when pigtailed macaques expanded was merely a chance event. However, pigtailed macaques are nowhere found on oceanic islands that were not connected to continental Asia during past sea lowerings. While most macaque populations of the *silenus* lineage kept ancestral characteristics in peripheral refuges, the Indochinese and Sundaic pigtailed macaques' progenitor would be the only one of its lineage that survived and expanded toward central ranges (Fig. 12.4c). This taxon could well have experienced stronger selective pressure due to repeated ecological disruptions and have thus emerged today as the most divergent of its lineage (see Thierry, this Chapter).

The longtailed macaques provide another example of how chance may have influenced the evolution of Asian macaques. After terrestrial colonization over exposed lands of Southeast Asia, longtailed macaques likely underwent a secondary expansion by sea rafting. Contrary to the views of some authors (e.g., Brandon-Jones, 1998) colonization by sea rafting is not common among Asian primates. Relatively large-sized mammals, including most primates, do not easily disperse to oceanic islands. Longtailed macaques stand as an exception. They managed to reach and stock the Philippines, Nicobar, Simeulue, Lasia, Nias, Maratua, and some of the Lesser Sunda Islands (Fooden, 1995). The only other medium-sized Asian primate that is thought to have dispersed once by rafting to a deep-water island is the progenitor of Sulawesi macaques. Longtailed macaques reached the Philippines and many oceanic islands that gibbon and langur species never populated. The habitat preferences of the longtailed macaques, often found near rivers, may well have favored its ability to disperse by raft (Abegg & Thierry, 2002). For an animal sleeping in a tree near the mouth of a river, the probability that the tree might fall and that some individuals might drift on it in the sea is certainly higher than elsewhere. The riverine habits of the longtailed macaque would have favored its dispersal by sea rafting, which was extremely rare for other primate species.

Contingent factors like the previous existence of a landbridge or the habitat preferences of each species, best explain the differing abilities of primate species to disperse over sea-barriers. Competition probably intervened in macaque evolution in a way heretofore never contemplated. Stocking by longtailed macaques using natural rafts would have only been successful in islands where no other macaque competitors were already present. On Sulawesi and Mentawai islands, earlier evolved macaque species belonging to the *silenus* lineage already populated these islands, which they had reached by a landbridge during maximal land exposure (Mentawai) or by sea rafting in spite of the odds (Sulawesi). Whatever the past climatic and sea level fluctuations, it is likely that longtailed ancestors could not occupy islands that had to be reached by raft whenever other macaques had already taken refuge in them (Abegg & Thierry, 2002). This might be because they were first eliminated by competitive exclusion for food-resources or then assimilated through interspecific reproduction. The integration of contingency events in the evolution of macaques along with competitive exclusion and adaptation to changing habitats accounts both for the fragmented distribution of macaques and their presence or absence on the islands of Southeast Asia.

13 The use of artificial-life models for the study of social organization

CHARLOTTE K. HEMELRIJK

Introduction

When observing differences between species, scientists tend to invoke a separate adaptive explanation for each difference. In this way, they explain the complexity that is observed at a higher level (such as the level of the relationship, the group, or the species) as if it resides in the individuals (Deneubourg & Goss, 1989; see Mason, Box 13). Accounting for each aspect of behavior separately, scientists have to invent theories to integrate these separate aspects. Now, however, a new method to arrive directly at integrative hypotheses has been introduced. This is the so-called "synthetic" or "bottom–up" approach, in which artificial individuals (robots and computer models) are designed that exclusively react to the information they receive from their nearby (local) environment (Braitenberg, 1984; Pfeifer & Scheier, 1999); the behavior of these virtual individuals is studied in detail. It then appears that many more behavioral patterns emerge than we expect to result from their simple behavioral rules. These patterns arise by self-organization from the interaction among these simple individuals as a consequence of the changes that take place in the agents and in their environment. Because we have complete knowledge of their behavioral rules, we can fully explain these patterns as arising from their integration, without resorting to additional and separate genetic or cognitive mechanisms. The aim of the present study is to show how this "bottom–up" method may be used to generate hypotheses that are simple in terms of the cognitive assumptions that are made and that can be tested in the study of macaques in the real world.

Group-life, in real animals, is supposed to serve the protection of group-members against predators (van Schaik, 1983); its main disadvantage is the competition for food it generates. Individual group-members vary in their competitive abilities and can be arranged in hierarchical order as regards their degree

Macaque Societies: A Model for the Study of Social Organization, ed. B. Thierry, M. Singh and W. Kaumanns. Published by Cambridge University Press. © Cambridge University Press 2004.

of dominance over others. How they acquire dominance is a point of discussion. While some argue for the importance of (genetic) inheritance of dominance (Ellis, 1991), others reject this for the following reasons: dominance is a relational phenomenon instead of an individual characteristic (Barette, 1993). Experimental results show that dominance depends on the order of introduction in a group (Bernstein & Gordon, 1980b). Dominance changes with experience, because the effects of victory and defeat in conflicts are self-reinforcing, the so-called winner–loser effect. This implies that winning a fight increases the probability of victory in the next fight and losing a fight increases the probability of defeat the next time. The effect has been established empirically in many animal species and is accompanied by psychological and physiological changes, such as hormonal fluctuations (Mazur, 1985; Chase *et al.*, 1994; Bonabeau *et al.*, 1996; Hemelrijk, 2000a). Macaques species differ in their competitive behavior, and in respect of this, Thierry (2000, Chapter 12) classifies them into different categories: among other things, a weak hierarchy classifies a group as an egalitarian society and a steep hierarchy as a despotic one. Clearly-despotic macaques display aggression of an intenser kind (e.g., biting) than egalitarian ones (which limit themselves to threatening) and are more nepotistic in their behavior than egalitarian ones (Thierry, 1985a,b). Despotic species differ also in other aspects from egalitarian ones, such as the degree to which they become reconciled after a fight and the degree of infant protection by mothers (Thierry, 1985a,b, 1990; de Waal & Luttrell, 1989). Instead of presenting a separate adaptive explanation for each single difference, Thierry believes that all differences in dominance style are brought about by only two internal differences: a higher intensity of aggression and a higher tendency to support related individuals (nepotism) among despotic macaques as compared to macaques that are egalitarian. In this chapter, I will use a model called "DomWorld" to explain how one of these internal differences at the individual level (intensity of aggression) may suffice to explain many of the differences observed at the level of social organization.

DomWorld is a so-called individual-based or multi-agent model. Such a model differs from models in which the "average" individual or strategy is studied (such as "hawks" versus "doves," Maynard Smith, 1983), because different individual agents are represented separately and each individual is supplied with rules to react to what it perceives locally. Therefore, it is possible to study social behavior in a more "natural" situation involving parts of the context of behavior. Representation of context is increased greatly in multi-agent models by the representation of space. In DomWorld, for instance, spatial proximity determines who meets whom (in contrast to a random draw of potential candidates for interaction). By means of the representation of the spatial distribution of food,

Deneubourg and co-authors have shown that in artificial ants, even when only a single set of behavioral rules is implemented, different environments may lead to extremely different behavioral patterns (Deneubourg & Goss, 1989). These kinds of models display a high potential for self-organization, which may or may not be accompanied by structural changes in the agents (Pfeifer & Scheier, 1999: 475). Whereas in most models structural changes are lacking (e.g., the fish-model of Huth & Wissel, 1992; for insects, see Deneubourg & Goss, 1989), in DomWorld individuals change their fighting capacity, as reflected in their changing dominance values.

The model DomWorld encompasses two aspects that are relevant for most animal societies: grouping behavior and competition. Although it reproduces grouping, the cause why agents group (e.g., to avoid predators or because of resources being clumped) is not specified and irrelevant to the model and all ecological characteristics are ignored. As regards competition in DomWorld, any "genetic" differences among individuals of one sex are absent, all are completely identical at the start. They even have the same capacity to win a fight. The hierarchy develops exclusively as a consequence of chance and the above-mentioned winner–loser effect because of its general occurrence (see above). In this way the model makes the working of self-organization clearly visible: despite its minimal rules and limited environmental properties, it leads to behavioral patterns in DomWorld that closely resemble certain behavioral phenomena found in macaque societies. Because of this correspondence, DomWorld can be used to infer a statistical method for measuring the gradient of the hierarchy. As it produces hypotheses for the interconnection of behavioral traits, we ought to ascertain whether such interconnections may also be found among real macaques. In order to establish this, these traits must be studied simultaneously in several groups of different species. Since DomWorld also produces other patterns that have not yet been studied in macaques, these should also be included (e.g., spatial structure and female dominance). In short, DomWorld generates simple explanations for social behavior and its evolution.

The model: Dominance World (DomWorld)

The model represents an artificial world that consists of a homogenous space inhabited by artificial agents that exclusively group and compete, but do nothing else (Fig. 13.1). Grouping arises as follows: as long as individuals are nearby (in NearView), "they are group-members" and therefore, continue to move straight ahead. However, in periods of sexual attraction, male behavior changes in that a male moves one step toward a female upon encountering her in his NearView.

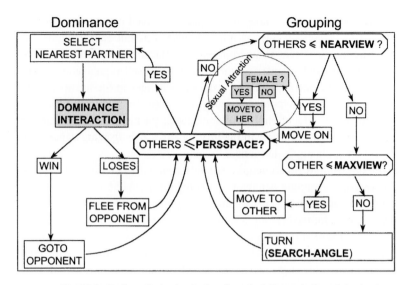

Fig. 13.1. Outline of behavioral rules. Grey shaded areas indicate behavioral interaction types that are experimented with. The encircled part of the diagram, labeled as sexual attraction, is activated for males in certain runs only. (See text for explanation.)

If individuals (of either sex) see no one close by, "group-members are too far away." If they notice an individual further off (in maxView), they move one step toward the other "in order to join a group;" but when they see no one, they turn over a specific angle in order to spot others (SearchAngle).

If, however, another agent is very near, in the personal space of "ego" (PersSpace), a competitive or dominance interaction may take place, though it is not specified what the cause of this competition is (food or the wish to mate). Therefore, encounters among agents are not random, but depend on their spatial proximity. After a fight, the winner chases the opponent (one unit) and the defeated agent flees (2 units).

The likelihood that an artificial individual undertakes an aggressive interaction (instead of remaining nonaggressively close by) increases with its chance to defeat its opponent (Hemelrijk, 2000a). The agent's capacity to be victorious is represented in its dominance value (i.e., body posture). This dominance value is displayed and observed among agents upon meeting each other. Initially, it is the same for all agents of one sex. Thus, during the first encounter, chance decides who wins. Subsequently, the higher-ranking agent has a greater chance to win. Changes in dominance values reflect the self-reinforcing effects of the victories and defeats in conflicts. In the model, this is implemented by an increase of the dominance value of the victorious agent after its victory (thus increasing the probability of victory the next time) and decreasing that of the defeated one (thus increasing the probability of defeat the next time) by the same amount. When, unexpectedly for us, an agent defeats a higher-ranking opponent, this outcome has a greater impact than when, as we would expect, the same agent conquers a lower-ranking opponent, and the dominance values of both opponents are changed by a greater amount (in accordance with detailed behavioral studies on bumble bees by Honk & Hogeweg, 1981). Thus, the impact of a conflict or the change in dominance values varies according to the expectation whether an outcome is likely or not. In this way we allow for rank reversals.

Besides, intensity of aggression is determined by a fixed scaling factor, called StepDom. This factor, StepDom, is multiplied with the degree of change in dominance value per fight. A high value of StepDom (of 1.0) implies that the impact of a single interaction may be high (such as in biting), a low value (of 0.1) allows only for low impact (such as threats and staring).

The sexes differ only in their attraction to the opposite sex and in their fighting capacity: "male" agents start with a higher initial dominance value (twice as high as that of females) and are characterized by a higher intensity of attack than "female" agents. To show the effects of winning and losing as clearly as possible, all agents of the same sex are identical at the start. Groups consist usually of four (or five) males and four (or five) females. Further, in those runs in which sexual attraction is operative, males have a greater inclination to approach females than individuals of their own sex, whereas females ignore the sexual identity of others.

Parameters are set so that the grouping of agents on the computer screen "looks like" a macaque group; the number of agents (8 or 10) is chosen so as to reflect the number of adults in groups of macaques; the high and low values of StepDom are chosen so as to create a large difference between both settings.

The behavior of the agents is analyzed by means of behavioral units and statistical methods similar to those used for observing real animals. For a more complete description, see Hemelrijk (1999a, 2000a).

Results of DomWorld

Hierarchical stability and the gradient of the hierarchy

At the beginning of each run, the dominance of all individuals of one sex is the same; in the course of time hierarchy develops and the dominance stabilizes

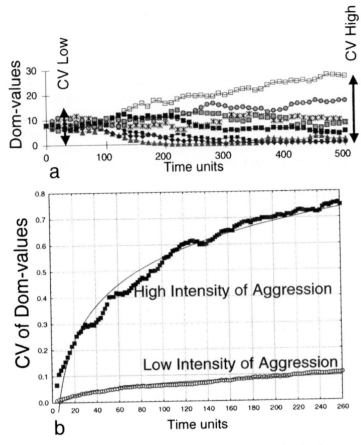

Fig. 13.2. Development of the dominance hierarchy. (a) Hierarchical development and the coefficient of variance (CV) of dominance values for one sex at a low and at a highly differentiated hierarchy; (b) average coefficient of variance of dominance values over time at a high and at a low intensity of aggression.

(Fig. 13.2a,b). The degree of hierarchical development is measured by the coefficient of variance of dominance values (Fig. 13.2a): at the beginning, during the weak development of hierarchy, the coefficient of variance is low, but later on, the coefficient of variance reaches a high value (reflecting a steep hierarchy). (From here on, the measurements of this chapter have been taken only during periods of clear hierarchical development.) Further, Figure 13.2b shows that the hierarchy develops more strongly at a high intensity of aggression than at a low one. Figure 13.3c shows that a steeper hierarchy is accompanied by more asymmetrical aggression. (Asymmetry of aggression is used as an indication of the gradient of the hierarchy in real animals.)

Positive feedback between spatial structure and dominance hierarchy

Figure 13.2b shows how, at a high intensity of aggression, a steep hierarchy develops. This comes about via a mutual feedback between the hierarchy and the spatial structure (with dominants in the center and subordinates at the periphery) (Fig. 13.3a) (Hemelrijk, 1999a,b, 2000a). In this process, pronounced rank-development causes low-ranking agents to be chased away by others continuously and thus the group spreads (1 in Fig. 13.3d and Distance in Fig. 13.3b). Consequently, the frequency of attack diminishes among the individuals (2 in Fig. 13.3d and Attack in Fig. 13.3b). Because of this reduction in the frequency of aggression, the hierarchy stabilizes (3 in Fig. 13.3d). While low-ranking individuals flee from everyone, this automatically leaves dominants in the center, and thus a spatial-social structure develops (Fig. 13.3a). Since individuals of adjacent dominance are treated by others in more or less the same way, similar agents remain close together; therefore, they interact mainly with others of adjacent rank. Further, if a rank reversal between two opponents occurs, it is only a minor one because opponents are often adjacent in dominance. In this way the spatial structure stabilizes the hierarchy and it maintains the hierarchical differentiation (4 and 5 in Fig. 13.3d). Also, the hierarchical differentiation and the hierarchical stability mutually increase each other (6 in Fig. 13.3d).

When at a high intensity of aggression a steep hierarchy develops (Fig. 13.2b), it is accompanied by aggression that is more asymmetrical (as measured by a negative τ_{Kr} – correlation for bidirectionality between aggression initiated with and received from other group-members, see Fig. 13.3c). This asymmetry is due to the great rank-distances, which inhibit low-ranking agents from initiating a dominance interaction with high-ranking ones. It also explains why at a low intensity, when a weaker hierarchy develops, aggression is more bidirectional. Further, increased unidirectionality (= decreased bidirectionality) strengthens the stability of the hierarchy, its differentiation and the spatial configuration (Fig. 13.3a,d).

Female dominance and the lowest-ranking male

How does the gradient of the hierarchy influence inter-sexual dominance relations? Unexpectedly, female dominance over males is greater at a high intensity of aggression than at a low one (Fig. 13.4a). This is a direct consequence of the stronger hierarchical development found at a high intensity, which automatically causes certain females to become very high- and some males very low-ranking. Consequently, certain females dominate some males and the dominance of the

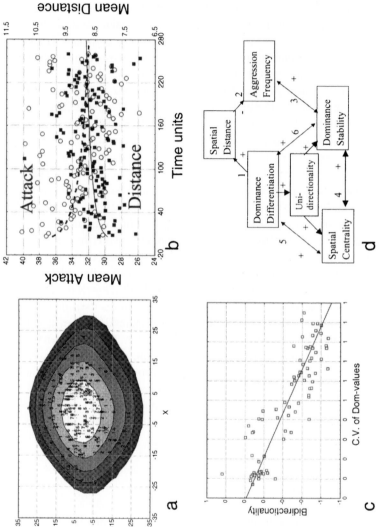

Fig. 13.3. Spatial–social structuring among agents at a high intensity of attack. (a) Spatial structure: different shades indicate areas occupied by individuals of different dominance; (b) decrease of aggression (open circles) and increase in average distance (closed blocks) over time; (c) bidirectionality (= symmetry) of dominance interactions measured by the τ Kr – statistic of the matrix-correlation between aggression initiated and received (Hemelrijk, 1990a,b) versus coefficient of variance (CV) of dominance values; (d) summary of interconnection between behavioral traits.

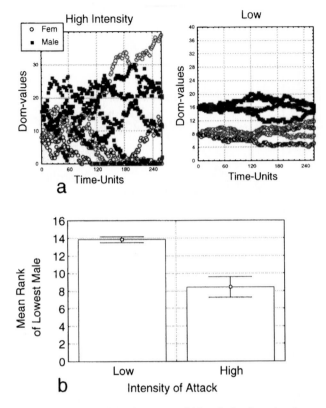

Fig. 13.4. Hierarchical development at a high and a low intensity of aggression. (a) A typical case for both sexes. Females, open circles. Males, closed blocks. (b) Average and standard error of dominance of the lowest-ranking males.

lowest-ranking male is lower than at a low intensity (Fig. 13.4b). In contrast, at a low intensity, nothing much happens to the hierarchy, and if females start out being lower than males, they remain so.

Sexual attraction

If we add social attraction of males to females to the model (as a preferential male orientation toward females rather than toward males), female dominance over males increases (Fig. 13.5a), but only if the intensity of aggression is high (Hemelrijk, 2000c), not if it is low (Hemelrijk, 2002)! Such an increase is due to the higher frequency of interactions between the sexes during sexual attraction in combination with the inbuilt mechanism that unexpected victories and defeats

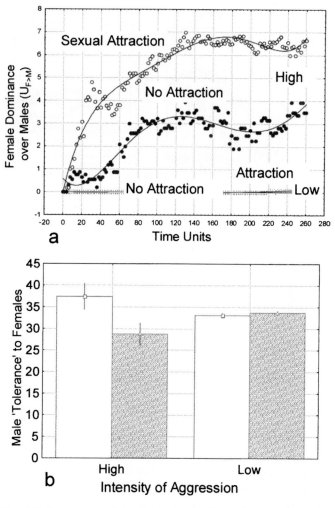

Fig. 13.5. Consequences of sexual attraction at a high and a low intensity of aggression. (a) Average female dominance over males (measured by the Mann–Whitney U statistic indicating the degree with which females rank above males); (b) average male "tolerance" (= non-aggressive proximity) towards females during sexual attraction (white bars) and without it (shaded bars).

cause a greater change in the dominance values of both opponents than expected outcomes do. Such an increase is absent at a low intensity, because their female dominance is very low anyway (Fig. 13.4a), and thus, even if the sexes interact more often during sexual attraction, there is almost no possibility for females to defeat males at all.

When during sexual attraction female dominance increases, at a high intensity of aggression, females run less risk if they attack males and thus, they display more frequent aggression against males. Males, in contrast, run a greater risk when they attack females and therefore, they approach females more often nonaggressively (Fig. 13.5b). It therefore seems as if males become more "tolerant" to females during sexual attraction, though, in fact, they become more "respectfully timid." Nothing of this happens at a low intensity, because there female dominance does not increase during sexual attraction.

Similarity of societies in DomWorld to real macaques

As shown above, in DomWorld, aggression declines more strongly over time when unfamiliar individuals are put together at a high intensity of aggression than at a low one. Further, at a high intensity, aggression is less frequent and less symmetrical, and the hierarchy is steeper. Also, approach- and attack-behavior are more clearly correlated with dominance. Besides, grouping is looser and spatial structure with dominants in the center and subordinates at the periphery is clearer. Compared to a low intensity of aggression, the dominance of the lowest male is lower, female dominance over males is higher and during sexual attraction the increase of female dominance over males is greater (Hemelrijk, 1999a, 2000a,c, 2002). As far as these patterns have been studied, the same is found when despotic societies of macaques are compared with egalitarian ones (Thierry, 1985a,b; de Waal & Luttrell, 1989). Thus, the result is an even simpler hypothesis than that of Thierry, namely that differences between both types of societies may be due to a difference in intensity of aggression only, ignoring all nepotistic tendencies.

Since in the model all differences are interconnected (Fig. 13.3d) and are due to a change in only one trait (intensity of aggression), this may also be true for real monkeys. Therefore, it is important to study in real monkeys the interconnection between all these variables: intensity of aggression, symmetry of aggression, cohesion, stability of the hierarchy, spatial structure, and female dominance (Hypothesis 1, Table 13.1). In addition, these results lead to several hypotheses for grooming and sexual behavior in real monkeys, as explained below.

Hypotheses for the study of real macaques

Grooming behavior

In DomWorld, spatial centrality of dominants is clearer at a high intensity of aggression than at a low one and, although spatial structure has not been studied systematically in macaques (or in any other primates), spatial centrality of

Table 13.1. *Hypotheses to be tested in macaques*

Hypothesis	Model-based hypotheses	That can be tested for macaques
1	Interconnection of variables	Spatial centrality, gradient and stability of the hierarchy, cohesion and unidirectionality are associated as in Fig. 13.3d
2	Spatial structure with dominants in center	Occurs more often in despotic groups of macaques than egalitarian ones
3	Spatial structure with dominants in center	Is mirrored in grooming distribution of despotic groups of macaques, but not of egalitarian ones
4	Female dominance over males	Is greater in despotic societies
5	Female dominance over males	Is greater if hierarchy is steeper (= more unidirectional, measured by τ_{Kr})
6	"	Leads to fewer mating partners for females
7	"	Leads to increased migration of subordinate males
8	"	Increases during periods of sexual attraction in species with some female dominance, and seemingly leads to "male tolerance"
9	"	Increases female aggression toward males, particularly during male attraction to females

dominants has, so far, been mentioned only for species that are despotic (e.g., Itani, 1954). If we assume that spatial structure influences the distribution of grooming, then we may expect that the distribution of grooming differs between egalitarian and despotic primate societies (Hypothesis 2 and 3, Table 13.1). If individuals groom others in proportion to their encounter rate, then spatial arrangement determines – through proximity – the grooming pattern at a group level. Then, at a high intensity of aggression, dominants should be groomed more often than subordinates: because they are more often in the center, and thus, meet others more frequently. Furthermore, individuals should groom more often those that are similar in dominance, because they are close to each other. Note that exactly the same patterns of grooming have been found by Seyfarth (1977) among female monkeys (but he does not distinguish between egalitarian and despotic species and does not expect a difference in the distribution of grooming between them): (1) high-ranking individuals receive more grooming than others; and (2) most grooming takes place between individuals that are adjacent in rank. Seyfarth gives, however, an explanation that is far more complex than those we derive from DomWorld. He assumes that two principles underlie these phenomena: (1) higher-ranking females are more attractive to groom, because potentially more benefits (such as more efficient support in fights) can be gained from them than from lower-ranking individuals; and (2) access to preferred

(= higher-ranking) grooming partners is restricted by competition. Consequently, in the end each female grooms most frequently close-ranking partners and is groomed herself most often by the female ranking just below her.

The hypothesis derived from DomWorld is simpler than Seyfarth's, in that it requires neither the occurrence of competition for grooming partners nor assumptions about exchanges for future social benefits, nor do individuals discern the relative rank of group members in order to groom higher-ranking partners more often than others (in contrast to assumptions by Seyfarth, 1981).

Another common hypothesis states that attraction among related individuals (which often are of similar rank) explains the high frequency of grooming among individuals of similar dominance. Note that the explanation presented by DomWorld holds without invoking kin-relations.

To establish the relevance of this explanation for real primates, it should be tested whether the patterns of grooming as described by Seyfarth occur particularly in groups with a spatial structure with individuals of adjacent rank close together and dominants in the center, and not in those with a weak spatial structure (Hypothesis 2 and 3, Table 13.1) and whether it occurs even in the absence of kin relations. This hypothesis is supported by the egalitarian society of Tonkean macaques where grooming is neither dominance- nor kin-oriented (Thierry *et al.*, 1990) and by several species of despotic macaques where grooming appears clearly dominance- and kin-oriented (rhesus macaques: Sade, 1972b; de Waal & Luttrell, 1986; Japanese macaques: Mehlman & Chapais, 1988).

Inter-sexual dominance, sexual behavior and migration

The close agreement of the results of DomWorld with the dominance styles of macaques inspires confidence for its further use for developing hypotheses regarding male–female dominance relationships among these monkeys.

When, for the sake of simplicity, the sexes in DomWorld are distinguished only in terms of an inferior fighting capacity of females as compared to that of males, then, surprisingly, at a high intensity of aggression, males appear to be less dominant over females than at a low intensity. This is due to the stronger hierarchical differentiation, which causes both sexes to develop their hierarchies in a more pronounced way and, thus, causes females to dominate over males to a higher degree. A similar thing is seen in monkeys. Circumstantial evidence shows that in despotic species of macaques, adolescent males have greater difficulty in outranking adult females than in egalitarian species (Thierry, 1990a). Thierry explains this as a consequence of the stronger cooperation to suppress males among related females of despotic macaques than of egalitarian ones, and van Schaik (1989) emphasizes the greater benefits associated with coalitions for females of despotic species than egalitarian ones. However, DomWorld explains female dominance, as we have seen, simply as a side effect of the more pronounced hierarchical differentiation (Hypothesis 4 and 5, Table 13.1). Differences in rank-overlap between the sexes may affect sexual behavior. In their study of male bonnet macaques, Rosenblum and Nadler (1971) discovered an ontogenetical effect: adult males ejaculate after a single mount, whereas young males need several mounts. These authors (and others, see Abernethy, 1974) suggest that males have difficulty in mating with females that outrank them. When we combine these observations with the patterns of female dominance over males found in DomWorld we come to the conclusion that at a high intensity of aggression, females have fewer males with which to mate than at a low intensity of aggression. In fact, observations of this kind have been reported for macaques. According to Caldecott (1986a) despotic females are observed to mate with fewer partners and almost exclusively with males of the highest ranks. He attributes this to the evolution of a more pronounced female preference in despotic than in egalitarian species of macaques. The explanation derived from the model, however, is simpler: the differences in sexual behavior between egalitarian and despotic macaques may directly arise from the difference in female dominance over males, which in turn arises from a difference in hierarchical development due to a different intensity of aggression. Therefore, we would expect in real animals, that in those groups in which females are dominant over some males, higher-ranking females have fewer mating partners than lower-ranking females (Hypothesis 6, Table 13.1).

Further, in real macaques, after a certain period of living in a group, males migrate to new groups whereas females usually remain in the same group for life (Gachot-Neveu & Ménard, Chapter 6). Males that emigrate are more often of low-rank. In the model, the bottom-ranking males are particularly low in dominance when the intensity of aggression is high (due to the stronger hierarchical differentiation). Thus, if emigration were possible in the model, we would expect that at a high intensity of aggression more males migrate than at a low one. Exactly this is described by Caldecott (1986a) for the rate of emigration of males of despotic primate species versus egalitarian ones. The explanation suggested by DomWorld is that the lower dominance of the bottom-ranking males implies that these males have mating problems with females and suffer competition for food from them and therefore benefit more from migration than those males that are dominant over females (Hypothesis 7, Table 13.1). Thus, it is not only competition with males (as is usually assumed), but also with females that may drive males from the group.

In the model, female dominance increases with sexual attraction as an automatic consequence of the more frequent encounters between the sexes. The explanation that more frequent interaction between individuals of two dominance classes makes them more alike may be a general process that holds for all natural species. This phenomenon is also observed in detailed behavioral studies of dominance interactions among bumblebees (Honk & Hogeweg, 1981): at the beginning there are two dominance classes of bumblebees, the high-ranking queen and the low-ranking "common"-workers. In due time, a third dominance class of high-ranking "elite"-workers develops. These elite-workers interact more often with the queen than do the common ones. Consequently, they come to resemble her.

However, instead of being associated with increased female dominance (as in Yerkes, 1939, 1940), sexual attraction in real animals is usually thought to be accompanied by reproductive strategies of exchange. For instance, chimpanzee males are described as exchanging sex for food with females (Tutin, 1980; Goodall, 1986; Stanford, 1996). Yet, in spite of detailed statistical studies, we have found no evidence that chimpanzee males obtain more copulations with, or more offspring from, those females who they allow more often to share their food (Hemelrijk *et al.*, 1992, 1999, 2001; Meier *et al.*, 2000). Thus, male tolerance seems to increase even without noticeable benefits. DomWorld provides us with the useful alternative hypothesis that males seem more tolerant at food sources to females when males are sexually attracted to them, because female dominance over males has increased. Whether the increase of female dominance is in fact greater in groups where there is already some female dominance, and whether the increase is greater among despotic macaques than among egalitarian ones, should be studied in real animals (Hypothesis 8, Table 13.1).

Further, DomWorld shows that artificial females become more aggressive when artificial males are attracted to them. Similarly, primate females are described as being more aggressive during their tumescent period (e.g., macaques: Michael & Zumpe, 1970; chimpanzees: Goodall, 1986). Though this may be due to their special hormonal state (as is traditionally supposed), the model suggests a more simple mechanism that may be operative: an increase of encounter-frequency with males and, consequently, an increase of female dominance over males (Hypothesis 9, Table 13.1).

Conclusion

DomWorld shows how many behavioral characteristics arise as side effects of other characteristics of individuals and of their interactions, and demonstrates how a number of behavioral aspects are interconnected. These are the gradient of the dominance hierarchy, bidirectionality, spatial centrality, rank-correlated behavior, the rank of the lowest male, and female dominance over males. The interconnection between, or the integration of, these traits causes many behavioral patterns to emerge and forms the basis of the alternative explanations generated by the model.

One of the emergent effects of a steeper hierarchy is the spatial structure with dominants in the center and subordinates at the periphery; another is the degree of female dominance over males. As regards spatial structure, DomWorld presents a hypothesis as to why it should be different in egalitarian and despotic societies and how it should be reflected in sociopositive behavior such as grooming. As regards female dominance over males, it may be pointed out that, so far, its variation in anthropoid primates has not been studied systematically. As a matter of course male dominance is regarded as the rule despite the frequent occasions on which females have been observed to be dominant over males (e.g., in talapoins, vervets and macaques, see Smuts, 1987: 407). Perhaps the lack of studies of this problem is due to the lack of a relevant theory. This gap may now be filled, as DomWorld presents us with the following hypotheses. Female dominance over males increases with the gradient of the hierarchy, and the gradient of the hierarchy in turn may increase from three causes: an increase in the intensity of aggression; in the frequency of interaction between the sexes; and in the degree of cohesion (as has been shown by Hemelrijk, 1999b). Cohesion increases female dominance by two processes (namely via the gradient of the hierarchy and via the frequency of inter-sexual encounters). First, as regards the hierarchical gradient, cohesion increases the frequency of interactions, and the degree of spatial limitation. Both combined cause a spatial structure to develop, which in turn strengthens the hierarchy and, thus, contributes to female dominance. Second, cohesion also increases the frequency of interaction between the sexes compared to what happens in loose groups and, thus, the opportunities for incidental victories of the weaker sex. Consequently, it increases female dominance too. Both effects may in part explain why female dominance is often reported to be stronger in bonobos than in common chimpanzees, in spite of their similar sexual dimorphism: groups of bonobos are more cohesive and groups include both sexes more often than groups of common chimpanzees (Hemelrijk, 2002).

Although individuals in the model represent certain key elements of real animals and their environment, they come, of course, nowhere near their real complexity. Therefore, as future additions to DomWorld, we intend to add offspring, youngsters, social positive behavior, and ecological conditions (see te Boekhorst & Hogeweg, 1994), and allow for immigration and emigration. This will permit us to study additional aspects that are often studied in primates, such as dominance acquisition by youngsters, post-conflict affiliation, etc., and how they are affected by the ecological environment (for preliminary studies on coalitions in DomWorld, see Hemelrijk, 1996, 1997). The remarkable resemblance of DomWorld (in its present form) with the behavior of real macaques, such as is typical of egalitarian and of despotic macaques, suggests that Dom-World sheds light on the processes that are essential to macaque societies. Since such resemblance occurs despite the omission in the model of ecological conditions, it may be suggested that social effects are probably also of crucial importance in macaques, overruling effects due to local ecological variation.

In distinguishing egalitarian macaques from despotic ones, it is difficult to decide which are their most important characteristics. Usually, the gradient of the hierarchy is taken as the key factor, because it is supposed to reflect the competitive regime and to cause variation in reproductive possibilities among group-members. The gradient of the hierarchy (the statistical variance of dominance values) cannot be measured in real animals, but in DomWorld it can be measured by the coefficient of variance of dominance values, and this coefficient of variance appears to be closely associated with the degree of bidirectionality (as measured by the τ_{Kr}-correlation between the frequency of aggression which each individual directs to, and receives from, each group-member) (see Hemelrijk, 2000a,b). Bidirectionality is a variable that is central in the interconnection of the behavioral traits (Fig. 13.3d); it is easily measurable in the real world (it can, for instance, be directly obtained from published matrices of aggression). In a comparative study of real primates, we have applied this measurement and we have shown that, as was to be expected, the degree of bidirectionality is greater during periods in which the hierarchy is unstable, and greater for typical egalitarian species than for despotic ones (Hemelrijk & Dübendorfer, unpublished data). Therefore, we regard it as a useful measure to grade the hierarchy. Yet, to scale species in regard of their degree of despotism is complicated by the fact that in real data the degree of bidirectionality varies greatly within groups over time and between groups of the same species. This implies that sometimes a group of typical egalitarian species may behave despotically and vice versa; for example, a group of typical egalitarian species may behave relatively despotically when its frequency of aggression has become very high because it is confined to a small space. In formulating hypotheses for real macaques it therefore seems appropriate to measure the degree of bidirectionality (as an indication of the gradient of the hierarchy) for each group separately and to determine per group whether it should be classified as despotic or egalitarian.

Cognitive and genetic traits

DomWorld provides us with general ideas regarding the minimum number of genetic and cognitive traits. As we have seen, changing only one trait (namely intensity of aggression) in the model leads to a great number of phenotypic differences at the level of the individual and of the group. Thus, the connection between the behavioral rules and the observed behavior (which respectively correspond loosely to the genotype and the phenotype) becomes nonlinear by self-organization. The results mentioned here not only bear a strong resemblance to primate societies (particularly of egalitarian and despotic macaques), but also to the behavior of fish as described in a selection experiment by Ruzzante and Doyle (1991, 1993). In this experiment, a decrease in aggression was accompanied by an increase in density of schooling and in social tolerance. This the authors explain by a so-called "threshold-hypothesis" for intensity of aggression, in which they assume that the selection they introduce results in a high genetic threshold for aggression (i.e., a reduced frequency of aggression) and that it genetically influences the other two aspects of social behavior, namely cohesion and social tolerance. To produce similar results in DomWorld, however, only the intensity of aggression has to be changed "genetically," and then the other changes of social behavior follow as side effects.

Further, DomWorld also produces hypotheses in which there is little need for assumptions of cognitive capabilities, because what may look like an exchange between the sexes (such as of female sex for male tolerance) appears to be a side effect of changes in female dominance and male timidity (Hemelrijk, 2000c). Further, the supposed exchange among females of grooming for something else (such as support), which is thought to underlie certain grooming patterns (see Seyfarth, 1977) and which should involve keeping track of the number of acts given and received, is according to DomWorld a superfluous supposition. There is no need for such cognitive book-keeping, nor for any genetic tendencies to exchange services, if such grooming patterns, as in our model, arise directly from a spatial configuration with dominants in the center and subordinates at the periphery.

It must be added, however, that on the other hand, DomWorld does not preclude the presence of such higher cognitive abilities in real animals. For instance, dominance perception of others and risk assessment may be based on direct perception, such as perceiving the body posture and size of an opponent (as reported here), but it may also be based on more complex cognitive processes, such as that agents recognize others individually and remember interactions they had with them in the past. Such memory-based dominance perception I have represented in the so-called "estimators" (Hemelrijk, 2000a). Whereas this leads to similar spatial-social structuring as in the case of direct perception, patterns were generally weaker, because the experiences each individual had with every other differed, and since different experiences by different individuals with the same partner cancel each other out, the dominance hierarchy becomes weaker.

Box 13 Proximate behaviors and natural selection 313

Evolutionary adaptations

As we have seen in respect of certain behavioral patterns, DomWorld generates hypotheses that require fewer adaptations by natural selection than are usually assumed. DomWorld makes clear how by changing the value of a single trait (that which represents the intensity of aggression), one may switch from an egalitarian to a despotic society. Because of the resemblance with societies of real macaques, natural selection may in the real world also have operated simply on intensity of aggression. One may imagine that in the distant past certain populations of the common ancestor of macaques (which was supposedly egalitarian, see Matsumura, 1999; Thierry *et al.*, 2000) may have lived under conditions of limited resources and that, therefore, a higher intensity of aggression developed as it was profitable to them. In such a case, individual selection would operate on only one trait (intensity of aggression) and this may have led, via self-organization, to a switch from the characteristics of an egalitarian society to those of a despotic one. Thus, we need not invoke a separate adaptation by natural selection for each single difference in social organization between an egalitarian and a despotic species. In this context, it may be mentioned that in a phylogenetic study of macaques by Thierry and collaborators (2000), greater male migration and female dominance over males (phenomena that are associated with despotism) display phylogenetic inertia. DomWorld makes this understandable, because both traits arise as side effects of the gradient of the hierarchy.

Box 13   Proximate behaviors and natural selection

William A. Mason

An evolutionary interpretation of the proximate behaviors contributing to macaque social organization requires a multi-level perspective that posits at least two major modes of natural selection. I call these modes *particulate selection* and *system selection*. The major contrasts between them are presented in Fig. 13.6. The heading: *phenotypic qualities* in Fig. 13.6 refers to differences in potentially observable attributes resulting from particulate selection versus system selection: *tempo of evolutionary change* (gradual versus punctuated), *units of behavioral change* (specific stimulus-response elements versus suites of behavior), *organization* of behavior (tight versus loose), behavioral *variability* (constrained versus permissive), behavioral *adaptability* (limited versus broad), and behavioral susceptibility to *environmental influences* (weak versus strong). Figure 13.6 also shows

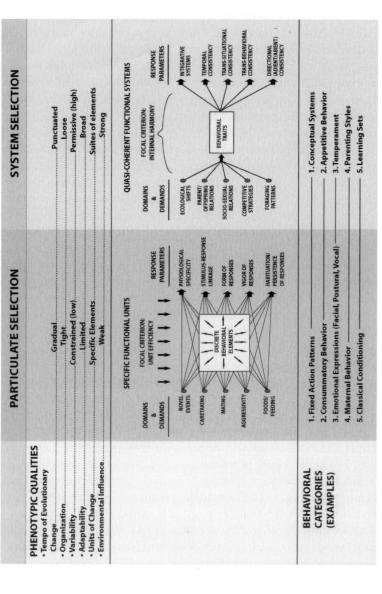

Fig. 13.6. Modes of natural selection.

Box 13 Proximate behaviors and natural selection   315

how the two modes are reflected in the relations between the environment (*domains-and-demands*) and behavior (*response parameters*), and provides examples of *behavioral categories* that are characteristic of each mode. The properties of each mode are discussed separately below.

Particulate selection

Results of particulate selection are most evident in a highly predictable relation between a stimulus and a specific response, as in the classic paradigms of the conditioned response and the fixed action pattern. A *focal selective criterion* in this mode is the reliability of the association between a "stimulus" and a "response." A general term for such stimulus-response units is *schema* (Piaget, 1971; Lorenz, 1981). Schemas cannot be dichotomized as "learned" or "innate," although they vary in the degree to which they are open to modification during ontogeny (Mason, 1984). Openness is generally greater on the stimulus (input) side than on the response (output) side of schemas. The adaptive value of schemas is as sources of organized behaviors that, with little or no practice, can become effective skills (knowing how).

Schemas are distinguished by the species-typical *form* of the response – a relatively stereotyped and coordinated pattern of motor elements. Schemas are abundant at all stages of development and serve many specific functions. Some of the more conspicuous and recurrent schemas in social interaction (social schemas) include grooming, facial expressions (e.g., threat, lipsmacking, fear, silent bared-teeth display), postures (e.g., sexual present, mounting, crouching, clasping, present-for-grooming), gait (e.g., brisk striding), and certain sounds (e.g., bark, coo, scream) (Hinde & Rowell, 1962; Rowell & Hinde, 1962; Altmann, 1963; Andrew, 1963; Kaufman & Rosenblum, 1966; van Hooff, 1967; Redican, 1975).

The forms of social schemas are fundamentally similar across individuals and species and they are also relatively impervious to environmental influences (Mason, 1985), suggesting they have been conserved in macaque evolution as *discrete behavioral elements* (Fig. 13.6), probably in a process of *mosaic evolution*. In addition, social schemas are readily observable categories, easily accommodated within the common "who-does-what-to-whom" formula, and serve communicative functions, features that may seem to qualify them as primary causes of social organization. More likely, however, they are surface manifestations of dynamic motivational and cognitive processes organized at a more fundamental level – the biobehavioral system (Mason, 1978b; Mendoza & Mason, 1989; Mendoza *et al.*, 2002; Capitanio, Chapter 2; Hemelrijk, this Chapter).

System selection

Selection at the level of the biobehavioral system operates on suites of behaviors within broad domains that constitute the basic functional dimensions of macaque social organization (Fig. 13.6). A suite consists of schemas that have been individually shaped by particulate selection and differ from other schemas within the domain in their forms and specific functions. Although they share an abiding affinity with the larger functions of the domain (e.g., sexual relations) and are often regarded as expressions of a unitary *behavioral trait*, they are not causally connected, but associated probabilistically in a *quasi-coherent functional system*. An important selective criterion in the evolution of these systems is *internal harmony* among constituent schemas.

A major adaptive value of quasi-coherent functional systems derives from the feature of openness within the configuration of schemas. In contrast to the functional emphasis of particulate selection on skill (knowing how), the emphasis of system selection is on knowledge (knowing that). This emphasis is supported by the evolved potential of macaques to form higher-order categories, concepts, rules, strategies, and the like. Such achievements are thoroughly documented and their development is known to depend on an individual's experience. Socialization influences are likely to be particularly influential and to contribute to reported variations within and between macaque species in social organization.

Evidence for the concept of quasi-coherent functional systems is abundant. Reliable differences exist between as well as within macaque species in such domains as foraging and feeding patterns, predator identification, relationships among kin, socio-sexual patterns, tolerance between and within the sexes, dispersal tendencies, characteristic stance toward the environment, and care of offspring.

A good illustration of the organization of a quasi-coherent functional system in macaques is maternal behavior (Mason, 2002). The principal schemas shown by a competent mother during the first few weeks following birth are clasping the infant to her ventral surface, carrying it as she goes about her normal routines, restraining it if it attempts to leave against her wishes, and retrieving it if she perceives it in danger or distress. These schemas appear in immature animals of both sexes and they are accessible throughout life (Breuggeman, 1973). To a casual observer the elements may appear as different expressions of a fully integrated unitary system of maternal care.

Contrary to this impression, the quality of maternal behavior is influenced by such factors as the mother's experience of being mothered and the quality of care she received, opportunities during development to interact with infants, to observe others caring for infants, and the experience of giving

Box 13 Proximate behaviors and natural selection   317

birth itself (see Chauvin & Berman, Chapter 10). Even when these opportunities are available, however, associations among the participating schemas may be tenuous. The same mother may show high levels of some positive aspects of maternal behavior (approach, making contact, restraining, grooming), while also displaying a strong tendency to reject, break contact, and leave the infant (Fairbanks, 1996). A mother may also be extremely abusive (dragging, stepping on, biting, hitting her infant), but also show high levels of proximity and protectiveness. Such contradictory behaviors have been noted in mothers raised in social groups and living in seemingly optimal conditions. Some effects may be modified by experience, whereas others apparently cannot (Maestripieri & Carroll, 1998).

The behavior of most mothers falls within the species-typical norm and the benefits to their infants of the inherent plasticity of the quasi-coherent maternal system are not noticed. There are cases, however, in which infants with serious developmental defects receive extensive compensatory maternal care that goes well beyond the normal pattern (Berkson, 1973; Fedigan & Fedigan, 1977; Nakamichi *et al.*, 1997).

Part V *An outside viewpoint*

Introduction

Primates represent a special order because this is the taxon to which *Homo* belongs – the "prime" order as we call it. Primatologists are also quite an odd scientific tribe, "Traditionally, primatologists have addressed questions raised by other primatologists in journals about primates read by other primatologists. Research on no other mammalian order has been so insular in its goals" (Whitten, 1988). The works of primatologists have had limited impact on related fields like ethology, behavioral ecology, and social anthropology (Hauser, 1993b). We need to widen our scope by submitting our findings and theories to outsiders (Chapter 14). Collaborating with conservation biology is another urgency (Strier, 1997). Habitat destruction and overhunting threaten the survival of nonhuman primate populations. In the twenty-first century primatologists may face the same event experienced by anthropologists in the twentieth century. They may witness the vanishing of a major part of the diversity they study.

14 An anthropologist among macaques

MAURICE GODELIER

It is unusual that a social anthropologist should be concerned by animal societies. In our field, it is commonly held that an impassable gap, a radical difference, separates humans from other primate species that live in societies. I do not
however think so. I have been relatively well-informed about the discoveries of
primatologists, in species like chimpanzees and bonobos in particular, and this
time I have learned a lot from the studies of macaques.

I am fascinated by the information available on the evolutionary history of
macaques. The knowledge that the 20 species of macaques recognized so far
belong to three different adaptive radiations, all originating from a common
ancestor, who appeared several millions years ago, provides a rich background
for inquiries. The divergence of most extant macaque species is thought to have
occurred more than 500 000 years ago, predating the appearance and dispersal of
Homo sapiens sapiens and even *Homo neanderthalensis*. A main achievement
of macaque studies is the finding that several components of their social life
bunch together and covary (Thierry, Chapter 12). Classifying the societies of
macaques on a continuum marked out by high power asymmetry at one extreme,
and more open and tolerant relationships at the other extreme, opens up the way
for the formulation of testable hypotheses. Yet, a few questions arise in my mind
regarding further issues.

First of all, how are kinship relationships defined in macaques? Some authors
use the terms "maternal lineage" and "matrilineage" to describe the descendants
stemming from an aged female, still present in the group, and surrounded by
her daughters and their offspring. These terms are inadequate. In anthropology,
matrilineage refers to a human group composed both of men and women who
descend from a common *ancestress* through the women only. In such groups,
political and economic authority is generally in the hands of men, that is of
the women's brothers, who exert their authority upon their sisters and their
offspring. The children of the men do not belong to the men's matrilineage, but
to the matrilineage of their wives, and they are under the authority of the wives'

Macaque Societies: A Model for the Study of Social Organization, ed. B. Thierry, M. Singh and
W. Kaumanns. Published by Cambridge University Press. © Cambridge University Press 2004.

321

brothers. No such structure is found among macaques. Instead what is described corresponds to matrilines as some authors use it, that is, lines of directly related females and their offspring. It is thus most interesting to learn that nieces do not have special relationships with the sisters of their mother, their maternal aunts, and that reciprocally aunts have no special relationships with their nieces and nephews. The experiments conducted by Chapais (Chapter 9) are especially interesting from this viewpoint, all the more so since they led to an observation which was probably unexpected on the part of the experimenters; whereas homosexual interactions did not occur between sisters or between grandmothers and granddaughters, they were observed between aunts and nieces (Chapais *et al.*, 1997).

We would like to know more about the behavior of macaque mothers toward their sons. Even if males leave their group when maturing, they live with their mothers, grandmothers, and sisters during their first years. Our knowledge is scant about them in most species. We know much more about dominance relationships between females, and in particular between the mother and her daughters. But the fact that the last-born daughter outranks her elders due to her mother's preference towards her is not really understood. We might propose an ultimate cause for such an outcome, though. By protecting her youngest daughter the mother would increase the offspring's chances of survival (Schulman & Chapais, 1980). Were there to be such a "natural reason," however, we should find the same privilege for the last-born daughters in all macaques, which in fact happens to be untrue (e.g., in Barbary macaques: Paul & Kuester, 1987; Prud'homme & Chapais, 1993a). We should also look at the behavior of a mother vis-à-vis her last offspring if it were a son. Biological protection should work as well in such instances. Finally, we should really conclude that there are no matrilineages in macaques. Matrilines exist but collateral kinship ties seem to be ignored. Since females are philopatric, immatures grow up in groups composed of females' relatives that often gather several sisters, each raising her own offspring herself. Therefore, some proximity and familiarity should link the offspring of several sisters living in the same local group. Since there is a preferential link for the mother, why is there no recognition of the mother's sister and no preferential link with her? When we address the degree of kinship through females only, it appears that female Japanese macaques display a preferential treatment of kin up to a relatedness threshold of 0.25. Beyond this value, the precedence given to relatives drops markedly, there are no more significant differences in the treatment of kin and nonkin (Chapais *et al.*, 1997; Bélisle & Chapais, 2001). If we introduce the role of males, however, the computation of the degree of kinship becomes quite tricky for individuals. Not only is the relatedness of males unknowable, since they usually come from neighboring groups, but kin discrimination and any computation of kinship relations involving both males and females are also beyond the cognitive abilities of macaques (see Chapais *et al.*, 2001).

I note that the study of macaques has mainly focused on females. We have little information on the dominance relationships between males or about the relative ranks of adult males and females. The fact that the adult males of a group generally come from other groups should allow the study of the processes of establishment of social dominance between males (e.g., van Noordwijk & van Schaik, 1985), who neither know each other nor the females of the group to which they immigrate. Here lies the question of the differences observed in dominance and subordination structures between macaque species, given the fact that data originate from wild, provisioned, or captive populations (Fig. 14.1). It is not clear how competitive and cooperative behaviors may vary according to these three different contexts. In particular, it is surprising to learn that the rule of the "youngest ascendancy" does not hold in some groups of wild Japanese macaques, and that some daughters can even outrank their mothers (Chapais, Chapter 9; Hill, Box 11).

Alliances between individuals are clearly critical in order to acquire and maintain social rank within the group hierarchy. This rank arises primarily from the support of the mother and relatives, but unrelated individuals may also bring help in conflicts. This demonstrates the importance of the social components of power, which allow individuals to acquire a social status that would be inaccessible for them through physical strength alone. This is especially true for the youngest daughters, whose rank obviously depends more on others than on themselves. But it appears to be a very general rule that alliances are indeed a social force that governs interactions between individuals. We are far from the image of the super-male primate. The correlation between social rank and physical strength is relatively limited, but this may vary according to individuals. Males in particular do not live among kin. The number of adult males in a group appears to be limited and they are often strangers, which may confer an advantage to females, who are organized in matrilines. Making alliances calls for an ability to manipulate conspecifics, which requires a fair amount of social intelligence (Call, Box 2).

We would like to know more about affiliative contacts, social grooming and "friendship" among macaques. In this volume, we learn that opponents may reconcile and groom each other after conflicts, but what about who does what to whom? What about variables such as the age and sex of partners, and the kind of reconciliatory behaviors they use? (see Call *et al.*, 2002). Affiliative interactions are genuine components of the survival strategies of individuals – they allow them to continue to live together. We also lack data about how macaques share resources. It seems that the "friendly" aspects of macaque social life still remain in the shadow. What are projected on our screen are mostly the dominance

a

b   c

Fig. 14.1. Weed macaques. Several Asian macaque species may live in commensal relationships with humans, they have been labeled as "weed" by Richard and collaborators (1989). (a) Urban rhesus macaques at rest (India); (b) a stumptailed macaque feeding from believers' offerings in a temple (Thailand); (c) longtailed macaques wandering near human settlements (Bali). (Photographs by R. Seitre.)

relationships of females and their ranking rules. In spite of the work of the last two decades (e.g., de Waal, 1986a, 1996a), it seems that competition more than cooperation remains the primary issue for primatologists (Flack & de Waal, Chapter 8).

The relation between a group and its home range raises another question. Macaque groups seem to have no real territory since there are wide overlaps between the home ranges of neighboring groups. Macaques do not defend territories as do chimpanzees. Encounters between groups necessarily occur if there are overlaps. In such circumstances, males are relatively aggressive. But we learn that adult females often do behave agressively (Cooper, Box 9). Obviously, distinct reasons might explain the agressiveness of males and females. Females would fight for control of the resources whereas the males would fight to keep a prior access to the females of the group. We would like to know whether the more aggressive individuals in these encounters between groups are higher-ranking males and females.

Let me say a few more words about the sexuality of macaques, either hetero or homo. It seems that mating success is weakly correlated with reproductive success in males. If this is statistically confirmed, part of the sociobiological thesis is stripped of its grounds. The discrepancy between mating and paternity records is usually accounted for by behaviors like sneaking by subordinate males. As an outcome, the female would be mainly responsible for mate choices; her matings with dominant and subordinate males would insure their own reproductive success. But this argument falls short of explaining why the females should prefer the subordinate males in some cases and not in others.

As for sexuality again, Vasey (Box 7) reports interactions between females that look like homosexuality. Such interactions, however, are not as intense as in bonobos (see de Waal & Lanting, 1997). No homosexual play is reported in macaques, but of course mounts do occur in the context of play. There seems to be no satisfactory explanation for the appearance and importance of homosexual behaviors from an evolutionary viewpoint (Vasey, Box 7). What is their adaptive function? Could it be that sexuality is both pleasure and reproduction for nonhuman primates as it is for humans? After all, one may seek pleasure without reproduction and homosexuality, like masturbation or fellatio, etc., gives pleasure. But if that is so, a large part of the sexual behavior of the primates is not driven only by the incentive of reproductive success and sexual selection.

Still, I would like to stress two major issues that emerge from the study of macaques: the importance of individual personalities in social relationships and the social transmission of behaviors. Differences in personalities may arise from a number of factors – for example, sex of the individual, birth rank, mother's personality, or ecological context. Critical though they are for the evolutionary process, these differences are not so important in the structuring of macaque societies. The individuals will carve out different places in a social hierarchy, but the hierarchy is an organization level that to some extent transcends interindividual variations. Putting it another way, the differences between individuals induce them to take up different places in a hierarchy for a certain time because, as we know, some months or years later these same places may be occupied by other individuals of the same band. An individual who is dominant today will be subordinated tomorrow. Through all these changes and permutations of place what remains constant is the need for a hierarchy. Because hierarchy (but also cooperation) is necessary both for the continuation of the group and for the evolutionary process. In human societies with neither class nor caste such as the so-called "primitive" societies of New Guinea, the social status of individuals depends on two processes (Godelier, 1986). Either the status is *ascribed*, as that of the ritual leader, whose function always belongs to the members of the same clan, or the status is *achieved*, as is the case of the Great Warrior, who must have killed enemies in single combat to deserve it.

Moreover, among these human societies, adults pay attention to the personality traits of children. They may predict for them a shamanic destiny if the child looks silent and thoughtful, or a warrior's destiny if the child appears aggressive and self-willed. This prediction often leads to an actualization of the destiny in question, but not always. Such predictions are made only for certain functions (shaman, warrior), which are accessible to everyone depending on the individual's gifts. They cannot stand for functions that are socially inherited, because in that case the functions are attached to a group but not to the individuals as such. In this case those liable to exert these functions are chosen by the group and not by their own will. This dialectic between the ascribed and achieved does not seem to exist among macaques. The fact that a female's social rank is dependent on that of her mother should not be seen as inheritance of a function, but rather as inheritance of a social power, which gives rise to nepotism, that is, a privileged protection.

Is there a social transmission of know-how in macaques? The answer is positive for the know-how related to material techniques, as for potato-washing in Japanese macaques. The innovation spread to the group-mates of the innovator, and then it was transmitted to the next generations, which learned it from their elders (Chauvin & Berman, Chapter 10). This was an innovation in the material way of life. But this change was not sufficient to transform the relation of macaques to their ecological environment. And there is no evidence that macaques transmit new social know-how, a new way of organizing social relationships. Certainly, mothers may influence their offspring by their dominance and protective style. Nonetheless, this does not change the operating rules of the group and the organization of the society.

I come to my last and main point. Macaques are social animals, they are primates who live in a society, and they must reproduce their social organization to perpetuate their life. Hence the need for cooperation, reconciliation, and hierarchies, and so on, because the society must exist for the individual to exist. It is a direct and a priori material reason for its being. Consequently, such statements as "investigating how societies arise from aggregates of individuals" (Flack & de Waal, Chapter 8) are difficult for me to understand since neither nonhuman primates nor humans "gave to themselves" the ability to live in society. It is nature, by its evolution, which has produced social species. A primary difference between humans and nonhumans is that the latter live in a society and must *reproduce* their society in order to sustain their life whereas humans do not merely live in society and cope with their environment to sustain their life: they themselves *produce* society – i.e., new forms of social organization, in order to live. Producing society is changing organization rules, creating new social groups, new classes or castes that fulfil distinct functions, it is creating towns separated from and opposed to the countryside, etc. Moreover, a substantial number of these new functions, which developed in human societies over the course of history, are "imaginary." These are, for instance, the different sorts of relationships with ancestors and with spirits, which are supposed to monitor the nature and the fate of human beings. These imaginary realities give rise to symbolic practices, and as a whole, have social consequences that are neither imaginary nor symbolic. For example, the fact that the Pharaoh was thought to be and feared as a god living among humans was not only an imaginary construct, it had real social consequences over the populations of ancient Egypt.

In this perspective, I believe that the four grades, along which macaque societies covary, represent a series of transformations of the same fundamental social organization (see Thierry, Chapter 12). Ultimate causes are usually called upon to explain such variation, in particular the patchy or dispersed character of food resources, and the more or less important impact of predation (see Ménard, Chapter 11). Most authors are not convinced that these ecological explanations can fully account for variations and they additionally resort to "phylogenetic inertia." If the capacities of one or another dominant macaque are not the source of a new form of organization in macaque groups, then we are driven back to the search of ultimate causes and these are still unknown to us.

I will close by stressing the need to study and protect nonhuman primates. As an anthropologist, I am sensitive to the threats upon wildlife all over the world (Muroyama & Eudey, Box 14). The expansion of the West and other dominant social systems has utterly disrupted the ways of life of hundreds of human societies that anthropologists study or have studied. Large portions of cultural invention have vanished with the disappearance of various human ways of life. It is not necessarily the individuals that disappeared, however, but their social systems, their customs and their values. North American Indians already lived in reservations in the nineteenth century. For nonhuman primates, it is the survival of populations and species which is at stake. There would be no more hope for understanding them were they to disappear. It would be unthinkable to dispense with the information held by our nonhuman fellows if we want to significantly improve our knowledge as regards the biological bases of human behavior.

Box 14   Do macaque species have a future?

Yasuyuki Muroyama and Ardith A. Eudey

Many primate species are in critical danger worldwide and threatened with extinction (Chapman & Peres, 2001). This is the case for most *Macaca* species. In the *2000 IUCN Red List of Threatened Species* (Hilton-Taylor, 2000), one macaque species is categorized as "Critically Endangered" (CR), three species are "Endangered" (EN), seven species are "Vulnerable" (VU), and the others are "Lower Risk" or "Data Deficient" (Table 14.1). Here, we describe the present status and population trends of macaque species, major threats to wild macaque populations, and prospects for their conservation and management.

Present status and population trends of macaque species

While macaques are likely to be considered as well known or common, the data on the present status of populations such as numbers, distribution, and population trends are deficient for most species, especially those that are widespread geographically. Based on available information, most of the macaque species appear to be experiencing reduction in numbers and/or distribution, although some populations may appear to be recovering (Table 14.1).

   The arboreal liontailed macaque, endemic to the Western Ghats in southern India (Fig. 6.1), is one of the most endangered macaque species (Easa *et al.*, 1997). The total number is estimated at 3550 individuals, spread over 41 subpopulations at nine locations, fragmented due to habitat destruction. Another endangered species is the crested macaque, endemic to the island of Sulawesi, Indonesia (Fig. 11.1b). Hunting is the primary threat to this species (Lee, 1997), for which the total number was estimated at less than 6000 in 1988 (Sugardjito *et al.*, 1989). Population numbers at the Tangkoko Reserve, the only area with effective protection for the species, have stabilized and in some cases even gone up, however (R. J. Lee, personal

Table 14.1. *Macaca spp. as listed in the 2000 IUCN Red List of Threatened Species, using categories and criteria adopted by the IUCN Council in 1994*[a]

Species	English name	Red List categories	Trends of populations	Estimated numbers	Sources
M. sylvanus	Barbary macaque	Vulnerable (VU)	Down	9000–24 000	1
M. silenus	liontailed macaque	Endangered (EN)	Down	3550	2
M. nigra	crested macaque	Endangered (EN)	Down	4000–6000[b]	3
M. nigrescens	Gorontalo macaque	Lower risk/conservation dependent	Down	60 000	4
M. hecki	Heck's macaque	Lower risk/not threatened	Down	NA[c]	
M. tonkeana	Tonkean macaque	Lower risk/not threatened	Down	NA[c]	
M. maurus	moor macaque	Endangered (EN)	Down	<5000[b]	3
M. ochreata	booted macaque	Data deficient (DD)	Down	NA[c]	
M. brunnescens	Muna-butung macaque	Vulnerable (VU)	Down	NA	
M. pagensis	Mentawai macaque	Critically endangered (CR)	Down	NA	5
M. nemestrina	pigtailed macaque	Vulnerable (VU)	Down	NA	
M. sinica	toque macaque	Vulnerable (VU)	Down	<300 000	6
M. radiata	bonnet macaque	Lower risk/least concern	Stable (forest), Up (urban)	>50 000	7
M. assamensis	Assamese macaque	Vulnerable (VU)	Down	NA	
M. thibetana	Tibetan macaque	Lower risk/conservation dependent	Stable	c. 20 000	8
M. arctoides	stumptailed macaque	Vulnerable (VU)	Down	NA	
M. fascicularis	longtailed macaque	Lower risk/not threatened	Down	>1 000 000	9
M. mulatta	rhesus macaque	Lower risk/not threatened	Unknown (forest), Up (urban)	>500 000	8, 10
M. fuscata	Japanese macaque	Data deficient (DD)	Unknown	NA	
M. cyclopis	Taiwanese macaque	Vulnerable (VU)	Unknown	250 000 ± 100 000	11

NA, not available.

[a] A reassessment of all primates using categories and criteria approved by the IUCN Council in 2000 and the taxonomic revisions proposed by the IUCN/SSC Primate Specialist Group (PSG) will be included in the 2003 Red List.

[b] The numbers of these species may have decreased because of habitat disturbance and loss, and increased pressure on populations in Sulawesi.

[c] The numbers of these species were extrapolated by Bynum et al. (1999) from estimated densities and estimated habitat area are as follows: M. hecki, 100 000; M. tonkeana, 150 000; M. ochreata, 50 000.

Sources: (1) IUCN, 1996; (2) Molur et al., 2003; (3) Bynum et al., 1999, pers. comm.; (4) R. J. Lee, pers. comm.; (5) Fuentes & Ray, 1997; (6) W. Dittus, pers. comm.; (7) M. Singh, pers. comm.; (8) Zhang et al., 2002; (9) Supriatna et al., 2001; (10) Southwick et al., 1995; (11) Fooden & Wu, 2001.

communication). The problematic Japanese macaque is reported to have been expanding its distribution for at least 40 years (Watanabe & Muroyama, 2004), while the change in its numbers is unknown. Distributional expansion is obvious in the eastern part of Japan, while both expansion and reduction (or disappearance) of local populations have been observed in western Japan (Watanabe & Muroyama, 2004). Rhesus macaque populations in India declined by 80–90% between 1960 and 1980 due to heavy trapping, especially of juveniles, for trade; the estimated number was 183 000 in 1978 but had recovered to over 500 000 in the 1990s following the activation of an Indian export ban on all primates in 1978 (Southwick & Siddiqi, 1995).

Threats to wild primate populations

The major threats to wild primate populations are habitat degradation and loss due to human encroachment, hunting for food and traditional medicine, eradication as agricultural pests, and live capture for biomedical research and pet trade (Erwin et al., 1995). In most cases, some of these threats are combined, resulting in more negative impacts on populations. Habitat destruction and loss are the most serious threats to primate populations and their ecosystems. Various human activities such as legal and illegal logging operations, encroachment and conversion of forests for agriculture and monoculture plantations, and economic development activities such as hydroelectric projects all contribute to habitat loss (Fuentes & Ray, 1997). Consumption as food and/or use for traditional medicine also may be threats to some macaque species. Hunting appears to be largely responsible for the decline of some species such as the crested macaque (Lee, 1997; O'Brien & Kinnaird, 2000), and Japanese macaques were hunted for food and traditional medicine until the 1940s, which may have resulted in the extermination of monkeys from much of their range in northeastern Honshu Island (Mito, 1992). Hunting of primates, including macaques, for food and medicine within Vietnam and for the trans-border trade between the Indochina region and China may rival the "bush meat" trade in Africa.

Many primate populations are virtually unable to survive outside primary forests (e.g., M. silenus), but some of the macaque species in Asia thrive in secondary forests and in commensal relationships with humans (Fig. 14.1). Conflict between human and nonhuman primates is inevitable, especially as the human population grows, and most frequently arises over crop-raiding, although conflicts between monkeys and humans are increasing in many urban areas in South Asia. Live capture of monkeys or the hunting of them as agricultural pests may be a means of reducing crop damage but also may lead to the extirpation of local populations.

Box 14 Do macaque species have a future? 331

The impact of live capture for biomedical research is usually insignificant in comparison with other influences on wild populations. With respect to Asian macaques, however, the trade in wild-caught primates has had one of the most negative impacts on wild populations (e.g., Kavanagh *et al.*, 1987; Eudey, 1995). It cannot be overemphasized that harvesting of wild monkeys as a means of reducing crop damage is ineffective unless some troops, or the local population, in a target area are eliminated; this could easily lead to the extirpation of the population.

Conservation and management of macaques

Some species or populations living mainly in evergreen forest and threatened by habitat destruction and loss, most notably the liontailed macaque, need appropriate, careful conservation programs for the metapopulation or ecosystem as even minor habitat degradation can be a significant threat to their survival. Isolated or fragmented habitats should be strictly protected and enriched by planting selected tree species, with the creation of corridors between habitats whenever feasible, so that local populations may be able to expand and exchange genes. Populations whose numbers are limited should be monitored carefully, and ex situ breeding programs might be considered if necessary.

Widespread species that may be very adaptable to different environments, such as rhesus, longtailed, and Japanese macaques, need appropriate management programs combining conservation and damage management of any populations in conflict with local people. Development of theories and techniques of damage management is critical for the conservation of these species because only programs that effectively address the needs of local people for relief from crop raiding have realistic prospects of success. Besides macaques, baboons and mangabeys in Africa also experience conflict with humans (e.g., Hill, 2000; Saj *et al.*, 2001).

Some researchers have argued that a sustained-yield management program could conserve self-sustaining, healthy primate populations, reduce crop damage and other human–macaque conflicts, promote a secure supply of primates for biomedical use and provide needed economic sustenance for local people (Erwin *et al.*, 1995; Crockett *et al.*, 1996), although it is highly questionable that primates are a renewable resource. This countermeasure would be very risky for conservation without personnel and systems to support and monitor management programs. With regard to Japanese macaques, for instance, many researchers have pointed out that live-capture or killing of crop raiding animals without appropriate management programs have proved to be ineffective for the reduction of crop damage (e.g., Agetsuma,

1999; Muroyama, 2000), and have even caused the extirpation of local populations (Muroyama *et al.*, 1999).

At present, most Asian countries appear to have insufficient economic bases for effective wildlife management. In addition, scientific data on primates and their habitats are usually too scarce to plan area-specific or species-specific conservation strategies. Effective conservation actions must concentrate on subspecies and/or local populations, because the extirpation at subspecies level and/or that of local populations can easily occur even in widespread common species such as longtailed and rhesus macaques.

Economic growth in Asia appears to enhance the rapid growth of human populations, accelerating habitat destruction, and local extinctions of wildlife. Under such conditions, the recruitment of young field biologists from within the habitat countries may be critical for the spread of knowledge, to government officials and on down to local people, on conservation and management of primates as well as other organisms and habitats. Such biologists should also be trained to survey and monitor wild populations, including macaques, where sufficient data are not available. The goal should be to empower people living in the habitat countries to plan and implement effective conservation and management programs on wildlife for themselves, in collaboration with non-governmental organizations (NGOs) and/or other countries if necessary. *Macaca* is a recent radiation (Pope *et al.*, 2000), of which not even the most threatened populations could be considered to be relict. Reductions in distribution and/or numbers are primarily the consequence of human threats: changes in human perception, policy, and action are necessary to conserve the genus.

Conclusion

15 Toward integrating the multiple dimensions of societies

BERNARD THIERRY, MEWA SINGH, AND WERNER
KAUMANNS

A society is a collective product made up of multiple dimensions. Each of the contributors in this volume has tackled one or the other dimension according to her or his own theoretical perspectives. This allowed them to go more deeply into the phenomenon. However, when attempting to integrate the knowledge supplied by various disciplines, a primary difficulty is to erase the boundaries between those disciplines and in fact to deconstruct part of the theoretical structures laboriously erected by each of them. In what follows, we point out the bridges we see between the various dimensions of macaque societies in the hope of paving the way toward an integration of our views regarding social organization.

A major fact emerging from the study of macaques is the link evidenced between the personality of individuals – their traits, emotions, and motives – and the species-typical social relationships (Capitanio, Chapter 2; Aureli & Schino, Chapter 3; Thierry, Chapter 12). Slight differences in the personality of individuals may give rise to significant differences in their behaviors and trigger different patterns of social relationships. In fine-tuning social styles, selective processes may act on tendencies and response thresholds rather than on particular behaviors (Capitanio, Chapter 2; Hemelrijk, Chapter 13). For example, an earlier dispersal is associated with a genetically-based reduced serotonin activity in rhesus macaques (Mehlman et al., 1995; Trefilov et al., 2000). Even if patterns of male dispersal are modulated by the social and demographic context, they consistently differ between macaque species and their variations follow the phyletic lines (Thierry et al., 2000).

Though macaques display a fair amount of variability in their time budget and the size and composition of groups according to ecological conditions (Dittus, Chapter 5; Ménard, Chapter 11; Hill, Box 11), fluctuations in social

Macaque Societies: A Model for the Study of Social Organization, ed. B. Thierry, M. Singh and W. Kaumanns. Published by Cambridge University Press. © Cambridge University Press 2004.

relationships remain limited, and overall grouping patterns remain constant in a given species. To take the example of the rhesus macaque, the diversity of habitats in which it lives covers most of the environmental range of the whole genus. Notwithstanding variations within the reaction norm of the species, we can recognize the intractable character of the rhesus macaque in any place where it has been studied. Given the acknowledged adaptability of the species, this weakens the common belief that niche breadth correlates positively with the extent of social variation (see Lott, 1991). Actually, the plasticity of individuals and the flexibility of their responses contrast with the relatively narrow range of intraspecific variation observed in macaque social organization (see Altmann & Altmann, 1979). This may be a consequence of the numerous epigenetic processes that intervene in the emergence of societies. The redundancy created by the cumulative behavioral effects of similarly built individuals would produce equifinal states at the level of the social organization (see Mendoza *et al.*, 2002).

If specific sociodemographic forms are associated with specific individual personalities, it would mean that interindividual differences clearly have their limits in macaques. There is no such thing as a temperament that can be associated with a given culture in human beings (Godelier, Chapter 14). Human groups are founded on the differentiation of roles, a process that permits a gainful exploitation of dissimilar characters and gifts of individuals. Human societies may both need and favor wide disparities in personalities. The extent of individual psychological variability could therefore stand as a major contrast between macaques and humans.

In the past decades, much of the scientific work focused on energy-flow systems. A wealth of data has been accumulated on how resource availability affects everyday life, patterns of growth, and reproductive effort (Bercovitch & Harvey, Chapter 4; Singh & Sinha, Box 4; Ménard, Chapter 11; Hill, Box 11), and how life-history traits in turn influence demographic dynamics (Dittus, Chapter 5; Okamoto, Box 5). We possess detailed accounts of the effects of the social context, and how nepotism, dominance asymmetry, and patterns of alliances channel the access of individuals to resources (Paul, Box 6; Flack & de Waal, Chapter 8; Butovskaya, Box 8; Chapais, Chapter 9; Cooper, Box 9). Whereas the study of proximate causes appears successful, our results regarding ultimate causes appear quite mixed in comparison. Socioecological models still fail to satisfactorily account for the various social styles observed among macaques (Cooper, Box 9; Ménard, Chapter 11; but see Okamoto, Box 5; Chapais, Chapter 9). Part of the problem may be methodological. Current modeling relies on factors that are notoriously difficult to assess. Debates on the role of predation and infanticide in the grouping patterns of primates are based on indirect evidence at best. And broad ecological classifications concerning food distribution may not be applicable in the field (Strier, 1994; Ménard,

Chapter 11): measuring the quality and patchiness of food resources, and separating the clumping of food from the amount of food may actually prove to be quite tricky.

Other limitations of the models are conceptual ones. The first problem is the neglect of the phylogenetic history of a species, which is a genuine Darwinian concern (Richard, 1981; Janson, 2000). It is recognized that "each species brings a different phylogenetic heritage into a particular ecological scene" (Struhsaker, 1969), but the conclusion that phylogeny is a major determinant is often viewed as a vexing one, only worth falling back on when adaptive hypotheses have failed. As Lott (1991: 173) notes, "phylogenetic inertia has become at once the salvation of particular socioecological hypotheses and the *bête noire* of socioecological theory in general." Common descent, nonetheless, accounts for a large amount of the variance found in the grouping patterns of nonhuman primates – e.g., most cercopithecine monkeys sharing traits like female philopatry or the formation of close bonds between females (di Fiore & Rendall, 1994). In macaques, it turns out that the overall social style, the temperament, the meaning of the bared-teeth display and the patterns of male dispersal all correlate with phylogeny, which points to a fair degree of conservatism in their evolution (Thierry *et al.*, 2000; Thierry, Chapter 12). One may argue that phylogenetic inertia and ecological adaptation represent complementary rather than mutually exclusive explanations (Sterck, 1999; Koenig, 2002; Singh & Sinha, Box 4; Chapais, Chapter 9).

A second problem lies in the nature of the selective processes acting on the social phenotype. Macaque societies are made of overlapping generations where the life history and social relationships of the individual are strongly influenced by the mother and other group members (Dittus, Chapter 5; Chauvin & Berman, Chapter 10; Maestripieri, Box 10). The transmission of phenotypic information from one generation to the next may give rise to indirect genetic effects and niche construction processes (Wolf *et al.*, 1998; Laland *et al.*, 1999). That may increase the covariation of characters and produce unusual evolutionary dynamics, either accelerating or impeding the rate of change. Macaque societies are complex systems, as they are built through multiple epigenetic processes that channel their evolutionary pathways (Thierry, Chapter 12; Hemelrijk, Chapter 13; Mason, Box 13). Future models will have to account for these facts. Correspondingly, the collapse of models that exclusively rely on functional trade-offs appears ineluctable.

It is acknowledged that social and demographic factors should affect the gene flow and evolutionary rates of populations (Gachot-Neveu & Ménard, Chapter 6). Many studies have focused on the sources of reproductive variance between individuals, particularly in the mating strategies pursued by both males and females (Bercovitch & Harvey, Chapter 4; Gachot-Neveu & Ménard, Chapter 6;

Paul, Box 6; Soltis, Chapter 7). For most species and populations, however, we lack data on the lifetime reproductive output of individuals. Population genetic analyses are scarce (e.g., Melnick, 1987; de Ruiter & Geffen, 1998). Patterns of male dispersal are not random, they vary according to social styles and this should affect the genetic structure of populations. Yet, there is no evidence that the genetic variability of populations differs according to the modes of social relationships (Gachot-Neveu & Ménard, Chapter 6). We do not know how the diverse social styles of macaques relate to their evolutionary dynamics. We cannot specify how the energy-flow system affects the information-flow system.

The expectation that the association between male dominance and number of offspring should be weaker in the more tolerant species than in the more hierarchical ones is not supported by the data (Paul, Box 6; Soltis, Chapter 7; Thierry, Chapter 12). The seasonality of reproduction has greater consequences on paternity than does the social style. If mating occurs throughout the year, there is usually no more than one receptive female at a time and the top-ranking male may have priority of access to her. When breeding is seasonal, several males may sire receptive females since no single male is able to monopolize all the females. If these findings were to be confirmed, an important outcome would be that the genetic structure of a group would differ depending on the climate. Whereas offspring may have different fathers in temperate regions, we would find a strong age-cohort effect among young individuals in the tropics – all members of the same generation being paternal half-siblings. Reproductive seasonality is a key element in the evolution of macaques. Over geological time, macaques may have switched several times between seasonality and aseasonality of reproduction following climatic changes, whereas the core of their societies underwent limited modifications during the same period (Bercovitch & Harvey, Chapter 4; Thierry, Chapter 12). In such a perspective, the occurrence of reproductive seasonality is a contingent event on the evolutionary trajectory of species. When the rules of the game come to be modified, winners and losers may not be the same as was the case before the change: individuals cannot help but cope with the new sorting rules. If seasonality determines mating behaviors more strongly than does the social style, the genetic representation of individuals in the next generations – and the ensuing genetic structure – should depend more on statistical sorting in the energy-flow system than on adaptive strategies (see Eldredge & Grene, 1992).

Many authors would disagree with the preceding conclusion. We come at this point to the Great Divide between biologists. The divide separates those who have faith in the power of natural selection to shape all characters of species from those who believe that the internal organization and imperfections of living beings channel the course of evolution. Both schools are represented among the contributors to this volume. The key word for the former scholars is

Fig. 15.1. Agonistic buffering in Barbary macaques (Kintzheim, France). Adult males carry infants so as to approach each other and engage in affiliative interactions (Deag, 1980). Males also display high rates of care toward infants who are not their own offspring (Paul *et al.*, 1996; Ménard *et al.*, 2001). Since there is no parental investment, an adaptive explanation may be that caring for infants represents a male's strategy to increase its chances of being chosen by the infant's mother as a mating partner (van Schaik & Paul, 1996). An alternative explanation is that the interest of males for infants merely represents a side effect of the use of infants in agonistic buffering (Paul *et al.*, 1996; Thierry, 2000). (Photograph by B. Thierry.)

"adaptation," as they focus on individuals and stress competition, kinship, repro-ductive strategies and the role of ecological environment in shaping functions (e.g., Singh & Sinha, Box 4; Dittus, Chapter 5; Paul, Box 6; Soltis, Chapter 7; Chapais, Chapter 9). The preferred term of the latter scholars is "constraint," as they emphasize systems and search for self-organizing processes, contin-gency, and nonadaptive correlates like homosexual behaviors or patterns of alloparental care (e.g., Vasey, Box 7; Thierry, Chapter 12; Abegg, Box 12; Hemelrijk, Chapter 13; Mason, Box 13) (Fig. 15.1).

We still have a long way to go in debating these issues and understanding why some beings are ruled by strong nepotism and power asymmetry whereas others live in a more tolerant and open social world. We wrote in the introduction of the book that there is no insuperable gap in our knowledge regarding the biology and societies of macaques. We must add that many of our conclusions rely on incomplete evidence. We have just begun to scratch the surface and we still lack comparative data for many characters. Though we have a fair knowledge of the social behaviors of macaques, we know little about temperament variations between species and almost nothing about their neurobiological and cognitive basis (Capitanio, Chapter 2; Call, Box 2; Aureli & Schino, Chapter 3). Moreover, our information regarding the ecology, demography, and population genetics of most macaques remains sketchy (Dittus, Chapter 5; Gachot-Neveu & Ménard, Chapter 6; Ménard, Chapter 11). A more complete synthesis requires more time and research.

The strength and weakness of this volume lies in its focus on macaque societies. We trust that it will also contribute to widening the scope of analysis of social organizations in other species. In order to answer our questions, to bring out new facts, and to continue to debate our differing views, we have to repeatedly go back to the animals themselves. We do hope this will remain possible for future generations of biologists. Macaques are at the forefront of providers of scientific knowledge, but many of them are on the verge of extinction in their natural environment. As is the case for many other animals, increased human pressures threaten their life and survival. Their habitats are rapidly shrinking to fragmented pockets that are no longer sustainable on a long-term basis. The wild populations of three species – liontailed, crested, and moor macaques – have already been reduced to fewer than 5000 individuals in their original habitats, i.e., the population of a village on the human scale (Muroyama & Eudey, Box 14). In most places, human beings see macaques as competitors who raid crops. Yet, we are indebted to them. Macaques have contributed to human welfare to an unexpected extent, if only for their involvement in the production of vaccines. Our most immediate challenge is to ensure that they have a future and preserve these unique representatives of the splendor of life.

References

Abegg, C. (1998). *Constance et Variabilité des Comportements Sociaux en Fonction de l'Environnement chez deux Espèces de Macaques* (Macaca fuscata, Macaca silenus). Thèse d'Université, Université Louis Pasteur, Strasbourg.

Abegg, C. & Thierry, B. (2002). Macaque evolution and dispersal in insular Southeast Asia. *Biological Journal of the Linnean Society*, **75**, 555–576.

Abegg, C., Thierry, B. & Kaumanns, W. (1996). Reconciliation in three groups of lion-tailed macaques. *International Journal of Primatology*, **17**, 803–816.

Abernethy, V. (1974). Dominance and sexual behavior: a hypothesis. *American Journal of Psychiatry*, **131**, 813.

Agetsuma, N. (1995a). Foraging strategies of Yakushima macaques (*Macaca fuscata yakui*). *International Journal of Primatology*, **16**, 595–609.

(1995b). Dietary selection by Yakushima macaques (*Macaca fuscata yakui*): the influence of food availability and temperature. *International Journal of Primatology*, **16**, 611–627.

(1999). Primatology and the protection and management of wildlife. In *For People Who Wish to Study Primatology* (in Japanese), ed. T. Nishida & S. Uehara, pp. 300–326. Kyoto: Sekaishiso-sha.

Agetsuma, N. & Nakagawa, N. (1998). Effects of habitat differences on feeding behaviors of Japanese monkeys: comparison between Yakushima and Kinkazan. *Primates*, **39**, 275–289.

Aggleton, J. P. & Young, A. W. (2000). The enigma of the amygdala: on its contribution to human emotion. In *Cognitive Neuroscience of Emotion*, ed. R. D. Lane & L. Nadel, pp. 106–128. New York: Oxford University Press.

Alexander, R. D. (1974). The evolution of social behavior. *Annual Review of Ecology and Systematics*, **5**, 325–383.

Allport, G. W. (1937). *Personality: A Psychological Interpretation*. New York: Holt.

Altmann, J. (1980). *Baboon Mothers and Infants*. Cambridge, MA: Harvard University Press.

Altmann, J. & Altmann, S. A. (1991). Models of status-correlated bias in offspring sex ratios. *American Naturalist*, **137**, 542–555.

Altmann, J., Alberts, S. C., Haines, S. A. *et al.* (1996). Behavior predicts genetic structure in wild primate group. *Proceedings of the National Academy of Sciences of the USA*, **93**, 5797–5801.

Altmann, J., Hausfater, G. & Altmann, S. A. (1985). Demography of Amboseli baboons. *American Journal of Primatology*, **8**, 113–125.

341

(1988). Determinants of reproductive success in savannah baboons, *Papio cyno-cephalus*. In *Reproductive Success*, ed. T. H. Clutton-Brock, pp. 403–418. Chicago: University of Chicago Press.

Altmann, S. A. (1962). A field study of the sociobiology of rhesus monkeys, *Macaca mulatta*. *Annals of the New York Academy of Sciences*, **102**, 338–435.

(1963). Social behavior of anthropoid primates: analysis of recent concepts. In *Roots of Behavior*, ed. E. L. Bliss, pp. 277–285. New York: Harper.

(1981). Dominance relationships: the Cheshire cat's grin? *Behavioral and Brain Sciences*, **4**, 430–431.

Altmann, S. A. & Altmann, J. (1970). *Baboon Ecology*. Chicago: University of Chicago Press.

(1979). Demographic constraints on behavior and social organization. In *Primate Ecology and Human Origins*, ed. I. S. Bernstein & E. O. Smith, pp. 47–63. New York: Garland Press.

Amaral, D. G. (2002). The primate amygdala and the neurobiology of social behavior: implications for understanding social anxiety. *Biological Psychiatry*, **51**, 11–17.

Anderson, C. O. & Mason, W. A. (1974). Early experience and complexity of social organization in groups of young rhesus monkeys (*Macaca mulatta*). *Journal of Comparative Psychology*, **87**, 681–690.

Anderson, J. R., Montant, M. & Schmitt, D. (1996). Rhesus monkeys fail to use gaze direction as an experimenter-given cue in an object-choice task. *Behavioural Processes*, **37**, 47–55.

Andersson, M. (1994). *Sexual Selection*. Princeton, NJ: Princeton University Press.

Andrew, R. J. (1963). The origin and evolution of the calls and facial expressions of the primates. *Behaviour*, **20**, 1–109.

Andrews, M. W. & Rosenblum, L. A. (1994). The development of affiliative and agonistic social patterns in differentially reared monkeys. *Child Development*, **65**, 1398–1404.

Angst, W. (1975). Basic data and concepts on the social organization of *Macaca fascicularis*. In *Primate Behavior: Developments in Field and Laboratory Research*, vol. 4, ed. L. A. Rosenblum, pp. 325–388. New York: Academic Press.

Antonovics, J. & van Tienderen, P. H. (1991). Ontoecogenophyloconstraints? The chaos of constraint terminology. *Trends in Ecology and Evolution*, **6**, 166–168.

Anzenberger, G., Mendoza, S. P. & Mason, W. A. (1986). Comparative studies of social behavior in *Callicebus* and *Saimiri*: behavioral and physiological responses of established pairs to unfamiliar pairs. *American Journal of Primatology*, **11**, 37–51.

Appleby, M. C. (1983). The probability of linearity in hierarchies. *Animal Behaviour*, **31**, 600–608.

Ashby, W. R. (1962). Principles of the self-organizing system. In *Principles of Organization*, ed. H. von Foerster & G. W. Zopf, pp. 255–278. New York: Pergamon Press.

Asquith, P. J. (1989). Provisioning and the study of free-ranging primates: history, effects, and prospects. *Yearbook of Physical Anthropology*, **32**, 129–158.

Aujard, F., Heistermann, M., Thierry, B. & Hodges, J. K. (1998). Functional significance of behavioral, morphological, and endocrine correlates across the ovarian cycle in semifree ranging female Tonkean macaques. *American Journal of Primatology*, **46**, 285–309.

Aureli, F. (1992). Post-conflict behaviour among wild long-tailed macaques (*Macaca fascicularis*). *Behavioral Ecology and Sociobiology*, **31**, 329–337.

(1997). Post-conflict anxiety in nonhuman primates: the mediating role of emotion in conflict resolution. *Aggressive Behavior*, **23**, 315–328.

Aureli, F. & Schaffner, C. M. (2002). Relationship assessment through emotional mediation. *Behaviour*, **139**, 393–420.

Aureli, F. & Smucny, D. A. (2000). The role of emotion in conflict and conflict resolution. In *Natural Conflict Resolution*, ed. F. Aureli & F. B. M. de Waal, pp. 199–224. Berkeley, CA: University of California Press.

Aureli, F. & van Schaik, C. P. (1991). Post-conflict behaviour in long-tailed macaques (*Macaca fascicularis*): II. Coping with the uncertainty. *Ethology*, **89**, 101–114.

Aureli, F. & Whiten, A. (2003). Emotions and behavioral flexibility. In *Primate Psychology: The Mind and Behavior of Human and Nonhuman Primates*, ed. D. Maestripieri, pp. 289–323. Cambridge, MA: Harvard University Press.

Aureli, F., Cords, M. & van Schaik, C. P. (2002). Conflict resolution following aggression in gregarious animals: a predictive framework. *Animal Behaviour*, **64**, 325–343.

Aureli, F., Cozzolino, R., Cordischi, C. & Scucchi, S. (1992). Kin-oriented redirection among Japanese macaques: an expression of a revenge system? *Animal Behaviour*, **44**, 283–292.

Aureli, F., Das, M. & Veenema, H. C. (1997). Differential kinship effect on reconciliation in three species of macaques (*Macaca fascicularis*, *M. fuscata*, and *M. sylvanus*). *Journal of Comparative Psychology*, **111**, 91–99.

Aureli, F., Preston, S. D. & de Waal, F. B. M. (1999). Heart rate responses to social interactions in free-moving rhesus macaques (*Macaca mulatta*): a pilot study. *Journal of Comparative Psychology*, **113**, 59–65.

Aureli, F., van Schaik, C. P. & van Hooff, J. A. R. A. M. (1989). Functional aspects of reconciliation among captive long-tailed macaques (*Macaca fascicularis*). *American Journal of Primatology*, **19**, 39–51.

Aureli, F., Veenema, H. C., van Panthaleon van Eck, C. J. & van Hooff, J. A. R. A. M. (1993). Reconciliation, consolation, and redirection in Japanese macaques (*Macaca fuscata*). *Behaviour*, **124**, 1–21.

Avital, E. & Jablonka, E. (2000). *Animal Traditions: Behavioural Inheritance in Evolution*. Cambridge: Cambridge University Press.

Axelrod, R. (1986). An evolutionary approach to norms. *American Political Science Review*, **80**, 1095–1111.

Bachorowski, J. A. & Owren, M. J. (1995). Vocal expression of emotion: acoustic properties of speech are associated with emotional intensity and context. *Psychological Science*, **6**, 219–224.

Bagemihl, B. (1999). *Biological Exuberance: Animal Homosexual and Natural Diversity*. New York: St. Martin's Press.

Baker, R. R. & Bellis, M. A. (1995). *Human Sperm Competition: Copulation, Masturbation and Infidelity*. London: Chapman & Hall.

Baker-Dittus, A. (1985). *Infant and Juvenile-directed Care Behaviors in Adult Toque Macaques, Macaca sinica*. Ph.D. dissertation, University of Maryland.

Baldellou, M. & Henzi, P. (1992). Vigilance, predator detection and the presence of supernumerary males in vervet monkey troops. *Animal Behaviour*, **43**, 451–461.

Barrett, G. M., Shimizu, K., Bardi, M., Asaba, S. & Mori, A. (2002). Endocrine correlates of rank, reproduction, and female-directed aggression in male Japanese macaques. *Hormones and Behavior*, **42**, 85–96.

Barrett, L. & Henzi, S. P. (1998). Epidemic deaths in a chacma baboon population. *South African Journal of Science*, **94**, 441.

Barrette, C. (1993). The "inheritance of dominance", or of an aptitude to dominate? *Animal Behaviour*, **46**, 591–593.

Barros, M., Boere, V., Huston, J. P. & Tomaz, C. (2000). Measuring fear and anxiety in the marmoset (*Callithrix penicillata*) with a novel predator confrontation model: effects of diazepam. *Behavioural Brain Research*, **108**, 205–211.

Barton, R. A. & Whiten, A. (1993). Female competition among female olive baboons, *Papio anubis*. *Animal Behaviour*, **46**, 777–789.

Barton, S. (1994). Chaos, self-organization, and psychology. *American Psychologist*, **49**, 5–14.

Basabose, K. & Yamagiwa, J. (1997). Predation on mammals by chimpanzees in the montane forest of Kahuzi, Zaïre. *Primates*, **38**, 45–55.

Bauer, R. M. (1998). Physiological measures of emotion. *Journal of Clinical Neurophysiology*, **15**, 388–396.

Bauers, K. A. & Hearn, J. P. (1994). Patterns of paternity in relation to male social rank in the stumptailed macaque, *Macaca arctoides*. *Behaviour*, **129**, 149–176.

Baulu, T. (1976). Seasonal sex skin coloration and hormonal fluctuations in free-ranging and captive monkeys. *Hormones and Behavior*, **7**, 481–494.

Beckerman, A., Benton, T. G., Ranta, E., Kaitala, V. & Lundberg, P. (2002). Population dynamic consequences of delayed life-history effects. *Trends in Ecology and Evolution*, **17**, 263–269.

Beckmann, F. (1997). *Vergleichende Untersuchungen zur Mimik bei Bartaffen* (*Macaca silenus L.*). Diplomarbeit, Freie Universität Berlin.

Beerli, P. & Felsenstein, J. (1999). Maximum-likelihood estimation of migration rates and effective population numbers in two populations using a coalescent approach. *Genetics*, **152**, 763–773.

Bélisle, P. & Chapais, B. (2001). Tolerated co-feeding in relation to degree of kinship in Japanese macaques. *Behaviour*, **138**, 487–509.

Bellarosa, A., Bedford, J. A. & Wilson, M. C. (1980). Sociopharmacology of d-amphetamine in *Macaca arctoides*. *Pharmacology, Biochemistry and Behavior*, **13**, 221–228.

Bennett, A. J., Lesch, K. P., Heils, A. *et al.* (2002). Early experience and serotonin transporter gene variation interact to influence primate CNS function. *Molecular Psychiatry*, **7**, 118–122.

Berard, J. D. (1999). A four-year study of the association between male dominance rank, residency status, and reproductive activity in rhesus macaques (*Macaca mulatta*). *Primates*, **40**, 159–175.

Berard J. D., Nürnberg, P., Epplen, J. T. & Schmidtke, J. (1993). Male rank, reproductive behavior, and reproductive success in free-ranging rhesus macaques. *Primates*, **34**, 481–489.

(1994). Alternative reproductive tactics and reproductive success in male rhesus macaques. *Behaviour*, **129**, 177–201.

Bercovitch, F. B. (1986). Male rank and reproductive activity in savanna baboons. *International Journal of Primatology*, **7**, 533–550.

(1987). Reproductive success in male savanna baboons. *Behavioral Ecology and Sociobiology*, **21**, 163–172.

(1988). Coalitions, cooperation and reproductive tactics among adult male baboons. *Animal Behaviour*, **36**, 1198–1209.

(1991). Social stratification, social strategies, and reproductive success in primates. *Ethology and Sociobiology*, **12**, 315–333.

(1992). Estradiol concentrations, fat deposits, and reproductive strategies in male rhesus macaaques. *Hormones and Behavior*, **26**, 272–282.

(1993). Dominance rank and reproductive maturation in male rhesus macaques (*Macaca mulatta*). *Journal of Reproduction and Fertility*, **99**, 113–120.

(1995). Female cooperation, consortship maintenance, and male mating success in savanna baboons. *Animal Behaviour*, **50**, 137–149.

(1997). Reproductive strategies of rhesus macaques. *Primates*, **38**, 247–263.

(1999). The physiology of male reproductive strategies. In *The Nonhuman Primates*, ed. P. Dolhinow & A. Fuentes, pp. 237–244. Mountain View, CA: Mayfield.

(2000). Behavioral ecology and socioendocrinology of reproductive maturation in cercopithecines. In *Old World Monkeys*, ed. P. F. Whitehead & C. J. Jolly, pp. 298–320. Cambridge: Cambridge University Press.

(2001). Reproductive ecology of Old World monkeys. In *Reproductive Ecology and Human Evolution*, ed. P. T. Ellison, pp. 369–396. New York: Aldine de Gruyter.

Bercovitch, F. B. & Berard, J. D. (1993). Life history costs and consequences of rapid reproductive maturation in female rhesus macaques. *Behavioral Ecology and Sociobiology*, **32**, 103–109.

Bercovitch, F. B. & Goy, R. W. (1990). The socioendocrinology of reproductive development and reproductive success in macaques. In *Socioendocrinology of Primate Reproduction*, ed. T. E. Ziegler & F. B. Bercovitch, pp. 59–93. New York: Wiley-Liss.

Bercovitch, F. B. & Nürnberg, P. (1996). Socioendocrine and morphological correlates of paternity in rhesus macaques. *Journal of Reproduction and Fertility*, **107**, 59–68.

(1997). Genetic determination of paternity and variation in male reproductive success in two populations of rhesus macaques. *Electrophoresis*, **18**, 1701–1705.

Bercovitch, F. B. & Rodriguez, J. F. (1993). Testis size, epididymis weight, and sperm competition in rhesus macaques. *American Journal of Primatology*, **30**, 163–168.

Bercovitch, F. B. & Ziegler, T. E. (2002). Current topics in primate socioendocrinology. *Annual Review of Anthropology*, **31**, 45–67.

Bercovitch, F. B., Lebron, M. R., Martinez, H. S. & Kessler, M. J. (1998). Primigravidity, body weight, and costs of rearing first offspring in rhesus macaques. *American Journal of Primatology*, **46**, 135–144.

Bercovitch, F. B., Saldky, K. K., Roy, M. M. & Goy, R. W. (1987). Intersexual aggression and male sexual activity in captive rhesus macaques. *Aggressive Behavior*, **13**, 347–358.

Bercovitch, F. B., Widdig, A. & Nürnberg, P. (2000). Maternal investment in rhesus macaques (*Macaca mulatta*): reproductive costs and consequences of raising sons. *Behavioral Ecology and Sociobiology*, **48**, 1–11.

Bercovitch, F. B., Widdig, A., Trefilov, A. *et al.* (2003). A longitudinal study of age-specific reproductive output and body condition among male rhesus macaques, *Macaca mulatta. Naturwissenschaften*, **90**, 309–312.

Berenstain, L. & Wade, T. D. (1983). Intrasexual selection and male mating strategies in baboons and macaques. *International Journal of Primatology*, **4**, 201–235.

Berkson, G. (1973). Social responses to abnormal infant monkeys. *American Journal of Physical Anthropology*, **38**, 583–586.

Berman, C. M. (1980). Early agonistic experience and rank acquisition among free-ranging infant rhesus monkeys. *International Journal of Primatology*, **1**, 153–170.

  (1982a). The ontogeny of social relationships with group companions among free-ranging infant rhesus monkeys: I. Social networks and differentiation. *Animal Behaviour*, **30**, 149–162.

  (1982b). The ontogeny of social relationships with group companions among free-ranging infant rhesus monkeys: II. Differentiation and attractiveness. *Animal Behaviour*, **30**, 163–170.

  (1984). Variation in mother-infant relationships: traditional and nontraditional factors. In *Female Primates*, ed. M. F. Small, pp. 17–36. New York: Alan R. Liss.

  (1990). Intergenerational transmission of maternal rejection rates among free-ranging rhesus monkeys. *Animal Behaviour*, **39**, 329–337.

  (1990a). Consistency in maternal behavior within families of free-ranging rhesus monkeys. *American Journal of Primatology*, **22**, 159–169.

  (1990b). Intergenerational transmission of maternal rejection rates among free-ranging rhesus monkeys. *Animal Behaviour*, **39**, 329–337.

  (1992). Immature siblings and mother-infant relationships among free-ranging rhesus monkeys on Cayo Santiago. *Animal Behaviour*, **44**, 247–258.

  (1996). An animal model for the intergenerational transmission of maternal style? *Family Systems*, **3**, 125–140.

  (2004). Developmental aspects of kin bias in behavior. In *Kinship and Behavior in Primates*, ed. B. Chapais & C. M. Berman, pp. 317–346. Oxford: Oxford University Press.

Berman, C. M. & Kapsalis, E. (1999). Development of kin bias among rhesus monkeys: maternal transmission or individual learning? *Animal Behaviour*, **58**, 883–894.

Berman, C. M., Rasmussen, K. L. R. & Suomi, S. J. (1993). Reproductive consequences of maternal care patterns during estrus among free-ranging rhesus monkeys. *Behavioral Ecology and Sociobiology*, **32**, 391–399.

  (1997). Group size, infant development and social networks in free-ranging rhesus monkeys. *Animal Behaviour*, **53**, 405–421.

Bernstein, I. S. (1969). Stability of the status hierarchy in a pigtail monkey group (*Macaca nemestrina*). *Animal Behaviour*, **17**, 452–458.

Bernstein, I. S. & Cooper, M. A. (1999). Dominance in Assamese macaques (*Macaca assamensis*). *American Journal of Primatology*, **48**, 283–289.

Bernstein, I. S. & Gordon, T. P. (1980a). Mixed taxa introductions, hybrids and macaque systematics. In *The Macaques: Studies in Ecology, Behavior and Evolution*, ed. D. G. Lindburg, pp. 125–147. New York: van Nostrand Reinhold.

(1980b). The social component of dominance relationships in rhesus monkeys (*Macaca mulatta*). *Animal Behaviour*, **28**, 1033–1039.

Bernstein, I. S., Gordon, T. P. & Rose, R. M. (1983a). The interaction of hormones, behavior, and social context in nonhuman primates. In *Hormones and Aggressive Behavior*, ed. B. B. Svaro, pp. 535–561. New York: Plenum Press.

Bernstein, I. S., Williams, L. & Ramsay, M. (1983b). The expression of aggression in Old World monkeys. *International Journal of Primatology*, **4**, 113–125.

Berntson, G. G., Boysen, S. T., Bauer, H. R. & Torello, M. S. (1989). Conspecific screams and laughter: cardiac and behavioral reactions of infant chimpanzees. *Developmental Psychobiology*, **22**, 771–787.

Berntson, G. G., Sarter, M. & Cacioppo, J. T. (1998). Anxiety and cardiovascular reactivity: the basal forebrain cholinergic link. *Behavioural Brain Research*, **94**, 225–248.

Bertrand, M. (1969). *The Behavioral Repertoire of the Stumptail Macaque*. Karger: Basel.

Biederman, J., Hirshfeld-Becker, D. R., Rosenbaum, J. F. *et al.* (2001). Further evidence of association between behavioral inhibition and social anxiety in children. *American Journal of Psychiatry*, **158**, 1673–1679.

Bielert, C. F., Vandenbergh, J. G., Scheffler, G., Robinson, J. A. & Goy, R. W. (1976). Mating in the rhesus monkey (*Macaca mulatta*) after conception and its relationship to oestradiol and progesterone levels throughout pregnancy. *Journal of Reproduction and Fertility*, **46**, 179–187.

Bierstedt, R. (1950). An analysis of social power. *American Sociological Review*, **15**, 161–184.

Bixler, R. H. (1981). Primate "mother–son" incest. *Psychological Reports*, **48**, 531–536.

Blakley, G. B., Beamer, T. W. & Dukelow, W. R. (1981). Characteristics of the menstrual cycle in nonhuman primates IV. Timed mating in *Macaca nemestrina*. *Laboratory Animals*, **15**, 351–353.

Blurton-Jones, N. G. & Trollope, J. (1968). Social behavior of stumptailed macaques in captivity. *Primates*, **9**, 365–394.

Boccia, M. L., Reite, M. & Laudenslager, M. (1989). On the physiology of grooming in a pigtail macaque. *Physiology and Behavior*, **45**, 667–670.

Boehm, C. (1993). Egalitarianism behavior and reverse dominance hierarchy. *Current Anthropology*, **34**, 227–254.

(1999). *Hierarchy in the Forest*. Cambridge, MA: Harvard University Press.

Boinski, S. (1987). Mating patterns in squirrel monkeys (*Saimiri oerstedi*): implications for seasonal sexual dimorphism. *Behavior Ecology and Sociobiology*, **21**, 13–21.

Boinski, S., Treves, A. & Chapman, C. A. (2000). A critical evaluation of the influence of predators on primates: effects on group travel. In *On the Move: How and Why Animals Travel in Groups*, ed. S. Boinski & P. A. Garber, pp. 43–72. Chicago: University of Chicago Press.

Bolig, R., Price, C. S., O'Neill, P. L. & Suomi, S. J. (1992). Subjective assessment of reactivity level and personality traits of rhesus monkeys. *International Journal of Primatology*, **13**, 287–306.

Bonabeau, E., Theraulaz, G. & Deneubourg, J. L. (1996). Mathematical models of self-organizing hierarchies in animal societies. *Bulletin of Mathematical Biology*, **58**, 661–717.

Bongaarts, J. (1980). Does malnutrition affect fecundity? A summary of evidence. *Science*, **208**, 564–569.

Borries, C. (1993). Ecology of female social relationships: Hanuman langurs and the van Schaik model. *Folia Primatologica*, **61**, 21–30.

Botchin, M. B., Kaplan, J. R., Manuck, S. B. & Mann, J. J. (1993). Low versus high prolactin responders to fenfluramine challenge: marker of behavioral difference in adult male cynomolgus macaques. *Neuropsychopharmacology*, **9**, 93–99.

Bourne, G. H., ed. (1975). *The Rhesus Monkey*, vols. 1 & 2. New York: Academic Press.

Bowman, J. E. & Lee, P. C. (1995). Growth and threshold weaning weights among captive rhesus macaques. *American Journal of Physical Anthropology*, **96**, 159–175.

Boyd, R. & Richerson, P. J. (1985). *Culture and Evolutionary Process*. Chicago: University of Chicago Press.

 (1992). Punishment allows the evolution of cooperation (or anything else) in sizable groups. *Ethology and Sociobiology*, **13**, 171–195.

Boyd, R. & Silk, J. (1983). A method for assigning cardinal dominance ranks. *Animal Behaviour*, **31**, 45–58.

Braitenberg, V. (1984). *Vehicles: Experiments in Synthetic Psychology*. Cambridge, MA: MIT Press.

Brandon-Jones, D. (1996). The Asian Colobinae (*Mammalia: Cercopithecidae*) as indicators of Quaternary climatic changes. *Biological Journal of the Linnean Society*, **59**, 327–350.

 (1998). Pre-glacial Bornean primate impoverishment and Wallace's line. In *Biogeography and Geological Evolution of SE Asia*, ed. R. Hall & J. D. Holloway, pp. 393–404. Leiden: Backhuys.

Breuggeman, J. A. (1973). Parental care in a group of free-ranging rhesus monkeys (*Macaca mulatta*). *Folia Primatologica*, **20**, 178–210.

Bronson, F. H. (1989). *Mammalian Reproductive Biology*. Chicago: University of Chicago Press.

Bronson, F. H. & Manning, J. M. (1991). The energetic regulation of ovulation: a realistic role for body fat. *Biology of Reproduction*, **44**, 945–950.

Brooks, D. R. & McLennan, D. A. (1991). *Phylogeny, Ecology, and Behavior*. Chicago: University of Chicago Press.

Brothers, L. (1990). The neural basis of primate social communication. *Motivation and Emotion*, **14**, 81–91.

Brown, J. L. (1997). A theory of mate choice based on heterozygosity. *Behavioral Ecology*, **8**, 60–65.

Bruce, K. E. & Estep, D. Q. (1992). Interruption of and harassment during copulation by stumptail macaques, *Macaca arctoides*. *Animal Behaviour*, **44**, 1029–1044.

Bruce, K. E., Estep, D. Q. & Baker, S. C. (1988). Social interactions following parturition in stumptail macaques. *American Journal of Primatology*, **15**, 247–261.

Brüggemann, S. & Dukelow, W. R. (1980). Characteristics of the menstrual cycle in nonhuman primates: III. Time mating in *Macaca arctoides*. *Journal of Medical Primatology*, **9**, 213–321.

Budnitz, N. & Dainis, K. (1975). *Lemur catta*: ecology and behavior. In *Lemur Biology*, ed. I. Tattersall & R. W. Sussman, pp. 219–235. New York: Plenum Press.

Burton, F. D. & Sawchuk, L. A. (1982). Birth intervals in *M. sylvanus* of Gibraltar. *Primates*, **23**, 140–144.

Bush, G. L. (1981). Stasipatric speciation and rapid evolution in animals. In *Evolution and Speciation: Essays in Honor of M. J. D. White*, ed. W. R. Atchley & D. S. Woodruff, pp. 201–218. Cambridge: Cambridge University Press.

Bush, G. L., Case, S. M., Wilson, A. C. & Patton, J. L. (1977). Rapid speciation and chromosomal evolution in mammals. *Proceedings of the National Academy of Sciences of the USA*, **74**, 3942–3946.

Buss, D., Haselton, M. G., Shackelford, T. K., Bleske, A. L. & Wakefield, J. C. (1998). Adaptations, exaptations, and spandrels. *American Psychologist*, **53**, 533–548.

Busse, C. D. (1977). Chimpanzee predation as a possible factor in the evolution of red colobus monkey social organization. *Evolution*, **31**, 907–911.

(1980). Leopard and lion predation upon chacma baboons living in the Moremi Wildlife Reserve. *Botswana Notes and Records*, **12**, 15–21.

Butynski, T. M., Werikhe, S. E. & Kalina, J. (1990). Status, distribution and conservation of the mountain gorilla in the Gorilla Game Reserve, Uganda. *Primate Conservation*, **11**, 31–41.

Butovskaya, M. (1993). Kinship and different dominance styles in groups of three species of the genus *Macaca* (*M. arctoides, M. mulatta, M. fascicularis*). *Folia Primatologica*, **60**, 210–224.

(1994). Towards a social psychology of personal behavior: attachments in four species of the genus *Macaca*. In *Current Primatology*, vol. 2, ed. J. J. Roeder, B. Thierry, J. R. Anderson & N. Herrenschmidt, pp. 119–126. Strasbourg: Université Louis Pasteur.

Butovskaya, M. L. & Kozintsev, A. G. (1996a). Group history and social style: the case of crab-eating monkeys. *Anthropologie*, **34**, 165–176.

(1996b). Gender-related factors affecting primate social behavior: grooming, rank, age and kinship in heterosexual and all-male groups of stumptail macaques. *American Journal of Physical Anthropology*, **101**, 39–54.

Butovskaya, M., Kozintsev, A. & Welker, C. (1996). Conflict and reconciliation in two groups of crab-eating monkeys differing in social status by birth. *Primates*, **37**, 261–270.

Butovskaya, M. L., Korotayev, A. & Kazankov, A. (2000). Variabilité des relations sociales chez les primates humains et non humains: à la recherche d'un paradigme général. *Primatologie*, **3**, 319–363.

Bygott, J. D. (1979). Agonistic behavior, dominance, and social structure in wild chimpanzees of the Gombe National Park. In *The Great Apes*, ed. D. A. Hamburg & E. R. McCown, pp. 405–427. Menlo Park, CA: Benjamin Cummings.

Bynum, E. V., Kohlhaas, A. K. & Pramono, A. H. (1999). Conservation status of Sulawesi macaques. *Tropical Biodiversity*, **6**, 123–144.

Byrne, R. W. (1997). Machiavellian intelligence. *Evolutionary Anthropology*, **5**, 172–180.

Byrne, R. & Whiten, A. (1988). *Machiavellian Intelligence: Social Expertise and the Evolution of Intellect in Monkeys, Apes, and Humans*. New York: Oxford University Press.

Byrne, R. W., Whiten, A. & Henzi, S. P. (1987). One-male groups and intergroup interactions of mountain baboons. *International Journal of Primatology*, **8**, 615–633.

Byrne, R. W., Whiten, A., Henzi, S. P. & McCullough, F. M. (1993). Nutritional
constraints on mountain baboons (*Papio ursinus*): implications for baboon socioecology. *Behavioral Ecology and Sociobiology*, **33**, 233–246.

Cacioppo, J. T. & Gardner, W. L. (1999). Emotion. *Annual Review of Psychology*, **50**,
191–214.

Cacioppo, J. T. & Tassinary, L. G. (1990). *Principles of Psychophysiology: Physical,
Social, and Inferential Elements*. Cambridge: Cambridge University Press.

Cacioppo, J. T., Klein, D. J., Berntson, G. G. & Hartfield, E. (1993). The psychophysiology of emotion. In *The Handbook of Emotions*, ed. M. Lewis & J. M. Haviland,
pp. 119–142. New York: Guilford Press.

Caine, N. G. & Mitchell, G. D. (1980). Species differences in the interest shown in
infants by juvenile female macaques (*Macaca mulatta* and *M. radiata*). *International Journal of Primatology*, **1**, 323–332.

Caine, N. G., Earle, H. & Reite, M. L. (1983). Personality traits of adolescent pig-tailed
monkeys (*Macaca nemestrina*): an analysis of social rank and early separation
experience. *American Journal of Primatology*, **4**, 253–260.

Caldecott, J. O. (1986a). Mating patterns, societies and the ecogeography of macaques.
Animal Behaviour, **34**, 208–220.

  (1986b). *An Ecological and Behavioral Study of the Pig-tailed Macaque*. Basel:
Karger.

Call, J. (2001). Chimpanzee social cognition. *Trends in Cognitive Sciences*, **5**, 388–393.

Call, J., Aureli, F. & de Waal, F. B. M. (2002). Post-conflict third-party affiliation in
stumptail macaques. *Animal Behaviour*, **63**, 209–216.

Camazine, S., Deneubourg, J. L., Franks, N. R., Sneyd, J., Theraulaz, G. & Bonabeau,
E. (2001). *Self-organization in Biological Systems*. Princeton, NJ: Princeton University Press.

Camperio Ciani, A. (1986). Intertroop agonistic behavior of a feral rhesus macaque troop
ranging in town and forest areas in India. *Aggressive Behavior*, **12**, 433–439.

Canli, T., Desmond, J. E., Zhao, Z., Glover, G. & Gabrieli, J. D. E. (1998). Hemispheric
asymmetry for emotional stimuli detected with fMRI. *NeuroReport*, **9**, 3233–3239.

Cant, J. G. (1980). What limits primates? *Primates*, **21**, 538–544.

Capitanio, J. P. (1985). Early experience and social processes in rhesus macaques
(*Macaca mulatta*): II. Complex social interaction. *Journal of Comparative Psychology*, **99**, 133–144.

  (1986). Behavioral pathology. In *Comparative Primate Biology*, vol. 2A, G. Mitchell
& J. Erwin, pp. 411–454. New York: Alan R. Liss.

  (1999). Personality dimensions in adult male rhesus macaques: prediction of behaviors
across time and situation. *American Journal of Primatology*, **47**, 299–320.

Capitanio, J. P., Mendoza, S. P. & Baroncelli, S. (1999). The relationship of personality
dimensions in adult male rhesus macaques to progression of simian immunodeficiency virus disease. *Brain, Behavior and Immunity*, **13**, 138–154.

Capitanio, J. P., Mendoza, S. P., Lerche, N. W. & Mason, W. A. (1998). Social stress
results in altered glucocorticoid regulation and shorter survival in simian acquired
immune deficiency syndrome. *Proceedings of the National Academy of Sciences
of the USA*, **95**, 4714–4719.

Carpenter, C. R. (1942). Sexual behavior of free-ranging rhesus monkeys (*Macaca mulatta*): II. Periodicity of estrus, homosexual, autoerotic and non-conformist behavior. *Journal of Comparative Psychology*, **33**, 143–162.

Cass, V. C. (1983/84). Homosexual identity: a concept in need of defense. *Journal of Homosexuality*, **9**, 105–126.

Castles, D. L., Aureli, F. & de Waal, F. B. M. (1996). Variation in conciliation tendency and relationship quality across groups of pigtail macaques. *Animal Behaviour*, **52**, 389–403.

Castles, D. L., Whiten, A. & Aureli F. (1999). Social anxiety, relationships and self-directed behaviour among wild female olive baboons. *Animal Behaviour*, **58**, 1207–1215.

Cattell, R. B. (1966). *The Scientific Analysis of Personality*. Chicago: Aldine.

Caughley, G. (1966). Mortality patterns in mammals. *Ecology*, **47**, 906–918.

(1977). *Analysis of Vertebrate Populations*. New York: John Wiley.

Cavalli-Sforza, L. L., Feldman, M. W., Chen, K. H. & Dornbusch, S. M. (1982). Theory and observation in cultural transmission. *Science*, **218**, 19–27.

Chaffin, C. L., Friedlen, K. & de Waal, F. B. M. (1995). Dominance style of Japanese macaques compared with rhesus and stumptail macaques. *American Journal of Primatology*, **35**, 103–116.

Chambers, K. C., Hess, D. L. & Phoenix, C. H. (1981). Relationship of free and bound testosterone to sexual behavior in old rhesus males. *Physiology and Behavior*, **27**, 615–620.

Chamove, A. S., Eysenck, H. J. & Harlow, H. F. (1972). Personality in monkeys: factor analyses of rhesus social behaviour. *Quarterly Journal of Experimental Psychology*, **24**, 496–504.

Champoux, M., Bennett, A. J., Lesch, K. P. *et al.* (1999). Serotonin transporter gene polymorphism and neurobehavioral development in rhesus monkey neonates. *Society for Neuroscience Abstracts*, **25**, 69.

Champoux, M., Bennett, A., Shannon, C., Higley, J. D., Lesch, K. P. & Suomi, S. J. (2002). Serotonin transporter gene polymorphism, differential early rearing, and behavior in rhesus monkey neonates. *Molecular Psychiatry*, **7**, 1058–1063.

Chapais, B. (1983). Dominance, relatedness and the structure of female relationships in rhesus monkeys. In *Primate Social Relationships: an Integrated Approach*, ed. R. A. Hinde, pp. 208–217. Oxford: Blackwell.

(1988a). Experimental matrilineal inheritance of rank in female Japanese macaques. *Animal Behaviour*, **36**, 1025–1037.

(1988b). Rank maintenance in female Japanese macaques: experimental evidence for social dependency. *Behaviour*, **104**, 41–59.

(1991). Primates and the origins of aggression, power, and politics among humans. In *Understanding Behavior: What Primates Studies Tell Us about Human Behavior*, ed. J. Loy & B. Peters, pp. 190–228. Oxford: Oxford University Press.

(1992a). The role of alliances in the social inheritance of rank among female primates. In *Coalitions and Alliances in Humans and Other Animals*, ed. A. H. Harcourt & F. B. M. de Waal, pp. 29–60. Oxford: Oxford University Press.

(1992b). The restoration of stability following dominance upheavals in Japanese macaques. *Paper presented at the 14th Congress of the International Primatological Society, Strasbourg.*

(1995). Alliances as a means of competition in primates: evolutionary, developmental and cognitive aspects. *Yearbook of Physical Anthropology*, **38**, 115–136.

(2001). Primate nepotism: what is the explanatory value of kin selection. *International Journal of Primatology*, **22**, 203–229.

Chapais, B. & Gates-St.-Pierre, C. (1997). Kinship bonds are not necessary for maintaining matrilineal rank in captive Japanese macaques. *International Journal of Primatology*, **18**, 375–385.

Chapais, B. & Gauthier, C. (1993). Early agonistic experience and the onset of matrilineal rank acquisition in Japanese macaques. In *Juvenile Primates: Life History, Development, and Behavior*, ed. M. E. Pereira & L. A. Fairbanks, pp. 245–258. New York: Oxford University Press.

Chapais, B. & Mignault, C. (1991). Homosexual incest avoidance among females in captive Japanese macaques. *American Journal of Primatology*, **23**, 171–183.

Chapais, B. & Schulman, S. (1980). An evolutionary model of female dominance relationships in primates. *Journal of Theoretical Biology*, **82**, 47–89.

Chapais, B., Gauthier, C., Prud'homme, J. & Vasey, P. (1997). Relatedness threshold for nepotism in Japanese macaques. *Animal Behaviour*, **53**, 1089–1101.

Chapais, B., Girard, M. & Primi, G. (1991). Non-kin alliances, and the stability of matrilineal dominance relations in Japanese macaques. *Animal Behaviour*, **41**, 481–491.

Chapais, B., Prud'homme, J. & Teijeiro, S. (1994). Dominance competition among siblings in Japanese macaques: constraints on nepotism. *Animal Behaviour*, **48**, 335–347.

Chapais, B., Savard, L. & Gauthier, C. (2001). Kin selection and the distribution of altruism in relation to degree of kinship in Japanese macaques (*Macaca fuscata*). *Behavioral Ecology and Sociobiology*, **49**, 493–502.

Chapman, C. A. & Peres, C. A. (2001). Primate conservation in the new millennium: the role of scientists. *Evolutionary Anthropology*, **10**, 16–33.

Chase, I. D., Bartelomeo, C. & Dugatkin, L. A. (1994). Aggressive interactions and inter-contest interval: how long do winners keep winning? *Animal Behaviour*, **48**, 393–400.

Cheney, D. L. (1977). The acquisition of rank and the development of reciprocal alliances among free-ranging immature baboons. *Behavioral Ecology and Sociobiology*, **2**, 303–318.

(1981). Intergroup encounters among free-ranging vervet monkeys. *Folia Primatologica*, **35**, 124–146.

(1983). Extra-familial alliances among vervet monkeys. In *Primate Social Relationships: an Integrated Approach*, ed. R. A. Hinde, pp. 278–286. Oxford: Blackwell.

(1987). Interactions and relationships between groups. In *Primate Societies*, ed. B. B. Smuts, D. L. Cheney, R. M. Seyfarth, R. W. Wrangham & T. T. Struhsaker, pp. 267–281. Chicago: University of Chicago Press.

Cheney, D. L. & Seyfarth, R. M. (1989). *How Monkeys See the World*. Chicago: University of Chicago Press.

(1990). Attending to behaviour versus attending to knowledge: examining monkeys' attribution of mental states. *Animal Behaviour*, **40**, 742–753.

Cheney, D. L. & Wrangham, R. W. (1987). Predation. In *Primate Societies*, ed. B. B. Smuts, D. L. Cheney, R. M. Seyfarth, R. W. Wrangham & T. T. Struhsaker, pp. 227–239. Chicago: University of Chicago Press.

Cheney, D. L., Lee, P. C. & Seyfarth, R. M. (1981). Behavioral correlates of non-random mortality among free-ranging female vervet monkeys. *Behavioral Ecology and Sociobiology*, **9**, 153–161.

Cheney, D. L., Seyfarth, R. M., Andelman, S. J. & Lee, P. C. (1988). Reproductive success in vervet monkeys. In *Reproductive Success*, ed. T. H. Clutton-Brock, pp. 384–402. Chicago: University of Chicago Press.

Chepko-Sade, B. D. & Olivier, T. J. (1979). Coefficient of genetic relationship and the probability of intragenealogical fission in *Macaca mulatta*. *Behavioral Ecology and Sociobiology*, **5**, 263–278.

Chepko-Sade, B. D. & Sade, D. S. (1979). Patterns of group splitting within matrilineal kinship group: study of social group structure in *Macaca mulatta* (*Cercopithecidae*, Primates). *Behavioral Ecology and Sociobiology*, **5**, 67–86.

Chesser, R. K. (1991a). Gene diversity and female philopatry. *Genetics*, **127**, 437–447.

(1991b). Influence of gene flow and breeding tactics on gene diversity within populations. *Genetics*, **129**, 573–583.

Chevalier-Skolnikoff, S. (1974). *The Ontogeny of Communication in the Stumptail Macaque (*Macaca arctoides*)*. Basel: Karger.

Cheverud, J. M. & Dow, M. M. (1985). An autocorrelation analysis of genetic variation due to lineal fission in social groups of rhesus macaques. *American Journal of Physical Anthropology*, **67**, 113–121.

Cheverud, J. M., Buettner-Janusch, J. & Sade, D. S. (1978). Social group fission and the origin of intergroup genetic differentiation among the rhesus monkeys of Cayo Santiago. *American Journal of Physical Anthropology*, **49**, 449–456.

Cheverud, J. M., Wilson, P. & Dittus, W. P. J. (1992). Primate population studies at Polonnaruwa. III. Somatometric growth in a natural population of toque macaques (*Macaca sinica*). *Journal of Human Evolution*, **23**, 51–77.

Chikazawa, D., Gordon, T. P., Bean, C. A. & Bernstein, I. S. (1979). Mother–daughter dominance reversals in rhesus monkeys (*Macaca mulatta*). *Primates*, **20**, 301–305.

Chism, J. (1986). Development and mother–infant relations among captive patas monkeys. *International Journal of Primatology*, **7**, 49–82.

Cilia, P. & Piper, D. C. (1997). Marmoset conspecific confrontation: an ethologically-based model of anxiety. *Pharmacology, Biochemistry and Behavior*, **58**, 85–91.

Clark, A. B. (1978). Sex ratio and local resource competition in a prosimian primate. *Science*, **201**, 163–165.

Clarke, A. S. & Boinski, S. (1995). Temperament in nonhuman primates. *American Journal of Primatology*, **37**, 103–125.

Clarke, A. S. & Mason, W. A. (1988). Differences among three macaque species in responsiveness to an observer. *International Journal of Primatology*, **9**, 347–364.

Clarke, A. S. & Schneider, M. L. (1993). Prenatal stress has long-term effects on behavioral responses to stress in juvenile rhesus monkeys. *Developmental Psychobiology*, **26**, 293–304.

Clarke, A. S., Czekala, N. M. & Lindburg, D. G. (1995). Behavioral and adrenocortical responses of male cynomolgus and lion-tailed macaques to social stimulation and group formation. *Primates*, **36**, 41–56.

Clarke, A. S., Harvey, N. C. & Lindburg, D. G. (1993). Extended postpregnancy estrous cycles in female lion-tailed macaques. *American Journal of Primatology*, **31**, 275–285.

Clarke, A. S., Mason, W. A. & Mendoza, S. P. (1994a). Heart rate patterns under stress in three species of macaques. *American Journal of Primatology*, **33**, 133–148.

Clarke, A. S., Mason, W. A. & Moberg, G. P. (1988). Differential behavioral and adrenocortical responses to stress among three macaque species. *American Journal of Primatology*, **14**, 37–52.

Clarke, A. S., Wittwer, D. J., Abbott, D. H. & Schneider, M. L. (1994b). Long-term effects of prenatal stress on HPA axis activity in juvenile rhesus monkeys. *Developmental Psychobiology*, **27**, 257–269.

Cloninger, C. R. (1986). A unified biosocial theory of personality and its role in the development of anxiety states. *Psychiatric Developments*, **3**, 167–226.

Clutton-Brock, T. H. & Harvey, P. H. (1976). Evolutionary rules and primate societies. In *Growing Points in Ethology*, ed. P. P. G. Bateson & R. A. Hinde, pp. 195–237. Cambridge: Cambridge University Press.

Clutton-Brock, T. H. & Iason, G. R. (1986). Sex ratio variation in mammals. *Quarterly Review of Biology*, **61**, 339–374.

Clutton-Brock, T. H., Harvey, P. & Rudder, B. (1977). Sexual dimorphism, socionomic sex ratio and body weights in primates. *Nature*, **269**, 797–800.

Coelho, A. M., Bramblett, C. A. & Quick, L. B. (1976). Resource availability and population density in primates: socio-bioenergetic analysis of energy budgets of Guatamalan howler and spider monkeys. *Primates*, **17**, 63–80.

Coelho, A. M., Coelho, L. S., Bramblett, C. A. & Quick, L. B. (1977). Ecology, population characteristics, and sympatric association in primates: a socio-bioenergetic analysis of howler and spider monkeys in Tikal, Guatemala. *Yearbook of Physical Anthropology*, **20**, 96–135.

Collias, N. & Southwick, C. H. (1952). A field study of population density and social organization in howling monkeys. *Proceedings of the American Philosophers Society*, **96**, 143–156.

Colishaw, G. & Dunbar, R. I. M. (1991). Dominance rank and mating success in male primates. *Animal Behaviour*, **41**, 1045–1056.

Colvin, J. D. (1986). Proximate causes of male emigration at puberty in rhesus monkeys. In *The Cayo Santiago Macaques*, ed. R. G. Rawlins & M. J. Kessler, pp. 131–157. Albany, NY: State University of New York Press.

Combes, S. L. & Altmann, J. (2001). Status change during adulthood: life-history by-product or kin selection based on reproductive value? *Proceedings of the Royal Society of London B*, **268**, 1367–1373.

Constable, J. L., Ashley, M. V., Goodall, J. & Pusey, A. E. (2001). Noninvasive paternity assignment in Gombe chimpanzees. *Molecular Ecology*, **10**, 1279–1300.

Cook, M. & Mineka, S. (1990). Selective associations in the observational conditioning of fear in rhesus monkeys. *Journal of Experimental Psychology*, **16**, 372–389.

Cooper, M. A. & Bernstein, I. S. (2000). Social grooming in Assamese macaques (*Macaca assamensis*). *American Journal of Primatology*, **50**, 77–85.

(2002). Counter aggression and reconciliation is Assamese macaques (*Macaca assamensis*). *American Journal of Primatology*, **56**, 215–230.

Cooper, M. A., Aureli, F. & Singh, M. (2004). Between-group encounters in bonnet macaques (*Macaca radiata*). *Behavioral Ecology and Sociobiology*. (In press.)

Coplan, J. D., Smith, E. L. P., Altemus, M., Scharf, B. A., Owens, M. J., Nemeroff, C. B., Gorman, J. M. & Rosenblum, L. A. (2001). Variable foraging demand rearing: sustained elevations in cisternal cerebrospinal fluid corticotropin-releasing factor concentrations in adult primates. *Biological Psychiatry*, **50**, 200–204.

Cords, M. (1992). Post-conflict reunions and reconciliation in long-tailed macaques. *Animal Behaviour*, **44**, 57–61.

Corradino, C. (1990). Proximity structure in a captive colony of Japanese monkeys (*Macaca fuscata*): an application of multidimensional scaling. *Primates*, **31**, 351–362.

Coss, R. G. & Ramakrishnan, U. (2000). Perceptual aspects of leopard recognition by wild bonnet macaques (*Macaca radiata*). *Behaviour*, **137**, 315–335.

Costa, P. T. & McCrae, R. R. (1995). Domains and facets: hierarchical personality assessment using the revised NEO Personality Inventory. *Journal of Personality Assessment*, **64**, 21–50.

Coussi-Korbel, S. & Fragaszy, D. M. (1995). On the relation between social dynamics and social learning. *Animal Behaviour*, **50**, 1441–1453.

Cowlishaw, G. (1994). Vulnerability to predation in baboon populations. *Behaviour*, **131**, 293–304.

(1995). Behavioral patterns in baboon group encounters: the role of resource competition and male reproductive strategies. *Behaviour*, **132**, 75–86.

Cowlishaw, G. & Dunbar, R. I. M. (1991). Dominance rank and mating success in male primates. *Animal Behaviour*, **41**, 1045–1056.

Crawford, M. P. (1938). A behavior rating scale for young chimpanzees. *Journal of Comparative Psychology*, **26**, 79–91.

Crockett, C. M., Kyes, R. C. & Sajuthi, D. (1996). Modeling managed monkey populations: sustainable harvest of longtailed macaques on a natural habitat island. *American Journal of Primatology*, **40**, 343–360.

Crook, J. H. (1989). Introduction: socioecological paradigms, evolution and history: perspectives for the 1990s. In *Comparative Socioecology: The Behavioural Ecology of Humans and Other Mammals*, ed. V. Standen & R. Foley, pp. 1–36. Oxford: Blackwell.

Crow, J. F. & Kimura, M. (1970). *An Introduction to Population Genetic Theory*. New York: Harper & Row.

Curie-Cohen, M., Yoshihara, D., Luttrell, L., Benforado, K., MacCluer, J. W. & Stone, W. H. (1983). The effects of dominance on mating behavior and paternity in a captive troop of rhesus monkeys (*Macaca mulatta*). *American Journal of Primatology*, **5**, 127–138.

Custance, D., Whiten, A. & Fredman, T. (1999). Social learning of an artificial fruit task in capuchin monkeys (*Cebus apella*). *Journal of Comparative Psychology*, **113**, 13–23.

Czaja, J. A., Robinson, J. A., Eisele, S. G., Scheffler, G. & Goy, R. W. (1977). Relation-ship between sexual skin colors of female rhesus monkeys and mid-cycle plasma levels of estradiol and progesterone. *Journal of Reproduction and Fertility*, **49**, 147–150.

Damasio, A. R. (1994). *Descartes' Error: Emotion, Reason, and The Human Brain.* New York: Grosset/Putnam.

(2000). *The Feeling of What Happens: Body, Emotion and the Making of Conscious-ness.* London: Vintage.

Dang, D. (1983). Female puberty in monkey *Macaca fascicularis* raised in laboratory: menarch–copulation–gestation–fertility. *Cahiers d'Anthropologie et Biométrie Humaine*, **1**, 33–45.

Dang, D. C. & Meussy-Desolle, N. (1984). Annual plasma testosterone cycle and ejac-ulatory ability in the laboratory-housed crab eating macaque (*Macaca fasicularis*). *Reproduction, Nutrition and Development*, **21**, 59–68.

Darwin, C. (1871). *The Descent of Man and Selection in Relation to Sex.* London: John Murray.

(1872). *The Expression of Emotions in Man and Animals.* London: John Murray.

(1999 [1859]). *The Origin of Species.* New York: Bantam Books.

Das, M. (2000). Conflict management via third parties: post-conflict affiliation of the aggressor. In *Natural Conflict Resolution*, ed. F. Aureli & F. B. M. de Waal, pp. 263–280. Berkeley, CA: University of California Press.

Das, M., Penke, Z. & van Hooff, J. A. R. A. M. (1997). Affiliation between aggressors and third parties following conflicts in long-tailed macaques (*Macaca fascicularis*). *International Journal of Primatology*, **18**, 159–181.

(1998). Post-conflict affiliation and stress-related behavior of long-tailed macaque aggressors. *International Journal of Primatology*, **19**, 53–71.

Dasser, V. (1988a). A social concept in Java monkeys. *Animal Behaviour*, **36**, 225–230.

(1988b). Mapping social concepts in monkeys. In *Machiavellian Intelligence. Social Expertise and the Evolution of Intellect in Monkeys, Apes, and Humans*, ed. R. W. Byrne & A. Whiten, pp. 85–93. New York: Oxford University Press.

Datta, S. B. (1983a). Relative power and the acquisition of rank. In *Primate Social Relationships: An Integrated Approach*, ed. R. A. Hinde, pp. 93–103. Oxford: Blackwell.

(1983b). Relative power and the maintenance or rank. In *Primate Social Relationships: An Integrated Approach*, ed. R. A. Hinde, pp. 103–112. Oxford: Blackwell.

(1988). The acquisition of dominance among free-ranging rhesus monkey siblings. *Animal Behaviour*, **36**, 754–772.

(1992). Effects of the availabilty of allies on female dominance structure. In *Coalitions and Alliances in Humans and Other Animals*, ed. A. H. Harcourt & F. B. M. de Waal, pp. 61–82. New York: Oxford University Press.

Datta, S. B. & Beauchamp, G. (1991). Effects of group demography on dominance rela-tionships among female primates. I. Mother–daughter and sister–sister relations. *American Naturalist*, **138**, 201–226.

Davidson, R. J. (1995). Cerebral asymmetry, emotion, and affective style. In *Brain Asymmetry*, ed. R. J. Davidson & K. Hugdahl, pp. 361–387. Cambridge, MA: MIT Press.

(2000). The neuroscience of affective style. In *The New Cognitive Neurosciences*, 2nd edn, ed. M. S. Gazzaniga, pp. 1149–1166. Cambridge, MA: MIT Press.

Davidson, R. J. & Sutton, S. K. (1995). Affective neuroscience: the emergence of a discipline. *Current Opinion in Neurobiology*, **5**, 217–224.

Davis, M. & Whalen, P. J. (2001). The amygdala: vigilance and emotion. *Molecular Psychiatry*, **6**, 13–34.

Dawkins, R. & Krebs, J. R. (1978). Animal signals: information or manipulation? In *Behavioural Ecology: An Evolutionary Approach*, ed. J. R. Krebs & N. B. Davies, pp. 282–309. Oxford: Blackwell.

Dazey, J. & Erwin, J. (1976). Infant mortality in *Macaca nemestrina*: neonatal and post-natal mortality at the Regional Primate Research Center Field Station, University of Washington, 1967–1974. *Theriogenology*, **5**, 267–279.

Deag, J. M. (1973). Intergroup encounters in the wild Barbary macaque (*Macaca sylvanus* L.). In *Comparative Ecology of Behavior of Primates*, ed. R. P. Michael & J. H. Crook, pp. 315–373. London: Academic Press.

(1974). *A Study of the Social Behavior and Ecology of the Wild Barbary Macaque, Macaca sylvanus.* Ph.D. dissertation, University of Bristol.

(1980). Interactions between males and unweaned Barbary macaques: testing the agonistic buffering hypothesis. *Behaviour*, **75**, 54–81.

Deag, J. M. & Crook, J. H. (1971). Social behaviour and "agonistic buffering" in the wild Barbary macaque *Macaca sylvana* L. *Folia Primatologica*, **15**, 183–200.

de Benedictis, T. (1973). The behavior of young primates during adult copulations: observations of a *Macaca irus* colony. *American Anthropologist*, **75**, 1469–1484.

Decker, B. S. (1994). Effects of habitat disturbance on the behavioral ecology and demo-graphics of the Tana River red colobus (*Colobus badius rufomitratus*). *International Journal of Primatology*, **15**, 703–737.

Deinard, A. & Smith, D. G. (2001). Phylogenetic relationships among the macaques: evidence from the nuclear locus *NRAMP1*. *Journal of Human Evolution*, **41**, 45–59.

de Jong, G., de Ruiter, J. R. & Haring, R. (1994). Genetic structure of a population with social structure and migration. In *Conservation Genetics*, ed. V. Loeschke, J. Tomiuk & S. K. Jain, pp. 147–164. Basel: Birkhauser.

Delson, E. (1980). Fossil macaques, phyletic relationships and a scenario of deployment. In *The Macaques: Studies in Ecology, Behavior, and Evolution*, ed. D. G. Lindburg, pp. 10–30. New York: van Nostrand Rheinhold.

Demaria, C. & Thierry, B. (1990). Formal biting in stumptailed macaques (*Macaca arctoides*). *American Journal of Primatology*, **20**, 133–140.

(2001). A comparative study of reconciliation in rhesus and Tonkean macaques. *Behaviour*, **138**, 397–410.

Deneubourg, J. L. & Goss, S. (1989). Collective patterns and decision-making. *Ethology, Ecology and Evolution*, **1**, 295–311.

de Ruiter, J. & Geffen, E. (1998). Relatedness of matrilines, dispersing males and social groups in long-tailed macaques (*Macaca fascicularis*). *Proceedings of the Royal Society of London B*, **265**, 79–87.

de Ruiter, J. R. & van Hooff, J. A. R. A. M. (1993). Male dominance rank and reproductive success in primate groups. *Primates*, **34**, 513–523.

de Ruiter, J. R., Scheffrahn, W., Trommelen, G. J. J. M., Uitterlinden, A. G., Martin, R. D. & van Hooff, J. A. R. A. M. (1992). Male social rank and reproductive success in wild long-tailed macaques. In *Paternity in Primates: Genetic Tests and Theories*, ed. R. D. Martin, A. F. Dixson & E. J. Wickings, pp. 175–191. Basel: Karger.

de Ruiter, J., van Hooff, J. A. R. A. M. & Scheffrahn, W. (1994). Social and genetic aspects of paternity in wild long-tailed macaques (*Macaca fascicularis*). *Behaviour*, **129**, 203–224.

de Silva, A. M., Dittus, W. P. J., Amerasinghe, P. H. & Amerasinghe, F. P. (1999). Serologic evidence for an epizootic dengue virus infecting toque macaques (*Macaca sinica*) at Polonnaruwa, Sri Lanka. *American Journal of Tropical Medicine and Hygiene*, **60**, 300–306.

Dettling, A., Pryce, C. R., Martin, R. D. & Döbeli, M. (1998). Physiological responses to parental separation and a strange situation are related to parental care received in juvenile Goeldi's monkeys (*Callimico goeldii*). *Developmental Psychobiology*, **33**, 31–31.

DeVinney, B. J., Berman, C. M. & Rasmussen, K. L. R. (2001). Changes in yearling rhesus monkeys' relationships with their mother after sibling birth. *American Journal of Primatology*, **54**, 193–210.

de Vries, H. (1995). An improved test of linearity in dominance hierarchies containing unknown or tied relationships. *Animal Behaviour*, **50**, 1375–1389.

    (1998). Finding a dominance order most consistent with a linear hierarchy: a new procedure and review. *Animal Behaviour*, **55**, 827–843.

de Vries H. & Appleby, M. C. (2000). Finding an appropriate order for a hierarchy: a comparison of the I&SI and the BBS methods. *Animal Behaviour*, **59**, 239–245.

de Waal, F. B. M. (1977). The organization of agonistic relations within two captive groups of Java monkeys (*Macaca fascicularis*). *Zeitschrift für Tierpsychologie*, **44**, 225–282.

    (1982). *Chimpanzee Politics: Power and Sex Among Apes*. London: Jonathan Cape.

    (1986a). The integration of dominance and social bonding in primates. *Quarterly Review of Biology*, **61**, 459–479.

    (1986b). The brutal elimination of a rival among captive male chimpanzees. *Ethology and Sociobiology*, **7**, 237–251.

    (1989a). Dominance 'style' and primate social organization. In *Comparative Socioecology: the Behavioral Ecology of Humans and Other Animals*, ed. V. Standen & R. A. Foley, pp. 243–263. Oxford: Blackwell.

    (1989b). *Peacemaking Among Primates*. Cambridge, MA: Harvard University Press.

    (1996a). *Good Natured: The Origins of Right and Wrong in Humans and Other Animals*. Cambridge, MA: Harvard University Press.

    (1996b). Macaque social culture: development and perpetuation of affiliative networks. *Journal of Comparative Psychology*, **110**, 147–154.

    (2000). Attitudinal reciprocity in food sharing among brown capuchin monkeys. *Animal Behaviour*, **60**, 253–261.

    (2001). *The Ape and the Sushi Master*. New York: Basic Books.

de Waal, F. B. M. & Aureli, F. (1997). Conflict resolution and distress alleviation in monkeys and apes. In *The Integrative Neurobiology of Affiliation*, ed. C. S. Carter, B. Kirkpatrick & I. Lenderhendler, pp. 317–328. New York: Annals of the New York Academy of Sciences.

de Waal, F. B. M. & Johanowicz, D. L. (1993). Modification of reconciliation behavior through social experience: an experiment with two macaque species. *Child Development*, **64**, 897–908.

de Waal, F. & Lanting, F. (1997). *Bonobo: The Forgotten Ape.* Berkeley, CA: University of California Press.

de Waal, F. B. M. & Luttrell, L. M. (1985). The formal hierarchy of rhesus monkeys: an investigation of the bared-teeth display. *American Journal of Primatology*, **9**, 73–85.

(1986). The similarity principle underlying social bonding among female rhesus monkeys. *Folia Primatologica*, **46**, 215–234.

(1989). Toward a comparative socioecology of the genus *Macaca*: different dominance styles in rhesus and stumptail macaques. *American Journal of Primatology*, **19**, 83–109.

de Waal, F. B. M., van Hooff, J. A. R. A. M. & Netto, W. J. (1976). An ethological analysis of types of agonistic interaction in a captive group of Java-monkeys (*Macaca fascicularis*). *Primates*, **17**, 257–290.

Dewit, I., Dittus, W. P. J., Vercruysse, J., Harris, E. A. & Gibson, D. I. (1991). Gastrointestinal helminths in a natural population of *Macaca sinica* and *Presbytis* spp at Polonnaruwa, Sri Lanka. *Primates*, **32**, 391–395.

Dewsbury, D. A. (1992). On the problems studied in ethology, comparative psychology, and animal behavior. *Ethology*, **92**, 89–107.

di Fiore, A. & Rendall, D. (1994). Evolution of social organization: a reappraisal for primates by using phylogenetic methods. *Proceedings of the National Academy of Sciences of the USA*, **91**, 9941–9945.

Diezinger, F. & Anderson, J. R. (1986). Starting from scratch: a first look at a "displacement activity" in group-living rhesus monkeys. *American Journal of Primatology*, **11**, 117–124.

Dittus, W. P. J. (1975). Population dynamics of the toque monkey (*Macaca sinica*). In *Socioecology and Psychology of Primates*, ed. R. H. Tuttle, pp. 124–153. The Hague: Mouton.

(1977). The social regulation of population density and age-sex distribution in the toque macaque. *Behaviour*, **63**, 281–322.

(1979). The evolution of behaviors regulating density and age-specific sex ratios in a primate population. *Behaviour*, **69**, 265–302.

(1980). The social regulation of primate populations: a synthesis. In *The Macaques: Studies in Ecology, Behavior and Evolution*, ed. D. G. Lindburg, pp. 263–286. New York: van Nostrand Reinhold.

(1982). Population regulation: the effects of severe environmental changes on the demography and behavior of wild toque macaques. *International Journal of Primatology*, **3**, 276.

(1985). The influence of cyclones on the dry evergreen forest of Sri Lanka. *Biotropica*, **17**, 1–14.

(1986). Sex differences in fitness following a group take-over among toque macaques: testing models of social evolution. *Behavioral Ecology and Sociobiology*, **19**, 257–266.

(1987). Group fusion among wild toque macaques: an extreme case of inter-group resource competition. *Behaviour*, **100**, 247–291.

(1988). Group fission among wild toque macaques as a consequence of female resource competition and environmental stress. *Animal Behaviour*, **36**, 1626–1645.

(1998). Birth sex ratios in toque macaques and other mammals: integrating the effects of maternal condition and competition. *Behavioral Ecology and Sociobiology*, **44**, 149–160.

Dixson, A. F. (1977). Observations on the displays, menstrual cycles and sexual behaviour of the "black ape" of Celebes (*Macaca nigra*). *Journal of Zoology (London)*, **182**, 63–84.

(1980). Androgens and aggressive behavior in primates: a review. *Aggressive Behavior*, **6**, 37–67.

(1983). Observations on the evolution and behavioral significance of "sexual skin" in female primates. *Advances in the Study of Behavior*, **13**, 63–106.

(1998). *Primate Sexuality: Comparative Studies of the Prosimians, Monkeys, Apes, and Human Beings*. Oxford: Oxford University Press.

Dobson, A. P. & Lyles, A. M. (1989). The population dynamics and conservation of primate populations. *Conservation Biology*, **5**, 362–380.

Dolan, R. J. & Morris, J. S. (2000). The functional anatomy of innate and acquired fear: perspectives from neuroimaging. In *Cognitive Neuroscience of Emotion*, ed. R. D. Lane & L. Nadel, pp. 225–241. New York: Oxford University Press.

Domb, L. G. & Pagel, M. (2001). Sexual swellings advertise female quality in wild baboons. *Nature*, **410**, 204–206.

Drapier, M. & Thierry, B. (2002). Social transmission of feeding techniques in Tonkean macaques? *International Journal of Primatology*, **23**, 105–122.

Drapier, M., Chauvin, C. & Thierry, B. (2002). Tonkean macaques (*Macaca tonkeana*) find food sources from cues conveyed by group-mates. *Animal Cognition*, **5**, 159–165.

Dücker, S. (1996). *Soziale Funktionen der affiliativen Gesichtsausdruecke bei* Theropithecus gelada. Diplomarbeit, Ruhr-Universität Bochum.

Ducoing, A. M. & Thierry, B. (2003). Withholding information in semifree-ranging Tonkean macaques (*Macaca tonkeana*). *Journal of Comparative Psychology*, **117**, 67–75.

(2004). Tool use learning in Tonkean macaques (*Macaca tonkeana*). *Animal Cognition*. (In press.)

Dum, J. & Herz, A. (1987). Opioids and motivation. *Interdisciplinary Science Reviews*, **12**, 180–190.

Dunbar, R. I. M. (1988). *Primate Social Systems*. London: Croom Helm.

(2000). Male mating strategies: a modeling approach. In *Primate Males. Causes and Consequences of Variation in Group Composition*, ed. P. M. Kappeler, pp. 259–268. Cambridge: Cambridge University Press.

Durkheim, E. (1933). *The Division of Labor in Society*. New York: Free Press.

Duvall, S. W., Bernstein, I. S. & Gordon, T. P. (1976). Paternity and status in a rhesus monkey group. *Journal of Reproduction and Fertility*, **47**, 25–31.

Dyke, B., Gage, T. B., Mamelka, P. M., Goy, R. W. & Stone, W. H. (1986). A demographic analysis of the Wisconsin Regional Primate Center rhesus colony, 1962–1982). *American Journal of Primatology*, **10**, 257–269.

Easa, P. S., Surendranathan Asari, P. K. & Chand Basha, S. (1997). Status and distribution of the endangered lion-tailed macaque *Macaca silenus* in Kerala, India. *Biological Conservation*, **80**, 33–37.

Eaton, G. G. (1973). Social and endocrine determinants of sexual behaviour in simian and prosimian females. In *Symposium of the IVth International Congress of Primatology*, vol. 2, ed. C. H. Phoenix, pp. 20–35. Basel: Karger.

(1978). Longitudinal studies of sexual behavior in the Oregon troop of Japanese macaques. In *Sex and Behavior*, ed. T. E. McGill, D. A. Dewsbury & B. D. Sachs, pp. 35–59. New York: Plenum Press.

Eaton, G. G., Johnson, D. F., Glick, B. B. & Worlein, J. M. (1985). Development in Japanese macaques (*Macaca fuscata*): sexually dimorphic behavior during the first year of life. *Primates*, **26**, 238–248.

Edwards, R. G. (1980). *Conception in the Human Female*. New York: Academic Press.

Ehardt, C. & Bernstein, I. (1992). Conflict intervention behavior by adult male macaques: structural and functional aspects. In *Coalitions and Alliances in Humans and Other Animals*, ed. S. Harcourt & F. B. M. de Waal, pp. 89–110. Oxford: Oxford University Press.

Eisenberg, J. F. (1977). The evolution of the reproductive unit in the class Mammalia. In *Reproductive Behavior and Evolution*, ed. J. S. Rosenblatt & B. R. Komisaruk, pp. 39–71. New York: Plenum Press.

(1981). *The Mammalian Radiations: an Analysis of Trends in Evolution, Adaptation, and Behavior*. Chicago: University of Chicago Press.

Ekman, P. (1984). Expression and the nature of emotion. In *Approaches to Emotion*, K. R. Scherer & P. Ekman, pp. 329–343. Hillsdale, NJ: Lawrence Erlbaum.

(1993). Facial expression and emotion. *American Psychologist*, **48**, 384–392.

Ekman, P., Levenson, R. W. & Friesen, W. V. (1983). Autonomic nervous system activity distinguishes among emotions. *Science*, **221**, 1208–1210.

Eldredge, N. (1985). *Unfinished Synthesis: Biological Hierarchies and Modern Evolutionary Thought*. Oxford: Oxford University Press.

Eldredge, N. & Grene, M. (1992). *Interactions: the Biological Context of Social Systems*. New York: Columbia University Press.

Eley, R. M., Strum, S. C., Muchemi, G. & Reid, G. D. (1989). Nutrition, body condition, activity patterns and parasitism if free-ranging troops of olive baboons (*Papio anubis*) in Kenya. *American Journal of Primatology*, **18**, 209–219.

Ellis, L. (1991). A biosocial theory of social stratification derived from the concepts of pro/antisociality and r/K selection. *Politics and Life Sciences*, **10**, 5–44.

(1995). Dominance and reproductive success among nonhuman animals: a cross-species comparison. *Ethology and Sociobiology*, **16**, 257–333.

Ellison, P. T. (2001). *On Fertile Ground*. Cambridge, MA: Harvard University Press.

Emery, N. J., Capitanio, J. P., Mason, W. A., Machado, C. J., Mendoza, S. P. & Amaral, D. G. (2001). The effects of bilateral lesions of the amygdale on dyadic social interactions in rhesus macaques (*Macaca mulatta*). *Behavioral Neuroscience*, **115**, 515–544.

Emery, N. J., Lorincz, E. N., Perrett, D. I. & Oram, M. W. (1997). Gaze following and joint attention in rhesus monkeys (*Macaca mulatta*). *Journal of Comparative Psychology*, **111**, 286–293.

Emlen, S. T. & Oring, L. W. (1977). Ecology, sexual selection, and the evolution of mating systems. *Science*, **197**, 215–223.

Emmons, L. H. (1984). Geographic variation in densitites and diversities of non-flying mammals in Amazonia. *Biotropica*, **16**, 210–222.

Emmons, R. A. (1989). Exploring the relations between motives and traits: the case of narcissism. In *Personality Psychology: Recent Trends and Emerging Directions*, ed. D. M. Buss & N. Cantor, pp. 32–44. New York: Springer.

(1997). Motives and life goals. In *Handbook of Personality Psychology*, ed. R. Hogan, J. Johnson & S. Briggs, pp. 485–512. San Diego: Academic Press.

Enomoto, T. (1974). The sexual behavior of Japanese monkeys. *Journal of Human Evolution*, **3**, 351–372.

(1981). Male aggression and the sexual behavior of Japanese monkeys. *Primates*, **22**, 15–23.

Enquist, M. & Leimar, O. (1983). Evolution of fighting behaviour: decision rules and assessment of relative strength. *Journal of Theoretical Biology*, **102**, 387–410.

Epstein, S. (1994). Trait theory as personality theory: can a part be as great as the whole? *Psychological Inquiry*, **5**, 120–122.

Erwin, J. M., Blood, B. D., Southwick, C. H. & Wolfle, T. L. (1995). Primate conservation. In *Nonhuman Primates in Biomedical Research: Biology and Management*, ed. B. T. Bennett, C. R. Abee & R. Henrickson, pp. 113–28. San Diego: Academic Press.

Estrada, A. (1984). Male–infant interactions among free-ranging stumptail macaques. In *Primate Paternalism*, ed. D. M. Taub, pp. 56–87. New York: van Nostrand Reinhold.

Estrada, A., Estrada, R. & Ervin, F. (1977). Establishment of a free-ranging colony of stumptail macaques (*Macaca arctoides*): social relations. *Primates*, **18**, 647–676.

Eudey, A. A. (1980). Pleistocene glacial phenomena and the evolution of Asian macaques. In *The Macaques: Studies in Ecology, Behavior, and Evolution*, ed. D. G. Lindburg, pp. 52–83. New York: van Nostrand Rheinhold.

(1995). Southeast Asian primate trade routes. *Primate Report*, **41**, 33–42.

Evans, B. J., Supriatna, J. & Melnick, D. J. (2001). Hybridization and population genetics of two macaque species in Sulawesi, Indonesia. *Evolution*, **55**, 1686–1702.

Eysenck, H. J. (1990). Biological dimensions of personality. In *Handbook of Personality Theory and Research*, ed. L. Pervin, pp. 244–276. New York: Guilford.

Fa, J. E. (1989). The genus *Macaca*: a review of taxonomy and evolution. *Mammal Reviews*, **19**, 45–81.

Fa, J. E. & Lindburg, D. G. (1996). *Evolution and Ecology of Macaque Societies*. Cambridge: Cambridge University Press.

Fa, J. E. & Southwick, C. H. (1988). *Ecology and Behaviour of Food-enhanced Primate Groups*. New York: Alan R. Liss.

Fairbanks, L. A. (1989). Early experience and cross-generational continuity of mother-infant contact in vervet monkeys. *Developmental Psychobiology*, **22**, 669–681.

(1996). Individual differences in maternal style: causes and consequences for mothers and offspring. *Advances in the Study of Behavior*, **25**, 579–611.

(2001). Individual differences in response to a stranger: social impulsivity as a dimension of temperament in vervet monkeys (*Cercopithecus aethiops sabaeus*). *Journal of Comparative Psychology*, **115**, 22–28.

Fairbanks, L. A. & McGuire, M. T. (1988). Long-term effects of early mothering behavior on responsiveness to the environment in vervet monkeys. *Developmental Psychobiology*, **21**, 711–724.

(1993). Maternal protectiveness and response to the unfamiliar in vervet monkeys. *American Journal of Primatology*, **30**, 119–129.

Fairbanks, L. A. & Pereira, M. E. (1993). Juvenile primates: dimensions for future research. In *Juvenile Primates: Life History, Development, and Behavior*, ed. M. E. Pereira & L. A. Fairbanks, pp. 359–366. New York: Oxford University Press.

Fairbanks, L. A., Melega, W. P., Jorgensen, M. J., Kaplan, J. R. & McGuire, M. T. (2001). Social impulsivity inversely associated with CSF 5-HIAA and fluoxetine exposure in vervet monkeys. *Neuropsychopharmacology*, **24**, 370–378.

Fashing, P. J. (2001). Male and female strategies during intergroup encounters in guerezas (*Colobus guereza*): evidence for resource defense mediated through males and a comparison with other primates. *Behavioral Ecology and Sociobiology*, **50**, 219–230.

Fedigan, L. M. (1983). Dominance and reproductive success in primates. *Yearbook of Physical Anthropology*, **26**, 91–129.

Fedigan, L. M. & Fedigan, L. (1977). The social development of a handicapped infant in a free-living troop of Japanese monkeys. In *Primate Bio-Social Development: Biological, Social, and Ecological Determinants*, ed. Chevalier-Skolnkoff, S. T. & Poirier, F. E., pp. 205–222. New York: Garland Press.

Fedigan, L. M. & Gouzoules, H. (1978). The consort relationship in a troop of Japanese monkeys. In *Recent Advances in Primatology*, vol. 1, ed. D. J. Chivers & J. Herbert, pp. 493–495. New York: Academic Press.

Fedigan, L. M. & Zohar, S. (1997). Sex differences in mortality of Japanese macaques: twenty-one years of data from the Arashiyama West population. *American Journal of Physical Anthropology*, **102**, 161–175.

Fedigan, L. M., Fedigan, L., Gouzoules, S., Gouzoules, H. & Koyama, N. (1986). Lifetime reproductive success in female Japanese macaques. *Folia Primatologica*, **47**, 143–157.

Fedigan, L. M., Gouzoules, H. & Gouzoules, S. (1983). Population dynamics of Arashiyama West Japanese macaques. *International Journal of Primatology*, **4**, 307–321.

Feibleman, J. K. (1954). Theory of integrative levels. *British Journal for the Philosophy of Science*, **5**, 59–66.

Feinman, S. (1982). Social referencing in infancy. *Merrill-Palmer Quarterly*, **28**, 445–470.

Ferrari, P. F., Kohler, E., Fogassi, L. & Gallese, V. (2000). The ability to follow eye gaze and its emergence during development in macaque monkeys. *Proceedings of the National Academy of Sciences of the USA*, **97**, 13997–14002.

Fichtel, C., Hammerschmidt, K. & Jürgens, U. (2001). On the vocal expression of emotion. A multi-parametric analysis of different states of aversion in the squirrel monkey. *Behaviour*, **138**, 97–116.

Figueredo, A. J., Cox, R. L. & Rhine, R. J. (1995). A generalizability analysis of subjective personality assessments in stumptail macaque and the zebra finch. *Multivariate Behavioral Research*, **30**, 167–197.

Fisher, R. A. (1958). *The Genetical Theory of Natural Selection*. New York: Dover Publications.

Fitch-Snyder, H., Harvey, N. C., Lindburg, D. G. & Tornatore, N. (1997). Adult/infant affiliative interactions in Tibetan macaques (*Macaca thibetana*) housed in a small social group. *American Journal of Primatology*, **42**, 110.

Flanagan, J. (1989). Hierarchy in simple "egalitarian" societies. *Annual Review of Anthropology*, **18**, 245–266.

Flesness, N. R. (1986). Captive status and genetic considerations. *Primates: the Road to Self-Sustaining Populations*, ed. W. K. Benirschke, pp. 845–856. New York: Springer.

Flynn, D., Clarke, A. S., Harvey, N. C. & Lindburg, D. L. (1989). Preliminary studies of rare Tibetan macaques. *Laboratory Primate Newsletter*, **28**, 4–5.

Fooden, J. (1964). Rhesus and crab-eating macaques: intergradation in Thailand. *Science*, **143**, 363–365.

(1969). *Taxonomy and Evolution of the Monkeys of Celebes* (*Primates: Cercopithecidae*). Basel: Karger.

(1971). Female genitalia and taxonomic relationships of *Macaca assamensis*. *Primates*, **12**, 63–73.

(1975). Taxonomy and evolution of liontail and pigtail macaques (Primates: *Cercopithecidae*). *Fieldiana: Zoology*, **67**, 1–168.

(1976). Provisional classification and and key to the living species of macaques (Primates: *Macaca*). *Folia Primatologica*, **25**, 225–236.

(1980). Classification and distribution of living macaques (*Macaca* Lacépède, 1799). In *The Macaques: Studies in Ecology, Behavior and Evolution*, ed. D. G. Lindburg, pp. 1–9. New York: van Nostrand Reinhold.

(1981). Taxonomy and evolution of the *sinica* group of macaques: 2. Species and subspecies accounts of the Indian bonnet macaque, *Macaca radiata*. *Fieldiana: Zoology*, **9**, 1–52.

(1982a). Ecogeographic segregation of macaque species. *Primates*, **23**, 574–579.

(1982b). Taxonomy and evolution of the *sinica* group of macaques: 3. Species and subspecies accounts of *Macaca assamensis*. *Fieldiana: Zoology*, **10**, 1–52.

(1990). The bear macaque, *Macaca arctoides*: a systematic review. *Journal of Human Evolution*, **19**, 607–686.

(1994). Malaria in macaques. *International Journal of Primatology*, **15**, 573–596.

(1995). Systematic review of Southeast Asian longtailed macaques, *Macaca fascicularis* (Raffles, [1821]). *Fieldiana: Zoology*, **81**: 1–206.

Fooden, J. & Wu, H. Y. (2001). Systematic review of the Taiwanese macaque, *Macaca cyclopis* Swinhoe, 1863. *Fieldiana: Zoology*, **98**, 1–70.

Fortes, M. & Evans-Pritchard, E. E., eds. (1940). *African Political Systems*. London: Oxford University Press.

Foucault, M. (1990). *The History of Sexuality: an Introduction*, vol. 1. New York: Vintage.

Fragaszy, D. M. & Perry, S. (2003). Towards a biology of traditions. In *The Biology of Traditions: Models and Evidence*, ed. D. M. Fragaszy & S. Perry, pp. 1–32. Cambridge: Cambridge University Press.

Frayn, K. N. (1986). *Metabolic Regulation*. London: Portland Press.

Frijda, N. H. (1986). *The Emotions*. New York: Cambridge University Press.

Frisch, R. E. (1984). Body fat, puberty, and fertility. *Biological Reviews*, **59**, 161–188.

Frisch, R. E. & McArthur, J. (1974). Menstrual cycles: fatness as a determinant of minimum weight for height necessary for their maintenance or onset. *Science*, **185**, 949–951.

Froehlich, J. W. & Supriatna, J. (1996). Secondary intergradation between *Macaca maurus* and *M. tonkeana* in South Sulawesi, and the species status of *M. togeanus*. In *Evolution and Ecology of Macaque Societies*, ed. J. E. Fa & D. G. Lindburg, pp. 43–70. Cambridge: Cambridge University Press.

Froehlich, J. W., Thorington, R. W. & Otis, J. S. (1981). The demography of howler monkeys (*Alouatta palliata*) on Barro Colorado Island, Panama. *International Journal of Primatology*, **2**, 207–236.

Fuentes, A. & Ray, E. (1997). Humans, habitat loss and hunting: the status of the Mentawai primates on Sipora and the Pagai islands. *Asian Primates*, **5**, 5–9.

Furuichi, T. (1983). Interindividual distance and influence of dominance on feeding in a natural Japanese macaque troop. *Primates*, **24**, 445–455.

Furuya, Y. (1968). On the fission of troops of Japanese monkeys. I. Five fissions and social changes between 1955 and 1966 in the Gagyusan troop. *Primates*, **9**, 323–350.

(1969). On the fission of troops of Japanese monkeys. II. General view of troop fission of Japanese monkeys. *Primates*, **10**, 47–69.

Futuyma, D. J. (1979). *Evolutionary Biology*. Sunderland, MA: Sinauer.

Futuyma, D. J. & Risch, S. J. (1984). Sexual orientation, sociobiology, and evolution. *Journal of Homosexuality*, **9**, 157–168.

Gachot-Neveu, H., Fazio, G. & Thierry, B. (2004). Evolution of the genetic structure in a group of Tonkean macaques: consequences of the mating system. *Folia Primatologica*, **75**. (In press.)

Gadgil, M. & Bossert, W. H. (1970). Life historical consequences of natural selection. *American Naturalist*, **104**, 1–24.

Galef, B. G. (1988). Imitation in animals: history, definition and interpretation of data from the psychological laboratory. In *Social Learning: Psychological and Biological Perspectives*, ed. T. R. Zentall & B. G. Galef, pp. 3–28. Hillsdale, NJ: Lawrence Erlbaum.

(1992). The question of animal culture. *Human Nature*, **3**, 157–178.

(1995). Why behaviour patterns that animals learn socially are locally adaptive? *Animal Behaviour*, **49**, 1325–1334.

(1998). Recent progress in studies of imitation and social learning in animals. In *Advances in the Psychological Sciences,* vol. 2, ed. M. Sbourin, F. Craig, & M. Robert, pp. 275–299. Hove, UK: Psychology Press/Erlbaum.

Galef, B. G. & Giraldeau, L. A. (2001). Social influences on foraging in vertebrates: causal mechanisms and adaptative functions. *Animal Behaviour*, **61**, 3–15.

Galef, B. G. & Whiskin, E. (1997). Effects of social and asocial learning on longevity of food-preference traditions. *Animal Behaviour*, **53**, 1313–1322.

Gerloff, U., Hartung, B., Fruth, B., Hohmann, G. & Tautz, D. (1999). Intracommunity relationships, dispersal pattern and paternity success in a wild living community of bonobos (*Pan paniscus*) determined from DNA analysis of faecal samples. *Proceedings of the Royal Society of London B*, **266**, 1189–1195.

Gibson, J. J. (1979). *The Ecological Approach to Visual Perception*. Boston: Houghton Mifflin.

Gilardi, K. V. K., Shideler, S. E., Valverde, C. R., Roberts, J. A. & Lasley, B. L. (1997). Characterization of the onset of menopause in the rhesus macaque. *Biology of Reproduction*, **57**, 335–340.

Glick, B. B. (1980). Ontogenetic and psychobiological aspects of the mating activities of male *Macaca radiata*. In *The Macaques: Studies in Ecology, Behavior, and Evolution*, ed. D. L. Lindburg, pp. 345–369. New York: van Nostrand Reinhold.

Glick, B. B., Eaton, G. G., Johnson, D. F. & Worlein, J. (1986a). Social behavior of infant and mother Japanese macaques (*Macaca fuscata*): effects of kinship, partner sex, and infant sex. *International Journal of Primatology*, **7**, 139–155.

(1986b). Development of partner preferences in Japanese macaques (*Macaca fuscata*): effects of gender and kinship during the second year of life. *International Journal of Primatology*, **7**, 467–479.

Godelier, M. (1986). *The Making of Great Men: Male Dominance and Power Among the New Guinea Baruya*. Cambridge: Cambridge University Press.

Goldfoot, D. A. (1981). Olfaction, sexual behavior, and the pheromone hypothesis in rhesus monkeys: a critique. *American Zoologist*, **21**, 153–164.

Goldsmith, H. H., Buss, A. H., Plomin, R. *et al.* (1987). Roundtable: What is temperament? Four approaches. *Child Development*, **58**, 505–529.

Goldstein, S. J. & Richard, A. F. (1989). Ecology of rhesus macaques (*Macaca mulatta*) in northwest Pakistan. *International Journal of Primatology*, **10**, 531–567.

Gomendio, M. (1989a). Differences in fertility and suckling patterns between primiparous and multiparous rhesus monkeys (*Macaca mulatta*). *Journal of Reproduction and Fertility*, **87**, 529–542.

(1989b). Suckling behaviour and fertility in rhesus macaques (*Macaca mulatta*). *Journal of Zoology (London)*, **217**, 449–467.

Goodall, J. (1983). Population dynamics during a fifteen year period in one community of free-living chimpanzees in the Gombe National Park. *Zeitschrift für Tierpsychologie*, **61**, 1–60.

(1986). *The Chimpanzees of Gombe: Patterns of Behavior*. Cambridge, MA: Belknap Press.

Goodman, S. M., O'Connor, S. & Langrand, O. (1993). A review of predation on lemurs: implications for the evolution of social behavior in small nocturnal primates. In *Lemur Social Systems and their Ecological Basis*, ed. P. M. Kappeler & J. U. Ganzhorn, pp. 51–66. New York: Plenum Press.

Gore, M. A. (1986). Mother–offspring conflict and interference at mother's mating in *Macaca fascicularis*. Primates, **27**, 205–214.

Gosling, S. D. (2001). From mice to men: what can we learn about personality from animal research? *Psychological Bulletin*, **127**, 45–86.

Gottlieb, G. (1992). *Individual Development and Evolution*. Oxford: Oxford University Press.

Gould, L., Sussman, R. W. & Sauther, M. L. (1999). Natural disasters and primate populations: the effects of a two-year drought on a naturally occurring population of ring-tailed lemurs (*Lemur catta*) in southwestern Madagascar. *International Journal of Primatology*, **20**, 69–84.

Gould, S. J. (1977). *Ontogeny and Phylogeny*. Cambridge, MA: Belknap Press.

Gould, S. J. & Lewontin, R. (1979). The spandrels of San Marco and the Panglossian paradigm: a critique of the adaptationist programme. *Proceedings of the Royal Society of London B*, **205**, 581–598.

Gould, S. J. & Vrba, E. S. (1982). Exaptation – a missing term in the science of form. *Paleobiology*, **8**, 4–15.

Gouzoules, H. & Goy, R. W. (1983). Physiological and social influences on mounting behavior of troop-living female monkeys (*Macaca fuscata*). *American Journal of Primatology*, **5**, 39–49.

Gouzoules, S. & Gouzoules, H. (1987). Kinship. In *Primate Societies*, ed. B. B. Smuts, D. L. Cheney, R. M. Seyfarth, R. W. Wrangham & T. T. Struhsaker, pp. 299–305. Chicago: University of Chicago Press.

Goy, R. W., Bridson, W. E. & Robinson, J. A. (1982). Puberty in the rhesus male. *International Journal of Primatology*, **3**, 288.

Grand, T. I. (1972). A mechanical interpretation of terminal branch feeding. *Journal of Mammalogy*, **53**, 198–201.

Gray, J. A. (1975). *Elements of a Two-Process Theory of Learning*. London: Academic Press.

Green, S. (1975). Dialects in Japanese monkeys: vocal learning and cultural transmission of locale-specific vocal behavior? *Zeitschrift für Tierpsychologie*, **38**, 304–314.

Greenwood, P. J. (1980). Mating systems, philopatry and dispersal in birds and mammals. *Animal Behaviour*, **28**, 1140–1162.

Groves, C. (2001). *Primate Taxonomy*. Washington, DC: Smithsonian Institution.

Gust, D. A. & Gordon, T. P. (1994). The absence of a matrilineally based dominance system in sooty mangabeys, *Cercocebus torquatus atys*. *Animal Behaviour*, **47**, 589–594.

Gust, D. A., Gordon, T. P., Gergits, W. F., Casna, N. J., Gould, K. G. & McClure, H. M. (1996). Male dominance rank and offspring-initiated affiliative behaviors were not predictors of paternity in a captive group of pigtail macaques (*Macaca nemestrina*). *Primates*, **37**, 271–278.

Gust, D. A., McCaster, T., Gordon, T. P., Gergits, W., Casna, N. & McClure, H. M. (1998). Paternity in sooty mangabeys. *International Journal of Primatology*, **19**, 83–94.

Gygax, L. (1995). Hiding behaviour of longtailed macaques (*Macaca fascicularis*). I. Theoretical background and data on mating. *Ethology*, **101**, 10–24.

Hadidian, J. & Bernstein, I. S. (1979). Female reproductive cycles and birth data from Old World monkey colony. *Primates*, **20**, 429–442.

Haffer, J. (1969). Speciation in Amazonian forest birds. *Science*, **165**, 131–137.

Hailman, J. P. (1982). Ontogeny: towards a general theoretical framework for ethology. In *Perspectives in ethology*, vol. 5, ed. P. P. G. Bateson & P. H. Klopfer, pp. 133–189. New York: Plenum Press.

Hall, K. R. L. (1965). Behaviour and ecology of the wild patas monkey, *Erythrocebus patas*, in Uganda. *Journal of Zoology (London)*, **148**, 15–87.

Halliday, T. R. (1983). The study of mate choice. In *Mate Choice*, ed. P. P. G. Bateson, pp. 3–32. Cambridge: Cambridge University Press.

Hamilton, W. D. (1964). The genetical theory of social behavior. *Journal of Theoretical Biology*, **7**, 1–52.

Hamilton, W. J. (1985). Demographic consequences of a food and water shortage to desert chacma baboons, *Papio ursinus*. *International Journal of Primatology*, **6**, 451–462.

Hanby, J. P., Robertson, L. T. & Phoenix, C. H. (1971). The sexual behavior of a confined troop of Japanese macaques. *Folia Primatologica*, **3**, 22–49.

Hand, J. L. (1986). Resolution of social conflicts: dominance, egalitarianism, spheres of dominance, and game theory. *Quarterly Review of Biology*, **61**, 201–220.

Harcourt, A. H. (1978). Strategies of emigration and transfer by primates, with particular reference to gorillas. *Zeitschrift für Tierpsychologie*, **48**, 401–420.

Harcourt, A. H., Harvey, P. H., Larson, S. G. & Short, R. V. (1981). Testis weight, body weight and breeding system primates. *Nature*, **293**, 55–57.

Harlow, H. F. (1958). The nature of love. *American Psychologist*, **13**, 673–685.

Harlow, H. F. & Harlow, M. K. (1965). The affectional systems. In *Behavior of Nonhuman Primates*, vol. 2, ed. A. M. Schrier, H. F. Harlow & F. Stollnitz, pp. 287–334. New York: Academic Press.

Harrison, M. J. S. (1983). Territorial behavior in the green monkey, *Cercopithecus sabaeus*: seasonal defense of local food supplies. *Behavioral Ecology and Sociobiology*, **12**, 85–94.

Harvey, N. C. & Rhine, R. J. (1983). Some reproductive parameters of stumptailed macaques (*Macaca arctoides*). *Primates*, **24**, 530–536.

Harvey, P. H. & Pagel, M. D. (1991). *The Comparative Method in Evolutionary Biology*. Oxford: Oxford University Press.

Harvey, P. H., Martin, R. D. & Clutton-Brock, T. H. (1987). Life histories in comparative perspective. In *Primate Societies*, ed. B. B. Smuts, D. L. Cheney, R. M. Seyfarth, R. W. Wrangham & T. T. Struhsaker, pp. 181–196. Chicago: University of Chicago Press.

Hauser, M. (1988). Invention and social transmission: new data from wild vervet monkeys. In *Machiavellian Intelligence. Social Expertise and the Evolution of Intellect in Monkeys, Apes and Humans*, ed. R. W. Byrne & A. Whiten, pp. 327–343. New York: Oxford University Press.

  (1993a). Right hemispheric dominance for the production of facial expression in monkeys. *Science*, **261**, 475–477.

  (1993b). Primatology: some lessons from and for related disciplines. *Evolutionary Anthropology*, **2**, 182–186.

Hauser, M. & Fairbanks, L. A. (1988). Mother–offspring conflict in vervet monkeys: variation in response to ecological conditions. *Animal Behaviour*, **36**, 802–813.

Hausfater, G. (1972). Intergroup behavior of free-ranging rhesus monkeys (*Macaca mulatta*). *Folia Primatologica*, **18**, 78–107.

Hausfater, G., Altmann, J. & Altmann, S. A. (1982). Long-term consistency of dominance relations among female baboons (*Papio cynocephalus*). *Science*, **217**, 752–755.

Hausfater, G. & Watson, D. F. (1976). Social and reproductive correlates of parasite ova emissions by baboons. *Nature*, **262**, 688–689.

Hayasaka, K., Kawamoto, Y., Shotake, T. & Nozawa, K. (1987). Population genetical study of Japanese macaques, *Macaca fuscata*, in the Shimokita A1 Troop, with special reference to genetic variability and relationships to Japanese macaques in other troops. *Primates*, **28**, 507–516.

Heils, A., Teufel, A., Petri, S., Stober, G., Riederer, P., Bengel, D. & Lesch, K. P. (1996). Allelic variation of human serotonin transporter gene expression. *Journal of Neurochemistry*, **66**, 2621–2624.

Heistermann, M., Mosti, E. & Hodges, J. K. (1995). Non-invasive endocrine monitoring of female reproductive status: methods and applications to captive breeding and conservation of exotic species. In *Research and Captive Propagation*, ed. U. Ganslosser, J. K. Hodges & W. Kaumanns, pp. 36–48. Erlanger: Filander.

Heistermann, M., Ziegler, T., van Schaik, C. P., Launhardt, K., Winkler, P. & Hodges, J. K. (2001). Loss of oestrus, concealed ovulation and paternity confusion in free-ranging Hanuman langurs. *Proceedings of the Royal Society of London B*, **268**, 2445–2451.

Hemelrijk, C. K. (1990a). A matrix partial correlation test used in investigations of reciprocity and other social interaction patterns at a group level. *Journal of Theoretical Biology*, **143**, 405–420.

(1990b). Models of, and tests for, reciprocity, unidirectional and other social interaction patterns at a group level. *Animal Behaviour*, **39**, 1013–1029.

(1996). Dominance interactions, spatial dynamics and emergent reciprocity in a virtual world. In *Proceedings of the Fourth International Conference on Simulation of Adaptive Behavior*, ed. P. Maes, M. J. Mataric, J. A. Meyer, J. Pollack & S. W. Wilson, pp. 545–552. Cambridge, MA: MIT Press.

(1997). Cooperation without genes, games or cognition. In *Fourth European Conference on Artificial Life*, ed. P. Husbands & I. Harvey, pp. 511–520. Cambridge, MA: MIT Press.

(1999a). An individual-oriented model on the emergence of despotic and egalitarian societies. *Proceedings of the Royal Society of London B*, **266**, 361–369.

(1999b). Effects of cohesiveness on intersexual dominance relationships and spatial structure among group-living virtual entities. In *Advances in Artificial Life: Fifth European Conference on Artificial Life*, ed. D. Floreano, J. D. Nicoud & F. Mondada, pp. 524–534. Berlin: Springer.

(2000a). Towards the integration of social dominance and spatial structure. *Animal Behaviour*, **59**, 1035–1048.

(2000b). Self-reinforcing dominance interactions between virtual males and females. Hypothesis generation for primate studies. *Adaptive Behavior*, **8**, 13–26.

(2000c). Sexual attraction and inter-sexual dominance. In *Second International Workshop on Multi Agent Based Simulation*, ed. S. Moss & P. Davidsson, pp. 167–180. Berlin: Springer.

(2002). Despotic societies, sexual attraction and the emergence of male 'tolerance': an agent-based model. *Behaviour*, **139**, 729–747.

Hemelrijk, C. K., Meier, C. M. & Martin, R. D. (1999). 'Friendship' for fitness in chimpanzees? *Animal Behaviour*, **58**, 1223–1229.

(2001). Social positive behaviour for reproductive benefits in primates? A response to comments by Stopka *et al.* (2001). *Animal Behaviour*, **61**, F22–F24.

Hemelrijk, C. K., van Laere, G. J. & van Hooff, J. A. R. A. M. (1992). Sexual exchange relationships in captive chimpanzees? *Behavioral Ecology and Sociobiology*, **30**, 269–275.

Higley, J. D. & Linnoila, M. (1997). Low central nervous system serotoninergic activity is traitlike and correlates with impulsive behavior. *Annals of the New York Academy of Sciences*, **836**, 39–56.

Higley, J. D. & Suomi, S. J. (1989). Temperamental reactivity in non-human primates. In *Temperament in Childhood*, ed. G. A. Kohnstamm, J. E. Bates & M. K. Rothbart, pp. 153–167. New York: Wiley.

Higley, J. D., King, S. T., Hasert, M. F., Champoux, M., Suomi, S. J. & Linnoila, M. (1996a). Stability of interindividual differences in serotonin function and its relationship to severe aggression and competent social behavior in rhesus macaque females. *Neuropsychopharmacology*, **14**, 67–76.

Higley, J. D., Mehlman, P. T., Higley, S. B. *et al.* (1996b). Excessive mortality in young free-ranging male nonhuman primates with low cerebrospinal fluid 5-hydroxyindoleacetic acid concentrations. *Archives of General Psychiatry*, **53**, 537–543.

Higley, J. D., Mehlman, P. T., Poland, R. E. *et al.* (1996c). CSF testosterone and 5-HIAA correlate with different types of aggressive behaviors. *Biological Psychiatry*, **40**, 1067–1082.

Higley, J. D., Suomi, S. J. & Linnoila, M. (1991). CSF monoamine metabolite concentrations vary according to age, rearing, and sex. *Psychopharmacology*, **103**, 551–556.

(1992). A longitudinal assessment of CSF monoamine metabolite and plasma cortisol concentrations in young rhesus monkeys. *Biological Psychiatry*, **32**, 127–145.

Higley, J. D., Thompson, W. W., Champoux, M. *et al.* (1993). Paternal and maternal genetic and environmental contributions to cerebrospinal fluid monoamine metabolites in rhesus monkeys (*Macaca mulatta*). *Archives of General Psychiatry*, **50**, 615–623.

Hikami, K., Hasegawa, Y. & Matsuzawa, T. (1990). Social transmission of food preferences in Japanese monkeys (*Macaca fuscata*) after mere exposure or aversion training. *Journal of Comparative Psychology*, **104**, 233–237.

Hill, C. M. (2000). Conflict of interest between people and baboons: crop raiding in Uganda. *International Journal of Primatology*, **21**, 299–315.

Hill, D. A. (1994). Affiliative behaviour between adult males of the genus *Macaca*. *Behaviour*, **130**, 293–308.

(1997). Seasonal variation in the feeding behavior and diet of wild Japanese macaques (*Macaca fuscata yakui*) in lowland forest of Yakushima. *American Journal of Primatology*, **43**, 305–322.

(1999). Effects of provisioning on the social behaviour of Japanese and rhesus macaques: implications for socioecology. *Primates*, **40**, 187–198.

Hill, D. A. & Okayasu, N. (1995). Absence of 'youngest ascendancy' in the dominance relations of sisters in wild Japanese macaques (*Macaca fuscata yakui*). *Behaviour*, **132**, 367–379.

(1996). Determinants of dominance among female macaques: nepotism, demography and danger. In *Evolution and Ecology of Macaque Societies*, ed. J. E. Fa & D. G. Lindburg, pp. 459–472. Cambridge: Cambridge University Press.

Hill, W. C. O. (1937). Longevity in a macaque. *Ceylon Journal of Science*, **20**, 255–256.

Hilton-Taylor, C., compiler (2000). *2000 IUCN Red List of Threatened Species*. Gland, Switzerland: IUCN.

Hinde, R. A. (1972). Concepts of emotion. In *Ciba Foundation Symposium N. 8 on Physiology, Emotion and Psychosomatic Illness*, pp. 3–13. Amsterdam: Elsevier.

(1974). *Biological Bases of Human Social Behaviour*. New York: Mc Graw-Hill.

(1976a). On describing relationships. *Journal of Child Psychology and Psychiatry*, **17**, 1–19.

(1976b). Interactions, relationships and social structure. *Man*, **11**, 1–17.

(1979). *Towards Understanding Relationships*. London: Academic Press.

Hinde, R. A. & Rowell, T. E. (1962). Communication by postures and facial expressions in the rhesus monkey (*Macaca mulatta*). *Proceedings of the Zoological Society of London*, **138**, 1–21.

Hinde, R. A. & Spencer-Booth, Y. (1967). The behaviour of socially living rhesus monkeys in their first two and a half years. *Animal Behaviour*, **15**, 169–196.

(1970). Individual differences in the responses of rhesus monkeys to a period of separation from their mothers. *Journal of Child Psychology and Psychiatry*, **11**, 159–176.

Hiraiwa, M. (1981). Maternal and alloparental care in a troop of free-ranging Japanese monkeys. *Primates*, **22**, 309–329.

Hirata, S., Watanabe, K. & Kawai, M. (2001). "Sweet-potato washing" revisited. In *Primate Origins of Human Cognition and Behavior*, ed. T. Matsuzawa, pp. 487–508. Tokyo: Springer.

Ho, M. W. & Saunders, P. T. (1979). Beyond neo-Darwinism: an epigenetic approach to evolution. *Journal of Theoretical Biology*, **78**, 573–591.

Hodgen, G. P., Goodman, A. L., O'Connor, A. & Johnson, D. K. (1977). Menopause in rhesus monkeys: model for study of disorders in the human climacteric. *American Journal of Obstetrics and Gynecology*, **127**, 581–584.

Hoelzer, G. A. & Melnick, D. J. (1996). Evolutionary relationships of the macaques. In *Ecology Evolution and Ecology of Macaque Societies*, ed. J. E. Fa & D. G. Lindburg, pp. 3–19. Cambridge: Cambridge University Press.

Hoelzer, G. A., Wallman, J. & Melnick, D. J. (1998). The effects of social structure, geographical structure, and population size on the evolution of mitochondrial DNA: II. Molecular clocks and the lineage sorting period. *Journal of Molecular Evolution*, **47**, 21–31.

Hooley, J. M. (1983). Primiparous and multiparous mothers and their infants. In *Primate Social Relationships: An Integrated Approach*, ed. R. A. Hinde, pp. 142–145. Sunderland, MA: Sinauer.

Horrocks, J. A. & Hunte, W. (1983). Maternal rank and offspring rank in vervet monkeys: an appraisal of the mechanism of rank acquisition. *Animal Behaviour*, **31**, 772–781.

Hrdy, S. B. (1976). Care and exploitation of nonhuman primate infants by conspecifics other than the mother. In *Advances in the Study of Animal Behaviour*, vol. 6, ed. J. S. Rosenblatt, R. A. Hinde, E. Shaw & C. Beer, pp. 101–158. London: Academic Press.

(1979). Infanticide among animals: a review, classification, and examination of the implications for the reproductive strategies of females. *Ethology and Sociobiology*, **1**, 13–40.

(1987). Sex-biased parental investment among primates and other mammals: a critical evaluation of the Trivers-Willard hypothesis. In *Child Abuse and Neglect: Biosocial Dimensions*, ed. R. Geller & J. Lancaster, pp. 97–147. New York: Aldine.

(1999). *Mother Nature*. New York: Pantheon Books.

Hrdy, S. B. & Hrdy, D. B. (1976). Hierarchical relations among female Hanuman langurs (Primates: Colobinae, *Presbytis entellus*). *Science*, **193**, 913–915.

Hrdy, S. B. & Whitten, P. L. (1987). Patterning of sexual activity. In *Primate Societies*, eds. B. B. Smuts, D. L. Cheney, R. M. Seyfarth, R. W. Wrangham & T. T. Struhsaker, pp. 370–384. Chicago: University of Chicago Press.

Hsu, M. J. & Lin, J. F. (2001). Troop size and structure in free-ranging Formosan macaques (*Macaca cyclopis*) at Mt Longevity, Taiwan. *Zoological Studies*, **40**, 49–60.

Hsu, M. J., Agoramoorthy, G. & Lin, J. F. (2001). Birth seasonality and interbirth intervals in free-ranging Formosan macaques (*Macaca cyclopis*) at Mt. Longevity, Taiwan. *Primates*, **42**, 15–25.

Huffman, M. A. (1984). Stone-play of *Macaca fuscata* in Arashiyama B troop: transmission of a non-adaptive behavior. *Journal of Human Evolution*, **13**, 725–735.

(1987). Consort intrusion and female mate choice in Japanese macaques (*Macaca fuscata*). *Ethology*, **75**, 221–234.

(1991). Mate selection and partner preferences in female Japanese macaques. In *The Monkeys of Arashiyama: Thirty-five Years of Research in Japan and the West*, ed. L. M. Fedigan & P. J. Asquith, pp. 101–122. Albany, NY: State University of New York Press.

(1996). Acquisition of innovative cultural behaviors in nonhuman primates: a case study of stone handling, a socially transmitted behavior in Japanese macaques. In *Social Learning in Animals: The Roots of Culture*, ed. C. M. Heyes & B. G. Galef, pp. 267–289. London: Academic Press.

Hunte, W. & Horrocks, J. A. (1986). Kin and non-kin interventions in the aggressive disputes of vervet monkeys. *Behavioral Ecology and Sociobiology*, **20**, 257–263.

Huth, A. & Wissel, C. (1992). The simulation of the movement of fish schools. *Journal of Theoretical Biology*, **156**, 365–385.

I'Anson, H., Foster, D. L., Foxcraft, G. R. & Booth, P. J. (1991). Nutrition and reproduction. *Oxford Reviews of Reproductive Biology*, **13**, 239–311.

Imanishi, K. (1952). Evolution of humanity. (In Japanese.) In *Man*, ed. K. Imanishi. Tokyo: Mainichi-Shinbunsha.

(1957). Social behavior in Japanese monkeys, *Macaca fuscata*. *Psychologia*, **1**, 47–54.

Inoue, M., Mitsunaga, F., Nozaki, M. *et al.* (1993). Male dominance rank and reproductive success in an enclosed group of Japanese macaques: with special reference to post-conception mating. *Primates*, **34**, 503–511.

Inoue, M., Mitsunaga, F., Ohsawa, H. *et al.* (1992). Paternity testing in captive Japanese macaques (*Macaca fuscata*) using DNA fingerprinting. In *Paternity in Primates: Genetic Tests and Theories*, ed. R. D. Martin, A. F. Dixson & E. J. Wickings, pp. 131–140. Basel: Karger.

Inoue, M., Mitsunaga, F., Ohsawa, H. *et al.* (1991). Male mating behavior and paternity discrimination by DNA fingerprinting in a Japanese macaque group. *Folia
Primatologica*, **56**, 202–210.

Inoue, M., Takenaka, A., Tanaka, S., Kominami, R. & Takenaka, O. (1990). Paternity
discrimination in a Japanese macaque group by DNA fingerprinting. *Primates*, **31**,
563–570.

Isbell, L. A. (1990). Sudden short-term increase in mortality in vervet monkeys (*Cercopithecus aethiops*) due to leopard predation in Amboseli National Park, Kenya.
American Journal of Primatology, **21**, 41–52.

Ishimoto, G. (1973). Blood protein variations in Asian macaques. III. Characteristics of
the macaque blood protein polymorphism. *Journal of Anthropology Society Nippon*,
81, 1–13.

Itani, J. (1954). *The Monkeys of Mt Takasaki*. Tokyo: Kobunsha.

(1959). Paternal care in wild Japanese monkey. *Primates*, **2**, 1–93.

(1975). Twenty years with Mount Takasaki monkeys. In *Primate Utilization and
Conservation*, ed. G. Berment & D. C. Lindburg, pp. 101–125. London: Wiley.

Itoigawa, N., Negayama, K. & Kondo, K. (1981). Experimental study on sexual behavior
between mother and son in Japanese monkeys (*Macaca fuscata*). *Primates*, **22**,
494–502.

Itoigawa, N., Tanaka, T., Ukai, N. *et al.* (1992). Demography and reproductive parameters
of a free-ranging group of Japanese macaques (*Macaca fuscata*) at Katsuyama.
Primates, **33**, 49–68.

IUCN (1996). *African Primates: Status Survey and Conservation Action Plan*, Revised
Edn. Gland, Switzerland: IUCN.

Iwamoto, T. (1982). Food and nutritional condition of free ranging Japanese monkeys
on Koshima islet during winter. *Primates*, **23**, 153–170.

Jablonski, N. G. (1998). The response of catharrhine primates to Pleistocene environmental fluctuations in East Asia. *Primates*, **39**, 29–37.

Jablonski, N. G., Whitfort, M., Roberts-Smith, N. & Qinqi, X. (2000). The influence of
life history and diet on the distribution of catarrhine primates during the Pleistocene
in East Asia. *Journal of Human Evolution*, **39**, 131–157.

Jack, K. M. & Pavelka, M. S. M. (1997). The behavior of peripheral males during the
mating season in *Macaca fuscata*. *Primates*, **38**, 369–377.

James, R. A., Leberg, P. L., Quattro, J. M. & Vrejenhoek, R. C. (1997). Genetic diversity
in lack howler monkeys (*Alouatta nigra*) from Belize. *American Journal of Physical
Anthropology*, **102**, 329–336.

Jameson, K. A., Appleby, M. C. & Freeman, L. C. (1999). Finding an appropriate order
for a hierarchy based on probabilistic dominance. *Animal Behaviour*, **57**, 991–998.

Jang, K. L., McCrae, R. R., Angleitner, A., Riemann, R., Livesley, W. J. (1998). Heritability of facet-level traits in a cross-cultural twin sample: support for a hierarchical model of personality. *Journal of Personality and Social Psychology*, **74**,
1556–1565.

Janson, C. H. (1984). Female choice and mating system of the brown capuchin monkey,
Cebus apella (Primates: *Cebidae*). *Zeitschrift für Tierpsychologie*, **65**, 177–200.

(2000). Primate socio-ecology: the end of a golden age. *Evolutionary Anthropology*,
9, 73–86.

Janson, C. H. & van Schaik, C. P. (1988). Recognizing the many faces of primate food competition: methods. *Behaviour*, **105**, 165–186.

John, O. P. & Robins, R. W. (1994). Traits and types, dynamics and development: no doors should be closed in the study of personality. *Psychological Inquiry*, **5**, 137–142.

John, O. P. & Srivastava, S. (1999). The big five trait taxonomy: history, measurement, and theoretical perspectives. In *Handbook of Personality*, 2nd edn, ed. L. A. Pervin & O. P. John, pp. 102–148. New York: Guilford Press.

John, O. P., Caspi, A., Robins, R. W., Moffitt, T. E. & Stouthamer-Loeber, M. (1994). The "little five": exploring the nomological network of the five-factor model of personality in adolescent boys. *Child Development*, **65**, 160–178.

Johnson, C. W. (1986). Sex-biased philopatry and dispersal in mammals. *Oecologia*, **69**, 626–627.

Johnson, R. L. & Kapsalis, E. (1995). Aging, infecundity and reproductive senescence in free-ranging female rhesus monkey. *Journal of Reproduction and Fertility*, **105**, 271–278.

Johnson-Laird, P. N. & Oatley, K. (1992). Basic emotions, rationality, and folk theory. *Cognition and Emotion*, **6**, 201–223.

Jones, M. L. (1962). Mammals in captivity – Primate longevity. *Laboratory Primate Newsletter*, **1**, 3–13.

Judge, P. G. (1982). Redirection of aggression based on kinship in a captive group of pigtail macaques. *International Journal of Primatology*, **3**, 301.

(1991). Dyadic and triadic reconciliation in pigtail macaques (*Macaca nemestrina*). *American Journal of Primatology*, **23**, 225–237.

Jürgens, U. (1998). Common features in the vocal expression of emotion in human and non-human primates. In *Oralité et Gestualité*, ed. S. Santi, J. Guaitella, C. Cavé & G. Konopczynski, pp. 153–158. Paris: L'Harmattan.

Kagan, J. & Snidman, N. (1999). Early childhood predictors of adult anxiety disorders. *Biological Psychiatry*, **46**, 1536–41.

Kagan, J., Reznick, J. S. & Snidman, N. (1988). Biological bases of childhood shyness. *Science*, **240**, 167–171.

Kagan, J., Snidman, N. & Arcus, D. M. (1998). Childhood derivatives of high and low reactivity in infancy. *Child Development*, **69**, 1483–1493.

Kalin, N. H. & Shelton, S. E. (1989). Defensive behaviors in infant rhesus monkeys: environmental cues and neurochemical regulation. *Science*, **243**, 1718–1721.

Kalin, N. H., Larson, C., Shelton, S. E. & Davidson, R. J. (1998). Asymmetric frontal brain activity, cortisol, and behavior associated with fearful temperament in rhesus monkeys. *Behavioral Neuroscience*, **112**, 286–292.

Kalin, N. H., Shelton, S. E. & Davidson, R. J. (2000). Cerebrospinal fluid corticotropin-releasing hormone levels are elevated in monkeys with patterns of brain activity associated with fearful temperament. *Biological Psychiatry*, **47**, 579–585.

Kalin, N. H., Shelton, S. E., Davidson, R. J. & Kelley, A. E. (2001). The primate amygdala mediates acute fear but not the behavioral and physiological component of anxious temperament. *Journal of Neuroscience*, **21**, 2067–2074.

Kaplan, J. R., Brent, L., Comuzzie, A. G. *et al.* (2001). Heritability of responses to novel objects among pedigreed baboons. *American Journal of Physical Anthropology*, **32** (Suppl.), 87–88.

Kaplan, J. R., Fontenot, M. B., Berard, J., Manuck, S. B. & Mann, J. J. (1995). Delayed dispersal and elevated monoaminergic activity in free-ranging hesus monkeys. *American Journal of Primatology*, **35**, 229–234.

Kaplan, J. R., Phillips-Conroy, J., Fontenot, M. B., Jolly, C. J., Fairbanks, L. A. & Mann, J. J. (1999). Cerebrospinal fluid monoaminergic metabolites differ in wild anubis and hybrid (*anubis hamadryas*) baboons: possible relationships to life history and behavior. *Neuropsychopharmacology*, **20**, 517–524.

Katsukake, N. (2000). Marilineal rank inheritance varies with absolute rank in Japanese macaques. *Primates*, **41**, 321–335.

Katz, N. S. (1995). *The Invention of Heterosexuality*. New York: Dutton.

Kaufman, I. C. & Rosenblum, L. A. (1966). A behavioral taxonomy for *Macaca nemestrina* and *Macaca radiata*: based on longitudinal observation of family groups in the laboratory. *Primates*, **7**, 205–258.

  (1969). The waning of the mother-infant bond in two species of macaque. In *Determinants of Infant Behaviour*, vol. 6, ed. B. M. Foss, pp. 41–59. London: Methuen.

Kavanagh, M., Eudey, A. A. & Mack, D. (1987). The effects of live trapping and trade on primate populations. In *Primate Conservation in the tropical rain forest*, ed. C. W. Marsh & R. A. Mittermeier, pp. 147–177. New York: Alan R. Liss.

Kawai, M. (1958). On the system of social ranks in a natural troop of Japanese monkeys: I & II. *Primates*, **1–2**, 111–148. (English translation. In *Japanese Monkey: a Collection of Translations*, ed. K. Imanishi & S. A. Altmann, pp. 66–104. Atlanta: Emory University, 1965.)

  (1965). Newly acquired pre-cultural behavior of a natural troop of Japanese monkeys on Koshima Island. *Primates*, **6**, 1–30.

Kawai, M., Watanabe, K. & Mori, A. (1992). Pre-cultural behaviors observed in free-ranging Japanese monkeys on Koshima Islet over the past 25 years. *Primate Report*, **32**, 143–153.

Kawamoto, Y. (1996). Population genetic study of Sulawesi macaques. In *Variations in the Asian Macaques*, ed. T. Shotake & K. Wada, pp. 37–65. Tokyo: Tokay University Press.

Kawamoto, Y., Ischak, Tb. M. & Supriatna, J. (1984). Genetic variations within and between troops of the crab-eating macaque (*Macaca fascicularis*) on Sumatra, Java, Bali, Lombok and Sumbawa, Indonesia. *Primates*, **25**, 131–159.

Kawamoto, Y., Ishida, T., Suzuki, J., Takenaka, O. & Varavudhi, P. (1989). A preliminary report on the genetic variations of crab-eating macaques in Thailand. *Kyoto University Overseas Research Report of Studies on Asian Non-human Primates*, **7**, 94–103.

Kawamoto, Y., Nozawa, K. & Ischak, T. M. (1981). Genetic variability and differentiation of local populations in the Indonesian crab-eating macaque (*Macaca fascicularis*). *Kyoto University Overseas Report of Studies on Indonesian Macaque*, **1**, 15–39.

Kawamoto, Y., Shotake, T. & Nozawa, K. (1982). Genetic differentiation among three genera of family *Cercopithecidae*. *Primates*, **23**, 272–286.

Kawamura, S. (1958). Matriarchal social ranks in the Minoo-B troop: a study of the rank system of Japanese macaques. *Primates*, **1**, 149–156. (English translation. In *Japanese Monkeys: a Collection of Translations*, ed. K. Imanishi & S. A. Altmann, pp. 105–112. Atlanta: Emory University, 1965.)

(1959). The process of sub-culture propagation among Japanese macaques. *Primates*, **2**, 43–60.

Kawamura, S., Norikoshi, K. & Azuma, N. (1991). Observation of Formosan monkeys (*Macaca cyclopis*) in Taipingshan Taiwan. In *Primatology Today*, ed. A. Ehara, T. Kimura, O. Takenaka & M. Iwamoto, pp. 97–100. Amsterdam: Elsevier.

Keane, B., Dittus, W. P. J. & Melnick, D. J. (1997). Paternity assessment in wild groups of toque macaques *Macaca sinica* at Polonnaruwa, Sri Lanka using molecular markers. *Molecular Ecology*, **6**, 267–282.

Keddy, A. C. (1986). Female mate choice in vervet monkeys (*Cercopithecus aethiops*). *American Journal of Primatology*, **10**, 405–416.

Keddy-Hector, A. C. (1992). Mate choice in non-human primates. *American Zoologist*, **32**, 62–70.

Kemnitz, J. W., Holston, K. A. & Colman, R. J. (1998). Nutrition, aging, and reproduction in rhesus monkeys. In *Nutrition and Reproduction*, ed. W. Hansel, G. A. Bray & D. H. Ryan, pp. 180–195. Baton Rouge, LA: Louisiana State University Press.

Ketterson, E. D. & Nolan, V. (1992). Hormones and life histories: an integrative approach. *American Naturalist*, **140**, S33–S62.

Keverne, E. B., Martensz, N. D. & Tuite, B. (1989). Beta-endorphin concentrations in cerebrospinal fluid of monkeys are influenced by grooming relationships. *Psychoneuroendocrinology*, **14**, 155–161.

Kleiber, M. (1975). *The Fire of Life: an Introduction to Animal Energetics*. Huntington, NY: Krieger Publishing.

Kling, A. (1972). Effects of amygdalectomy on social-affective behavior in non-human primates. In *The Neuroanatomy of the Amygdala*, ed. B. E. Eleftheriou, pp. 511–536. New York: Plenum Press.

Knobbout, D. A., Ellenbroek, B. A. & Cools, A. R. (1996). The influence of social structure on social isolation in amphetamine-treated Java monkeys (*Macaca fascicularis*). *Behavioural Pharmacology*, **7**, 417–429.

Koenig, A. (2000). Competitive regimes in forest-dwelling Hanuman langur females (*Semnopithecus entellus*). *Behavioral Ecology and Sociobiology*, **48**, 93–109.

(2002). Competition for resources and its behavioral consequences among female primates. *International Journal of Primatology*, **23**, 759–783.

Koyama, N. (1967). On dominance rank and kinship of a wild Japanese monkey troop in Arashiyama. *Primates*, **8**, 105–112.

(1970). Changes in dominance rank and division of a wild Japanese monkey troop at Arashiyama. *Primates*, **11**, 335–90.

(1973). Dominance, grooming and clasped-sleeping relationships among bonnet macaques in India. *Primates*, **14**, 225–254.

Koyama, N., Takahata, Y., Huffman, M. A., Norikoshi, K. & Suzuki, H. (1992). Reproductive parameters of female Japanese macaques: thirty years data from the Arashiyama troops, Japan. *Primates*, **33**, 33–47.

Kraemer, G. W., Ebert, M. H., Schmidt, D. E. & McKinney, W. T. (1989). A longitudinal study of the effects of different social rearing conditions on cerebrospinal fluid norepinephrine and biogenic amine metabolites in rhesus monkeys. *Neuropsychopharmacology*, **2**, 175–189.

Krebs, J. R. & Dawkins, R. (1984). Animal signals: mind reading and manipulation. In *Behavioural Ecology*, ed. J. R. Krebs & N. B. Davies, pp. 380–402. Oxford: Blackwell.

Kroeber, A. & Parsons, T. (1958). The concept of culture and social system. *American Sociological Review*, **23**, 582–583.

Kuester, J. & Paul, A. (1989). Reproductive strategies of subadult Barbary macaque males at Affenberg Salem. In *The Sociobiology of Sexual and Reproductive Strategies*, ed. A. E. Rasa, C. Vogel & E. Voland, pp. 93–109. London: Chapman & Hall.

(1992). Influence of male competition and female mate choice on male mating success in Barbary macaques (*Macaca sylvanus*). *Behaviour*, **120**, 192–217.

(1996). Female–female competition and male mate choice in Barbary macaques (*Macaca sylvanus*). *Behaviour*, **133**, 763–790.

(1997). Group fission in Barbary macaques (*Macaca sylvanus*) at Affenberg Salem. *International Journal of Primatology*, **18**, 941–966.

Kuester, J., Paul, A. & Arnemann, J. (1994). Kinship, familiarity and mating avoidance in Barbary macaques, *Macaca sylvanus*. *Animal Behaviour*, **48**, 1183–1194.

(1995). Age-related and individual differences of reproductive success in male and female Barbary macaques (*Macaca sylvanus*). *Primates*, **36**, 461–476.

Kumar, A. (1987). *The Ecology and Population Dynamics of the Lion-tailed Macaque (*Macaca silenus*) in South India*. Ph.D. dissertation, Cambridge University.

Kumar, A. & Kurup, G. U. (1981). Infant development in the lion-tailed macaque, *Macaca silenus* (Linnaeus): the first eight weeks. *Primates*, **22**, 512–522.

(1985). Inter-troop interactions in the lion-tailed macaque, *Macaca silenus*. In *The Lion-tailed Macaque: Status and Conservation*, ed. P. G. Heltne, pp. 91–107. New York: Wiley-Liss.

Kummer, H. (1971). *Primate Societies*. Chicago: Aldine.

(1975). Rules of dyad and group formation among captive gelada baboons (*Theropithecus gelada*). In *Proceedings from the Symposia of the Fifth Congress of the International Primatological Society*, ed. S. Kondo, M. Kawai, A. Ehara & S. Kawamura, pp. 129–159. Tokyo: Japan Science Press.

(1984). From laboratory to desert and back: a social system of hamadryas baboons. *Animal Behaviour*, **32**, 965–971.

Kummer, H., Anzenberger, G. & Hemelrijk, C. K. (1996). Hiding and perspective taking in long-tailed macaques (*Macaca fascicularis*). *Journal of Comparative Psychology*, **110**, 97–102.

Kurland, J. A. (1977). *Kin Selection in the Japanese Monkey*. Basel: Karger.

Kurup, G. U. & Kumar, A. (1993). Time budget and activity patterns of the lion-tailed macaque (*Macaca silenus*). *International Journal of Primatology*, **14**, 27–39.

Küster, J. & Paul, A. (1984). Female reproductive characteristics in semi-free ranging Barbary macaques (*M. sylvanus* L. 1753). *Folia Primatologica*, **43**, 68–83.

Kutsukake, N. & Castles, D. L. (2001). Reconciliation and variation in post-conflict stress in Japanese macaques (*Macaca fuscata fuscata*): testing the integrated hypothesis. *Animal Cognition*, **4**, 259–268.

Laland, K., Odling-Smee, J. & Feldman, M. (2000). Niche construction, biological evolution and cultural change. *Behavioral and Brain Sciences*, **23**, 131–146.

Laland, K., Richerson, P. J. & Boyd, R. (1996). Developing a theory of animal social learning. In *Social Learning in Animals: the Roots of Culture*, ed. C. M. Heyes & B. G. Galef, pp. 129–154. London: Academic Press.

Laland, K. N., Odling-Smee, F. J. & Feldman, M. W. (1999). Evolutionary consequences of niche construction and their implications for ecology. *Proceedings of the National Academy of Sciences of the USA*, **96**, 10242–10247.

Lancaster, J. B. (1971). Play mothering: the relations between juvenile females and young infants among free ranging vervet monkeys (*Cercopithecus aethiops*). *Folia Primatologica*, **15**, 161–182.

Lathuillière, M. (2002). Influence de la Fragmentation des Populations et des Stratégies de Reproduction et de Dispersion sur la Diversité Génétique des Populations de Magots (*Macaca sylvanus*). Thèse d'Université, Université François Rabelais, Tours.

Lathuillière, M., Crouau-Roy, B., Petit, E., Scheffrahn, W. & Ménard, N. (2004). Influence of group fission on gene dispersion in the Barbary macaque (*Macaca sylvanus*). *Folia Primatologica*, **75**. (In press.)

Launhardt, K., Borries, C., Hardt, C., Epplen, J. T. & Winkler, P. (2001). Paternity analysis of alternative male reproductive routes among the langurs (*Semnopithecus entellus*) of Ramnagar. *Animal Behaviour*, **61**, 53–64.

Lazarus, R. S. (1991). *Emotion and Adaptation*. New York: Oxford University Press.

LeDoux, J. E. (1995). In search of an emotional system in the brain: leaping from fear to emotion and consciousness. In *The Cognitive Neurosciences*, ed. M. S. Gazzaniga, pp. 1049–1061. Cambridge, MA: MIT Press.

(1996). *The Emotional Brain*. New York: Simon & Schuster.

Lee, L. L. & Lin, Y. S. (1991). Status of Formosan macaques in Taiwan. In *Primatology Today*, ed. A. Ehara, T. Kimura, O. Takenaka & M. Iwamoto, pp. 33–36. Amsterdam: Elsevier.

Lee, P. C. & Bowman, J. E. (1995). Influence of ecology and energetics on primate mothers and infants. In *Motherhood in Human and Nonhuman Primates*, ed. C. R. Pryce, R. D. Martin & D. Skuse, pp. 47–58. Basel: Karger.

Lee, P. C. & Hauser, M. D. (1998). Long-term consequences of changes in territory quality on feeding and reproductive strategies of vervet monkeys. *Journal of Animal Ecology*, **67**, 347–358.

Lee, R. J. (1997). Impact of Hunting and Habitat Disturbance on the Population Dynamics and Behavioral Ecology of the Crested Black Macaque (*Macaca nigra*). Ph.D. dissertation, University of Oregon.

Lefebvre, L. (1995). Culturally-transmitted feeding behaviour in primates: evidence for accelerating learning rates. *Primates*, **36**, 227–239.

Leinfelder, I., de Vries, H., Deleu, R. & Nelissen, M. (2001). Rank and grooming reciprocity among females in a mixed-sex group of captive hamadryas baboons. *American Journal of Primatology*, **55**, 25–42.

Lévi-Strauss, C. (1955). *Tristes Tropiques*. Plon: Paris.

Lewis, M. H., Gluck, J. P., Beauchamp, A. J., Keresztury, M. F. & Mailman, R. B. (1990). Long-term effects of early social isolation in *Macaca mulatta*: changes in dopamine receptor function following apomorphine challenge. *Brain Research*, **513**, 67–73.

Lewis, R. J. (2002). Beyond dominance: the importance of leverage. *Quarterly Review of Biology*, **77**, 149–164.

Lewontin, R. C. (1974). *The Genetic Basis of Evolutionary Change*. New York: Columbia University Press.

Li, J., Wang, Q. & Han, D. (1996). Fission in a free-ranging Tibetan macaque troop at Huangshan Mountain, China. *Chinese Science Bulletin*, **41**, 1377–1381.

Lindburg, D. G. (1971). The rhesus monkey in North India: an ecological and behavior study. In *Primate Behavior: Developments in Field and Laboratory Research*, vol. 2, ed. L. A. Rosenblum, pp. 1–106. New York: Academic Press.

(1983). Mating behavior and estrus in the Indian rhesus monkey. In *Perspectives in Primate Biology*, ed. P. K. Seth, pp. 45–61. New Delhi: Today & Tomorrow.

Lindburg, D. G. & Harvey, N. C. (1996). Reproductive biology of captive lion-tailed macaques. In *Evolution and Ecology of Macaque Societies*, ed. J. E. Fa & D. G. Lindburg, pp. 318–341. Cambridge: Cambridge University Press.

Lindburg, D. G. & Lasley, B. L. (1985). Strategies for optimizing the reproductive potential of lion-tailed macaque colonies in captivity. In *The Lion-Tailed Macaques: Status and Conservation*, ed. P. G. Heltne, pp. 343–356, New York: Alan R. Liss.

Lindell, S. G., Maestripieri, D., Megna, N. L. & Higley, J. D. (2000). CSF 5-HIAA and MHPG predict infant abuse and rejection by rhesus macaque mothers. *American Journal of Primatology*, **51** (Suppl. 1), 70.

Lockard, J. S., Fahrenbruch, C. E., Smith, J. L & Morgan, C. J. (1977). Smiling and laughter: different phylogenetic origins. *Bulletin of the Psychonomic Society*, **10**, 183–186.

Lorenz, K. (1963). *Das sogenannte Böse. Zur Naturgeschichte der Aggression*. Wien: Borotha-Schoeler.

(1981). *The Foundations of Ethology*. New York: Springer.

Lott, D. F. (1991). *Intraspecific Variation in the Social Systems of Wild Vertebrates*. Cambridge: Cambridge University Press.

Low, B. S. (1978). Environmental uncertainty and the parental strategies of marsupials and placentals. *American Naturalist*, **112**, 197–213.

Loy, J. (1971). Estrous behavior of free-ranging rhesus monkeys (*Macaca mulatta*). *Primates*, **12**, 1–31.

(1988). Effects of supplementary feeding on maturation and fertility in primate groups. In *Ecology and Behavior of Food-Enhanced Primate Groups*, ed. J. Fa & C. H. Southwick, pp. 153–166. New York: Alan R. Liss.

Lucotte, G., Hazout, S. & Smith, D. G. (1984). Distinction électrophorétique des groupes naturels d'espèces dans le genre *Macaca*. *Biochemical Systematics and Ecology*, **12**, 339–347.

Lunardini, A. (1989). Social organization in a confined group of Japanese macaques (*Macaca fuscata*): an application of correspondence analysis. *Primates*, **30**, 175–185.

Lyles, A. M. & Dobson, A. P. (1988). The population biology of provisioned and unprovisioned primate populations. In *Ecology and Behavior of Food-Enhanced Primate Groups*, ed. J. E. Fa & C. H. Southwick, pp. 167–198. New York: Alan R. Liss.

MacDonald, G. J. (1971). Reproductive patterns of three species of macaques. *Fertility and Sterility*, **22**, 373–377.

Machida, S. (1990). Standing and climbing a pole by members of a captive group of Japanese monkeys. *Primates*, **31**, 291–298.

MacLean, P. D. (1952). Some psychiatric implications of physiological studies on frontotemporal portion of limbic system (visceral brain). *Electroencephalography and Clinical Neurophysiology*, **4**, 407–418.

Maestripieri, D. (1993). Maternal anxiety in rhesus macaques (*Macaca mulatta*). II. Emotional basis of individual differences in mothering style. *Ethology*, **95**, 32–42.

(1994a). Mother-infant relationships in three species of macaques (*Macaca mulatta*, *M. nemestrina, M. arctoides*). I. Development of the mother-infant relationship in the first three months. *Behaviour*, **131**, 75–96.

(1994b). Mother–infant relationships in three species of macaques (*Macaca mulatta*, *M. nemestrina, M. arctoides*). II. The social environment. *Behaviour*, **131**, 97–113.

(1994c). Social structure, infant handling, and mothering styles in group-living Old World monkeys. *International Journal of Primatology*, **15**, 531–553.

(1995). First steps in the macaque world: do rhesus mothers encourage their infants' independent locomotion? *Animal Behaviour*, **49**, 1541–1549.

(1996a). Maternal encouragement of infant locomotion in pigtail macaques (*Macaca nemestrina*). *Animal Behaviour*, **51**, 603–610.

(1996b). Gestural communication and its cognitive implications in pigtail macaques (*Macaca nemestrina*). *Behaviour*, **133**, 997–1022.

(1998). The evolution of male–infant interactions in the tribe Papionini (*Primates: Cercopithecidae*). *Folia Primatologica*, **69**, 247–251.

(1999). Fatal attraction: interest in infants and infant abuse in rhesus macaques. *American Journal of Physical Anthropology*, **110**, 17–25.

(2001). Intraspecific variability in parenting styles of rhesus macaques (*Macaca mulatta*): the role of the social environment. *Ethology*, **107**, 237–248.

Maestripieri, D. & Carroll, K. A. (1998). Risk factors for infant abuse and neglect in group-living rhesus monkeys. *Psychological Science*, **9**, 143–145.

Maestripieri, D., Martel, F. L., Nevison, C. M., Simpson, M. J. A. & Keverne, E. B. (1992b). Anxiety in rhesus monkey infants in relation to interactions with their mother and other social companions. *Developmental Psychobiology*, **24**, 571–581.

Maestripieri, D., Schino, G., Aureli, F. & Troisi, A. (1992a). A modest proposal: displacement activities as indicators of emotions in primates. *Animal Behaviour*, **44**, 967–979.

Maestripieri, D., Wallen, K. & Carroll, K. A. (1997). Infant abuse runs in families of group-living pigtail macaques. *Child Abuse and Neglect*, **21**, 465–471.

Malik, I. & Southwick, C. H. (1988). Feeding behavior and activity patterns of rhesus monkeys (*Macaca mulatta*) at Tughlaqabad, India. In *Ecology and Behavior of Food-enhanced Primate Groups*, ed. J. E. Fa & C. H. Southwick, pp. 95–111. New York: Alan R. Liss.

Malik, I., Seth, P. K. & Southwick, C. H. (1984). Population growth of free-ranging rhesus monkeys at Tughlaqabad. *American Journal of Primatology*, **7**, 311–321.

(1985). Group fission in free-ranging rhesus monkeys of Tughlaqabad, Northern India. *International Journal of Primatology*, **6**, 411–422.

Manson, J. H. (1996). Male dominance and mount series duration in Cayo Santiago rhesus macaques. *Animal Behaviour*, **51**, 1219–1231.

(1992). Measuring female mate choice in Cayo Santiago rhesus macaques. *Animal Behaviour*, **44**, 405–416.

(1993). Sons of low-ranking female rhesus macaques can attain high dominance rank in their natal groups. *Primates*, **34**, 285–288.

(1994). Male aggression: a cost of female mate choice in Cayo Santiago rhesus macaques. *Animal Behaviour*, **48**, 473–475.

(1995). Do female rhesus monkeys choose novel males? *American Journal of Primatology*, **37**, 285–296.

Manson, J. H. & Perry, S. (1993). Inbreeding avoidance in rhesus macaques. Whose choice? *American Journal of Physical Anthropology*, **90**, 335–344.

Manuck, S. B., Kaplan, J. R. & Clarkson, T. B. (1986). Atherosclerosis, social dominance, and cardiovascular reactivity. In *Biological and Psychological Factors in Cardiovascular Disease*, ed. T. H. Schmidt, T. M. Dembrosky & G. Blumchen, pp. 461–475. Berlin: Springer.

Martel, F. L., Nevison, C. M., Simpson, M. J. A. & Keverne, E. B. (1995). Effects of opioid receptor blockade on the social behavior of rhesus monkeys living in large family groups. *Developmental Psychobiology*, **28**, 71–84.

Martin, L. J., Spicer, D. M., Lewis, M. H., Gluck, J. P. & Cork, L. C. (1991). Social deprivation of infant rhesus monkeys alters the chemoarchitecture of the brain: I. Subcortical regions. *Journal of Neuroscience*, **11**, 3344–3358.

Maruhashi, T. (1982). An ecological study of troop fissions of Japanese monkeys (*Macaca fuscata yakui*) on Yakushima Island, Japan. *Primates*, **23**, 317–337.

(1992). Fission, takeover, and extinction of a troop of Japanese monkeys (*Macaca fuscata yakui*) on Yakushima Island, Japan. In *Topics in Primatology*, vol. 2, ed. N. Itoigawa, Y. Sugiyama, G. Sackett & R. K. R. Thompson, pp. 47–56. Tokyo: University of Tokyo Press.

Maruhashi, T., Saito, C. & Agetsuma, N. (1998). Home range structure and inter-group competition for land of Japanese macaques in evergreen and deciduous forests. *Primates*, **39**, 291–301.

Masataka, N. & Thierry, B. (1993). Vocal communication of Tonkean macaques in confined environments. *Primates*, **34**, 169–120.

Maslow, A. H. (1940). Dominance-quality and social behavior in infra-human primates. *Journal of Social Psychology*, **11**, 313–324.

Mason, W. A. (1978a). Social experience and primate cognitive development. In *The Development of Behavior: Comparative and Evolutionary Aspects*, ed. G. M. Burghardt & M. Bekoff, pp. 233–251. New York: Garland Press.

(1978b). Ontogeny of social systems. In *Recent Advances of Primatology*, vol. 1, ed. D. J. Chivers & J. Herbert, pp. 5–14. London: Academic Press.

(1984). Animal learning: experience, life modes and cognitive style. *Verhandlungen deutsche Zoologische Gesellschaft*, **45**, 45–56.

(1985). Experimental influences on the development of expressive behaviors in rhesus monkeys. In *The Development of Expressive Behavior: Biology–Environment Interactions*, ed. G. Zivin, pp. 117–152. New York: Academic Press.

(2002). The natural history of primate development: an organismic perspective. In *Conceptions of Development: Lessons from the Laboratory*, ed. D. J. Lewkowicz & R. Lickliter, pp. 105–134. New York: Psychology Press.

Mason, W. A., Long, D. D. & Mendoza, S. P. (1993). Temperament and mother-infant conflict in macaques: a transactional analysis. In *Primate Social Conflict*, ed. W. A. Mason & S. P. Mendoza, pp. 205–227. Albany, NY: State University of New York Press.

Mason, W. A., Mendoza, S. P. & Moberg, G. P. (1991). Persistent effects of early social experience on physiological responsiveness. In *Primatology Today*, ed. A. Ehara, pp. 469–471. Amsterdam: Elsevier.

Masui, K., Sugiyama, Y., Nishimura, A. & Ohsawa, H. (1975). The life table of Japanese monkeys at Takasakiyama: a preliminary report. In *Proceedings of the Fifth Congress of the International Primatological Society*, ed. M. Kawai, S. Kondo & A. Ehara, pp. 401–406. Basel: Karger.

Matsubayashi, K. & Mochuzuki, K. (1982). Growth of male reproductive organs with observation of their seasonal morphologic changes in the Japanese monkey (*Macaca fuscata*). *Japanese Journal of Veterinary Science*, **44**, 891–902.

Matsumura, S. (1991). A preliminary report on the ecology and social behavior of moor macaques (*Macaca maurus*) in Sulawesi, Indonesia. *Kyoto University Overseas Research Report Studies in Asian Non-human Primates*, **8**, 27–41.

(1993). Female reproductive cycles and sexual behavior of moor macaques (*Macaca maura*) in their natural habitat, South Sulawesi, Indonesia. *Primates*, **34**, 99–103.

(1995). Affiliative mounting interference in *Macaca maurus*. *Kyoto University Overseas Research Report Studies in Asian Non-human Primates*, **9**, 1–5.

(1996). Postconflict affiliative contacts between former opponents among wild moor macaques (*Macaca maurus*). *American Journal of Primatology*, **38**, 211–219.

(1998). Relaxed dominance relations among female moor macaques (*Macaca maurus*) in their natural habitat, South Sulawesi, Indonesia. *Folia Primatologica*, **69**, 346–356.

(1999). The evolution of "egalitarian" and "despotic" social systems among macaques. *Primates*, **40**, 23–31.

(2001). The myth of despotism and nepotism: dominance and kinship in matrilineal societies of macaques. In *Primate Origins of Human Cognition and Behavior*, ed. T. Matsuzawa, pp. 441–462. Tokyo: Springer.

Matsumura, S. & Okamoto, K. (1997). Factors affecting proximity among members of a wild group of moor macaques during, feeding, moving, and resting. *International Journal of Primatology*, **18**, 929–940.

Matsuzawa, T., Diro, D., Humle, T., Inoue-Nakamura, N., Tonooka, R. & Yamakoshi, G. (2001). Emergence of culture in wild chimpanzees: education by master-apprenticeship. In *Primate Origins of Human Cognition and Behavior*, ed. T. Matsuzawa, pp. 557–574. Tokyo: Springer.

Maynard Smith, J. (1983). *Evolution and Theory of Games*. Cambridge: Cambridge University Press.

Maynard Smith, J., Burian, S., Kaufmann S. *et al.* (1985). Developmental constraints and evolution. *Quarterly Review of Biology*, **6**, 265–287.

Mayr, E. (1961). Cause and effect in biology. *Science*, **134**, 1501–1506.

(1963). *Animal Species and Evolution*. Cambridge, MA: Harvard University Press.

Mazur, A. (1985). A biosocial model of status in face-to face primate groups. *Social Forces*, **64**, 377–402.

McFarland, D. (1966). On the causal and functional significance of displacement activities. *Zeitschrift für Tierpsychologie*, **23**, 217–235.

McFarland Symington, M. (1988). Environmental determinants of population density in *Ateles*. *Primate Conservation*, **9**, 74–79.

McGue, M. & Bouchard, T. J. (1998). Genetic and environmental influences on human behavioral differences. *Annual Review of Neuroscience*, **21**, 1–24.

McKenna, J. J. (1979). The evolution of allomothering among colobine monkeys: function and opportunism in evolution. *American Anthropologist*, **81**, 818–840.

McMillan, C. A. & Duggleby, C. R. (1981). Interlineage genetic differentiation among rhesus macaques on Cayo Santiago. *American Journal of Physical Anthropology*, **56**, 305–312.

Medawar, P. B. & Medawar, J. S. (1983). *Aristotle to Zoos: a Philosophical Dictionary of Biology*. Cambridge, MA: Harvard University Press.

Mehlman, P. (1986). Male intergroup mobility in a wild population of the Barbary macaque (*Macaca sylvanus*), Ghomaran Rif Mountains, Morocco. *American Journal of Primatology*, **10**, 67–81.

Mehlman, P. T. & Chapais, B. (1988). Differential effects of kinship, dominance and the mating season on female allogrooming in a captive group of *Macaca fuscata*. *Primates*, **29**, 195–217.

Mehlman, P. T. & Parkhill, R. S. (1988). Intergroup interactions in wild Barbary macaques (*Macaca sylvanus*), Ghomaran Rif Mountains, Morocco. *American Journal of Primatology*, **15**, 31–44.

Mehlman, P. T., Higley, J. D., Faucher, I. *et al.* (1994). Low CSF 5-HIAA concentrations and severe aggression and impaired impulse control in nonhuman primates. *American Journal of Psychiatry*, **151**, 1485–1491.

(1995). Correlation of CSF 5-HIAA concentration with sociality and the timing of emigration in free-ranging primates. *American Journal of Psychiatry*, **152**, 907–913.

Meier, C., Hemelrijk, C. K. & Martin, R. D. (2000). Paternity determination, genetic characterization and social correlates in a captive group of chimpanzees (*Pan troglodytes*). *Primates*, **41**, 175–183.

Meikle, D. B. & Vessey, S. H. (1981). Nepotism among rhesus monkey brothers. *Nature*, **294**, 160–161.

Melnick, D. J. (1983). Genetic diversity in wild rhesus monkeys. *American Journal of Primatology*, **4**, 348–349.

(1987). The genetic consequences of primate social organization: a review of macaques, baboons and vervet monkeys. *Genetica*, **73**, 117–135.

(1988). The genetic structure of a primate species: rhesus macaques and other Cercopithecine monkeys. *International Journal of Primatology*, **9**, 195–231.

Melnick, D. J. & Hoelzer, G. A. (1996). The population genetic consequences of macaque social organisation and behaviour. In *Evolution and Ecology of Macaque Societies*, ed. J. E. Fa & D. G. Lindburg, pp. 413–443. Cambridge: Cambridge University Press.

Melnick, D. J. & Kidd, K. K. (1983). The genetic consequences of social group fission in a wild population of rhesus monkeys (*Macaca mulatta*). *Behavioral Ecology and Sociobiology*, **12**, 229–236.

Melnick, D. J. & Pearl, M. C. (1987). Cercopithecines in multimale groups: genetic diversity and population structure. In *Primate Societies*, ed. B. B. Smuts, D. L. Cheney, R. M., Seyfarth, R. W. Wrangham & T. T. Struhsaker, pp. 121–134. Chicago: University of Chicago Press.

Melnick, D. J., Jolly, C. J. & Kidd, K. K. (1984b). The genetics of a wild population of rhesus monkeys (*Macaca mulatta*). I. Genetic variability within and between social groups. *American Journal of Physical Anthropology*, **63**, 341–360.

(1986). The genetics of a wild population of rhesus monkeys (*Macaca mulatta*). II. The Dunga Gali population in species-wide perspective. *American Journal of Physical Anthropology*, **71**, 129–140.

Melnick, D. J., Pearl, M. C. & Richard, A. F. (1984a). Male migration and inbreeding avoidance in wild rhesus monkeys. *American Journal of Primatology*, **7**, 229–243.

Ménard, N. (1985). Le régime alimentaire de *Macaca sylvanus* dans différents habitats d'Algérie: 1. Régime en chênaie décidue. *Revue d'Ecologie (Terre Vie)*, **40**, 451–466.

(2002). Ecological plasticity of Barbary macaques (*Macaca sylvanus*). *Evolutionary Anthropology*, **11** (Suppl. 1), 95–100.

Ménard, N. & Qarro, M. (1999). Bark stripping and water availability: a comparative study between Moroccan and Algerian Barbary macaques (*Macaca sylvanus*). *Revue d'Ecologie (Terre et Vie)*, **54**, 123–132.

Ménard, N. & Vallet, D. (1986). Le régime alimentaire de *Macaca sylvanus* dans différents habitats d'Algérie. 2. Régime en forêt sempervirente et sur les sommets rocheux. *Revue d'Ecologie (Terre et Vie)*, **41**, 173–192.

(1988). Disponibilités et utilisation des ressources par le magot (*Macaca sylvanus*) dans différents milieux en Algérie. *Revue d'Ecologie (Terre et Vie)*, **43**, 201–250.

(1993a). Population dynamics of *Macaca sylvanus* in Algeria: an 8-year study. *American Journal of Primatology*, **30**, 101–118.

(1993b). Dynamics of fission in a wild Barbary macaque group (*Macaca sylvanus*). *International Journal of Primatology*, **14**, 479–500.

(1996). Demography and ecology of Barbary macaques (*Macaca sylvanus*) in two different habitats. In *Evolutionary Ecology and Behaviour of Macaques*, ed. J. E. Fa & D. G. Lindburg, pp. 106–131. Cambridge: Cambridge University Press.

(1997). Behavioral responses of Barbary macaques (*Macaca sylvanus*) to variations in environmental conditions in Algeria. *American Journal of Primatology*, **43**, 285–304.

Ménard, N., von Segesser, F., Scheffrahn, W. *et al.* (2001). Is male-infant caretaking related to paternity and/or mating activities in wild Barbary macaques (*Macaca sylvanus*)? *Comptes Rendus de l'Académie des Sciences III*, **324**, 601–610.

Mendoza, S. P. & Mason, W. A. (1989). Primate relationships: social dispositions and physiological responses. In *Perspectives in Primate Biology*, vol. 2, ed. P. K. Seth & S. Seth, pp. 129–143. New Delhi: Today & Tomorrow.

Mendoza, S. P., Reeder, D. M. & Mason, W. A. (2002). Nature of proximate mechanisms underlying primate social systems: simplicity and redundancy. *Evolutionary Anthropology*, 1 (Suppl.), 112–116.

Menon, S. & Poirier, F. E. (1996). Lion-tailed macaques (*Macaca silenus*) in a disturbed forest fragment: activity patterns and time budget. *International Journal of Primatology*, **17**, 969–985.

Michael, R. P. & Zumpe, D. (1970). Aggression and gonadal hormones in captive rhesus monkeys (*Macaca mulatta*). *Animal Behaviour*, **18**, 1–10.

(1993). A review of hormonal factors influencing the sexual and aggressive behavior of macaques. *American Journal of Primatology*, **30**, 213–241.

Michels, R. (1911 [1962]). *Political Parties: A Sociological Study of the Oligarchical Tendencies of the Modern Democracy*. New York: Collier Books.

Miller, N. E. (1959). Liberalisation of basic S-R concepts: extension to conflict, motivation and social learning. In *Psychology: A Study of a Science. Study 1*, vol. 2, ed. S. Koch, pp. 196–292. New York: McGraw-Hill.

Miller, P. B., Charleston, J. S., Battaglia, D. E., Klein, N. A. & Soules, M. (1999). Morphometric analysis of primordial follicle number in pigtailed monkey ovaries: symmetry and relationship with age. *Biology of Reproduction*, **61**, 553–556.

Miller, R. E. (1971). Experimental studies of communication in the monkey. *Primate Behavior: Developments in Field and Laboratory Research*, vol. 2, ed. L. A. Rosenblum, pp. 131–175. New York: Academic Press.

Milton, K. (1982). Dietary quality and demographic regulation in a howler monkey population. In *The Ecology of a Tropical Forest*, ed. E. G. Leigh, A. S. Rand & D. M. Windsor, pp. 273–289. Washington, DC: Smithsonian Institution Press.

(1996). Effects of bot fly (*Alouattamyia baeri*) parasitism on a free-ranging howler monkey (*Alouatta palliata*) population in Panama. *Journal of Zoology, London*, **239**, 39–63.

Mineka, S., Davidson, M., Cook, M. & Keir, R. (1984). Observational conditioning of snake fear in rhesus monkeys. *Journal of Abnormal Psychology*, **93**, 335–372.

Missakian, E. A. (1972). Genealogical and cross-genealogical dominance relations in a group of free-ranging rhesus monkeys (*Macaca mulatta*) on Cayo Santiago. *Primates*, **13**, 169–180.

(1973a). The timing of fission among free-ranging rhesus monkeys. *American Journal of Physical Anthropology*, **36**, 621–624.

(1973b). Genealogical mating activity in free-ranging groups of rhesus monkeys (*M. mulatta*) in Cayo Santiago. *Behaviour*, **45**, 225–241.

Mito, Y. (1992). Why is Japanese monkey distribution so limited in the northern Tohoku Region? (In Japanese.) *Biological Science*, **44**, 141–158.

Molur, S., Brandon-Jones, D., Dittus, W. *et al.*, eds. (2003). *Status of South Asian Primates*. Conservation Assessment and Management Plan (CAMP) Workshop Report. Coimbatore, India: Zoo Outreach Organisation and CBSG-South Asia.

Moore, J. (1984). Female transfer in primates. *International Journal of Primatology*, **5**, 537–589.

Moore, J. & Ali, R. (1984). Are dispersal and inbreeding avoidance related? *Animal Behaviour*, **32**, 94–112.

Morales, J. C. & Melnick, D. J. (1998). Phylogenetic relationships of the macaques (*Cercopithecidae: Macaca*), as revealed by high resolution restriction site mapping of mitochondrial ribosomal genes. *Journal of Human Evolution*, **34**, 1–23.

Morgan, D., Grant, K. A., Gage, H. D. *et al.* (2002). Social dominance in monkeys: dopamine D2 receptors and cocaine self-administration. *Nature Neuroscience*, **5**, 169–174.

Mori, A. (1977). The social organization of the provisioned Japanese monkey troops which have extraordinary large population sizes. *Anthropological Society of Nippon*, **84**, 325–345.

(1979a). Analysis of population changes by measurement of body weight in the Koshima troop of Japanese monkeys. *Primates*, **20**, 371–397.

(1979b). Intra-troop spacing mechanism of the wild Japanese monkeys of the Koshima troop. *Primates*, **18**, 331–357.

Mori, A., Watanabe, K. & Yamaguchi, N. (1989). Longitudinal changes of dominance rules among the females of the Koshima group of Japanese monkeys. *Primates*, **30**, 147–173.

Mori, A., Yamaguchi, N., Watanabe, K. & Shimizu, K. (1997). Sexual maturation of female Japanese macaques under poor nutritional conditions and food enhanced perineal swelling in the Koshima troop. *International Journal of Primatology*, **18**, 553–579.

Morley, R. J. (2000). Origin and evolution of tropical rain forest. Chichester, NY: Wiley.

Moschos, S., Chan, J. L. & Mantzoros, C. S. (2002). Leptin and reproduction: a review. *Fertility and Sterility*, **77**, 433–444.

Muckenhirn, N. A. & Eisenberg, J. F. (1973). Home ranges and predation of the Ceylon leopard. In *The World's Cats*, vol. 1, ed. R. L. Eaton, pp. 142–175. Winston, OR: World Wildlife Safari.

Muroyama, Y. (2000). Monkeys living near villages: Japanese macaques that have started a new life. (In Japanese.) In *Primate Ecology: Dynamics of Environments and Behaviours*, ed. Y. Sugiyama, pp. 225–247. Kyoto: Kyoto-Gakujutsu-shuppankai.

Muroyama, Y., Torii, H. & Maekawa, S. (1999). Status of wild Japanese macaques (*Macaca fuscata fuscata*) in Kinki district, Japan. (In Japanese.) *Wildlife Forum*, **5**, 1–15.

Murray, H. A. (1938). *Explorations in Personality*. New York: Oxford University Press.

Murray, R. D. & Smith, E. O. (1982). The role of dominance and intrafamilial bonding in the avoidance of close inbreeding. *Journal of Human Evolution*, **12**, 481–486.

Murray, R. D., Bour, E. S. & Smith, E. O. (1985). Female menstrual cyclicity and sexual behavior in stumptail macaques (*Macaca arctoides*). *International Journal of Primatology*, **6**, 101–113.

Nadler, R. D. (1990). Homosexual behavior in nonhuman primates. In *Homosexuality/ Heterosexuality: Concepts of Sexual Orientation*, ed. D. P. McWhirter, S. A. Sanders & J. M. Reinisch, pp. 138–170. New York: Oxford University Press.

Nakamichi, M. (1988). Aging and behavioral changes of female Japanese macaques. *Research Reports of the Arashiyama West and East Groups of Japanese Monkeys*, 87–97.

(1996). Proximity relationships within a birth cohort of immature Japanese monkeys (*Macaca fuscata*) in a free-ranging group during the first four years of life. *American Journal of Primatology*, **40**, 315–325.

Nakamichi, M., Itoigawa, N., Imakawa, S. & Machida S. (1995). Dominance relations among adult females in a free-ranging group of Japanese monkeys at Katsuyama. *American Journal of Primatology*, **37**, 241–251.

Nakamichi, M., Nobuhara, H., Nobuhara, T., Nakahashi, M. & Nigi, H. (1997). Birth rate and mortality rate of infants with congenital malformations of the limbs in the

Awajishima free-ranging group of Japanese monkeys (*Macaca fuscata*). *American Journal of Primatology*, **42**, 225–234.

Nanda, S. (2000). *Gender Diversity: Cross-cultural Variations*. Prospect Heights, IL: Waveland Press.

National Research Council (1981). *Techniques for the Study of Primate Population Ecology*. Washington, DC: National Academy Press.

Nawar, M. M. & Hafez, E. S. E. (1972). The reproductive cycle of the crab-eating macaque (*Macaca fascicularis*). *Primates*, **13**, 43–56.

Nei, M. (1973). Analysis of gene diversity in subdivided populations. *Proceedings of the National Academy of Sciences of the USA*, **70**, 3321–3323.

(1975). *Molecular Population Genetics*. Amsterdam: North-Holland.

(1977). F-statistics and analysis of gene diversity in subdivided populations. *Annals of Human Genetics*, **41**, 225–233.

Netto, J. W. & van Hooff, J. A. R. A. M. (1986). Conflict interference and the development of dominance relationships in immature *Macaca fascicularis*. In *Primate Ontogeny, Cognition and Social Behaviour*, ed. J. G. Else & P. C. Lee, pp. 291–300. Cambridge: Cambridge University Press.

Niemeyer, C. L. & Anderson, J. R. (1983). Primate harassment of matings. *Ethology and Sociobiology*, **4**, 205–220.

Niemeyer, C. L. & Chamove, A. S. (1983). Motivation of harassment of matings in stumptailed macaques. *Behaviour*, **87**, 298–323.

Nieuwenhuijsen, K., Bonke-Jansen, M., de Neef, K. J, van der Werfff ten Bosch, J. J. & Slob, A. K. (1987). Physiological aspects of puberty in group-living stumptail monkeys (*Macaca arctoides*). *Physiology and Behavior*, **41**, 37–45.

Nieuwenhuijsen, K., Bonke-Jansen, M., Broekhuijzen, E. *et al.* (1988). Behavioral aspects of puberty in group-living stumptail monkeys (*Macaca arctoides*). *Physiology and Behavior*, **42**, 255–264.

Nieuwenhuijsen, K., de Neef, K. J. & Slob, A. K. (1986). Sexual behavior during ovarian cycles, pregnancy and lactation in group living stumptail macaques (*Macaca arctoides*). *Human Reproduction*, **1**, 159–169.

Nieuwenhuijsen, K., Lammers, A. J. J. C., de Neef, K. & Slob, A. K. (1985). Reproduction and social rank in female stumptail macaques (*Macaca arctoides*). *International Journal of Primatology*, **6**, 77–98.

Nigi, H. (1975). Menstrual cycle and some other related aspects of Japanese monkeys (*Macaca fuscata*). *Primates*, **16**, 207–216.

(1976). Some aspects related to conception of the Japanese monkey (*Macaca fuscata*). *Primates*, **17**, 81–87.

Nigi, H., Tiba, T., Yamamoto, S., Floescheim, Y. & Ohsawa, N. (1980). Sexual maturation and seasonal changes in reproductive phenomenon of male Japanese monkeys (*Macaca fuscata*) at Takasakiyama. *Primates*, **21**, 230–240.

Ninan, P. T., Insel, T. M., Cohen, R. M., Cook, J. M., Skolnik, P. & Paul, S. M. (1982). Benzodiazepine receptor-mediated experimental 'anxiety' in primates. *Science*, **218**, 1332–1334.

Nishida, T. (1963). Intertroop relationships of the Formosan monkeys (*Macaca cyclopsis*) relocated on Nojima Island. *Primates*, **4**, 121–122.

(1979). The social structure of chimpanzees of the Mahale mountains. In *The Great Apes*, ed. D. A. Hamburg & E. R. McCown, pp. 73–121. Menlo Park, CA: Benjamin Cummings.

(1987). Local traditions and cultural transmission. In *Primate Societies*, ed. B. B. Smuts, D. L. Cheney, R. M. Seyfarth, R. W. Wrangham & T. T. Struhsaker, pp. 462–474. Chicago: University of Chicago Press.

Nishimura, A. (1973). The third fission of a Japanese monkey group at Takasakiyama. In *Behavioral Regulators of Behavior in Primates*, ed. C. R. Carpenter, pp. 115–123. Lewisburg: Bucknell University Press.

Noë, R. (1992). Alliance formation among male baboons: shopping for profitable partners. In *Coalitions and Alliances in Humans and Other Animals*, ed. A. H. Harcourt & F. B. M. de Waal, pp. 285–322. Oxford: Oxford University Press.

Nöe, R., de Waal, F. B. M. & van Hooff, J. A. R. A. M. (1980). Types of dominance in a chimpanzee colony. *Folia Primatologica*, **34**, 90–110.

Nomura, T, Ohsawa, N., Tajima, Y. *et al.* (1972). Reproduction of Japanese monkeys. In *The Use of Non-Human Primates on Human Reproduction*, ed. E. Diczfalusy, E. & C. C. Standley, pp. 473–482. Copenhagen: Bogtrykkeriet Foraum.

Noordwijk, M. & van Schaik, C. P. (1999). The effects of dominance rank and group size on female lifetime reproductive success in wild long-tailed macaques, *Macaca fascicularis*. *Primates*, **40**, 105–130.

Nozaki, M., Mitsunaga, F. & Shimizu, K. (1995). Reproductive senescence in female Japanese monkeys (*Macaca fuscata*): age-and season-related changes in hypothalamic–pituitary–ovarian functions and fecundity rates. *Biology of Reproduction*, **52**, 1250–1257.

Nozaki, M., Yamashita, K. & Shimizu, K. (1997). Age-related changes in ovarian morphology from birth to menopause in the Japanese monkey (*Macaca fuscata fuscata*). *Primates*, **38**, 89–100.

Nozawa, K., Shotake, T., Kawamoto, Y. & Tanabe, Y. (1982). Population genetics of Japanese monkeys: II. Blood protein polymorphisms and population structure. *Primates*, **23**, 252–271.

Nozawa, K., Shotake, T., Minezawa, M. *et al.* (1991). Population genetics of Japanese monkeys: III. Ancestry and differentiation of local populations. *Primates*, **32**, 411–435.

Nunn, C. L. (1999). The evolution of exaggerated sexual swellings in primates and the graded-signal hypothesis. *Animal Behaviour*, **58**, 229–246.

O'Brien, T. G. & Kinnaird, M. F. (1997). Behavior, diet, and movements of the Sulawesi crested black macaque (*Macaca nigra*). *International Journal of Primatology*, **18**, 321–351.

(2000). Differential vulnerability of large birds and mammals to hunting in North Sulawesi, Indonesia and the outlook for the future. In *Hunting for Sustainability in Tropical Forests*, ed. J. G. Robinson & E. L. Bennett, pp. 199–213. New York: Columbia University Press.

O'Connor, K. A., Holman, D. J. & Wood, J. W. (2001). Menstrual cycle variability and the perimenopause. *American Journal of Human Biology*, **13**, 465–478.

Ober, C., Olivier, T. J. & Buettner-Janusch, J. (1980). Genetic aspects of migration in a rhesus monkey population. *Journal of Human Evolution*, **9**, 197–203.

Ober, C., Olivier, T. J., Sade, D. S., Schneider, J. M., Cheverud, J. & Buettner-Janusch, J. (1984). Demographic components of gene frequency change in free-ranging macaques on Cayo Santiago. *American Journal of Physical Anthropology*, **64**, 223–231.

Odling-Smee, F. J. (1988). Niche-constructing phenotypes. In *The Role of Behavior in Evolution*, ed. H. C. Plotkin, pp. 73–132. Cambridge, MA: MIT Press.

Ogawa, H. (1995). Bridging behavior and other affiliative interactions among male Tibetan macaques (*Macaca thibetana*). *International Journal of Primatology*, **16**, 707–729.

Öhman, A. (1993). Fear and anxiety as emotional phenomena: clinical phenomenology, evolutionary perspectives, and information-processing mechanisms. In *Handbook of Emotions*, ed. M. Lewis & J. M. Haviland, pp. 511–536. New York: Guilford Press.

Ohsawa, H. & Sugiyama, Y. (1996). Population dynamics of Japanese monkeys at Takasakiyama: trends in 1985–1992. In *Variations in the Asian Macaques*, ed. T. Shotake & K. Wada, pp. 163–179. Tokyo: Tokai University Press.

Oi, T. (1988). Sociological study on the fission of wild Japanese monkeys (*Macaca fuscata yakui*) on Yakushima Island. *Primates*, **29**, 1–19.

(1990). Population organization of wild pig-tailed macaques (*Macaca nemestrina nemestrina*) in West Sumatra. *Primates*, **31**, 15–31.

(1996). Sexual behaviour and mating system of the wild pig-tailed macaque in West Sumatra. In *Evolution and Ecology of Macaque Societies*, ed. J. E. Fa & D. G. Lindburg, pp. 342–368. Cambridge: Cambridge University Press.

Okamoto, K. (2004). Change in inter-birth intervals prior to a group fission in moor macaques (*Macaca maurus*). *Bulletin of Tokai Gauken University*, **9**. (In press.)

Okamoto, K. & Matsumura, S. (2001). Group fission in moor macaques (*Macaca maurus*). *International Journal of Primatology*, **22**, 481–493.

(2002). Intergroup encounters in wild moor macaques (*Macaca maurus*). *Primates*, **43**, 119–125.

Okamoto, K., Matsumura, S. & Watanabe, K. (2000). Life history and demography of wild moor macaques (*Macaca maurus*): summary of 10 years of observation. *American Journal of Primatology*, **52**, 1–11.

Olivier, T. J., Ober, C., Buettner-Janusch, J. & Sade, D. S. (1981). Genetic differentiation among matrilines in social groups of rhesus monkeys. *Behavioral Ecology and Sociobiology*, **8**, 279–285.

Oswald, M. & Erwin, J. (1976). Control of intragroup aggression by male pigtail monkeys (*Macaca nemestrina*). *Nature*, **262**, 686–688.

Otis, J. S., Froehlich, J. W. & Thorington, R. W. (1981). Seasonal and age-related differential mortality by sex in the mantled howler monkey, *Alouatta palliata*. *International Journal of Primatology*, **2**, 197–205.

Owren, M. J. & Rendall, D. (1997). An affect-conditioning model of nonhuman primate vocal signaling. *Perspectives in Ethology*, **12**, 299–346.

Owren, M. J., Dieter, J. A., Seyfarth, R. M. & Cheney, D. L. (1992). Food calls produced by adult female rhesus (*Macaca mulatta*) and Japanese (*M. fuscata*) macaques, their normally raised infants, and offspring cross-fostered between species. *Behaviour*, **120**, 218–231.

Oyama, S. (1985). *The Ontogeny of Information*. Cambridge: Cambridge University Press.

Packer, C. (1979). Inter-troop transfer and inbreeding avoidance in *Papio anubis. Animal Behaviour*, **27**, 1–36.

Packer, C. & Pusey, A. (1979). Female aggression and male membership in troops of Japanese macaques and olive baboons. *Folia Primatologica*, **31**, 212–218.

Packer, C., Collins, D. A., Sindimwo, A. & Goodall, J. (1995). Reproductive constraints on aggressive competition in female baboons. *Nature*, **373**, 60–63.

Packer, C., Tatart, M. & Collins, A. (1998). Reproductive cessation in female mammals. *Nature*, **392**, 807–811.

Panksepp, J. (1989). The psychobiology of emotions: the animal side of human feelings. In *Emotions and the Dual Brain*, ed. G. Gainotti & C. Caltagirone, pp. 31–55. Berlin: Springer.

(1998). *Affective Neuroscience: the Foundation of Human and Animal Emotions*. New York: Oxford University Press.

Panksepp, J., Nelson, E. & Bekkedal, M. (1997). Brain systems for the mediation of social separation-distress and social-reward. *Annals of the New York Academy of Sciences*, **807**, 78–100.

Parker, G. A. (1974). Assessment strategy and the evolution of fighting behavior. *Journal of Theoretical Biology*, **47**, 223–243.

Parker, R. & Hendrickx, A. (1975). The temporal relationship between preovulatory estradiol peak and the optimal mating period in rhesus and the bonnet monkey. *Biology of Reproduction*, **13**, 617–622.

Parr, L. A. (2001). Cognitive and physiological markers of emotional awareness in chimpanzees (*Pan troglodytes*). *Animal Cognition*, **4**, 223–229.

Parr, L. A. & Hopkins, W. D. (2000). Brain temperature asymmetries and emotional perception in chimpanzees, *Pan troglodytes. Physiology and Behavior*, **71**, 363–371.

Parsons, T. (1963). On the concept of influence. *Public Opinion Quarterly*, **27**, 37–62.

Paul, A. (1997). Breeding seasonality affects the association between dominance and reproductive success in non-human male primates. *Folia Primatologica*, **68**, 344–349.

(1999). The sociecology of infant handling in primates: is the current model convincing? *Primates*, **40**, 33–46.

(2002). Sexual selection and mate choice. *International Journal of Primatology*, **23**, 877–904.

Paul, A. & Kuester, J. (1985). Intergroup transfer and incest avoidance in semifree-ranging Barbary macaques (*Macaca sylvanus*) at Salem (FRG). *American Journal of Primatology*, **8**, 317–322.

(1987). Dominance, kinship and reproductive value in female Barbary macaques (*Macaca sylvanus*) at Affenberg Salem (FRG). *Behavioral Ecology and Sociobiology*, **21**, 323–331.

(1996a). Differential reproduction in male and female Barbary macaques. In *Evolution and Ecology of Macaque Societies*, ed. J. E. Fa & D. G. Lindburg, pp. 293–317. Cambridge: Cambridge University Press.

(1996b). Infant handling by female Barbary macaques (*Macaca sylvanus*) at Affenberg Salem: testing functional and evolutionary hypotheses. *Behavioral Ecology and Sociobiology*, **39**, 133–145.

Paul, A. & Thommen, D. (1984). Timing of birth, female reproductive success and infant sex ratio in semi-free ranging Barbary macaques (*Macaca sylvanus*). *Folia Primatologica*, **42**, 2–16.

Paul, A., Kuester, J. & Arnemann, J. (1992a). DNA fingerprinting reveals that infant care by male Barbary macaques (*Macaca sylvanus*) is not paternal investment. *Folia Primatologica*, **58**, 93–98.

(1992b). Maternal rank affects reproductive success of male Barbary macaques (*Macaca sylvanus*): evidence from DNA fingerprinting. *Behavioral Ecology and Sociobiology*, **30**, 337–341.

(1996). The sociobiology of male–infant interactions in Barbary macaques, *Macaca sylvanus*. *Animal Behaviour*, **51**, 155–170.

Paul, A., Kuester, J. & Podzuweit, D. (1993a). Reproductive senescence and terminal investiment in female Barbary macaques (*Macaca sylvanus*) at Salem. *International Journal of Primatology*, **14**, 105–124.

Paul, A., Kuester, J., Timme, A. & Arnemann, J. (1993b). The association between rank, mating effort, and reproductive success in male Barbary macaques (*Macaca sylvanus*). *Primates*, **34**, 491–502.

Pavani, S., Maestripieri, D., Schino, G., Turillazzi, P. G. & Scucchi, S. (1991). Factors influencing scratching behaviour in long-tailed macaques (*Macaca fascicularis*). *Folia Primatologica*, **57**, 34–38.

Pavelka, M. S. M. & Fedigan, L. M. (1999). Reproductive termination in female Japanese monkeys: a comparative life history perspective. *American Journal of Physical Anthropology*, **109**, 455–464.

Pavelka, M. S. M., Fedigan, L. M. & Zohar, S. (2002). Availability and adaptive value of reproductive and postreproductive Japanese macaque mothers and grandmothers. *Animal Behaviour*, **64**, 407–414.

Peccei, J. C. (2001). A critique of the grandmother hypotheses: old and new. *American Journal of Human Biology*, **13**, 434–452.

Peng, M. T., Lai, Y. L., Yang, C. S., Chiang, H. S., New, A. E. & Chang, C. P. (1973). Reproductive parameters of the Taiwan monkey (*Macaca cyclopis*). *Primates*, **14**, 201–213.

Pereira, M. E. (1989). Agonistic interactions of juvenile savanna baboons. II. Agonistic support and rank acquisition. *Ethology*, **80**, 152–171.

(1992). The development of dominance relations before puberty in cercopithecine societies. In *Aggression and Peacefulness in Humans and Other Primates*, ed. P. Gray & J. Silverberg, pp. 117–149. Oxford: Oxford University Press.

Peres, C. A. (1994). Primate responses to phenological changes in an Amazon terra firme forest. *Biotropica*, **26**, 98–112.

Perloe, S. I. (1989). *Monkeys of Minoo*. Produced by S. I. Perloe, Dept. of Psychology, Haverford College [VHS, 45 min].

(1992). Male mating competition, female choice and dominance in a free-ranging group of Japanese macaques. *Primates*, **33**, 289–304.

Perrett, D. I. & Mistlin, A. J. (1990). Perception of facial characteristics by monkeys. In *Comparative perception*, ed. W. C. Stebbins & M. A. Berkley, pp. 187–215. New York: John Wiley.

Perry, S. (1996). Intergroup encounters in wild white-faced capuchins (*Cebus capucinus*). *International Journal of Primatology*, **17**, 309–330.

(2003). Conclusions and research agendas. In *The Biology of Traditions: Models and Evidence*, ed. D. M. Fragaszy & S. Perry. Cambridge: Cambridge University Press.

Pervin, L. A. (1994a). A critical analysis of current trait theory. *Psychological Inquiry*, **5**, 103–113.

(1994b). Further reflections on current trait theory. *Psychological Inquiry*, **5**, 169–178.

Petit, O. & Thierry, B. (1992). Affiliative function of the silent bared-teeth display in moor macaques (*Macaca maurus*): further evidence for the particular status of Sulawesi macaques. *International Journal of Primatology*, **13**, 97–105.

(1994). Aggressive and peaceful interventions in conflicts in Tonkean macaques. *Animal Behaviour*, **48**, 1427–1436.

Petit, O., Abegg, C. & Thierry, B. (1997). A comparative study of aggression and conciliation in three cercopithecine monkeys (*Macaca fuscata*, *Macaca nigra*, *Papio papio*). *Behaviour*, **134**, 415–432.

Pfeifer, R. & Scheier, C. (1999). *Understanding Intelligence*. Cambridge, MA: MIT Press.

Phoenix, C. H. & Chambers, K. C. (1988). Old age and sexual exhaustion in male rhesus macaques. *Physiology and Behavior*, **44**, 157–163.

Phoenix, C. H., Goy, R. W., Gerall, A. A. & Young, W. C. (1959). Organizing action of prenatally administered testosterone propionate on the tissues mediating mating behavior in the female Guinea pig. *Endocrinology*, **65**, 369–382.

Piaget, J. (1971). *Biology and Knowledge*. Chicago: University of Chicago Press.

Pickering, A. D. & Gray, J. A. (1999). The neuroscience of personality. In *Handbook of Personality*, ed. L. A. Pervin & O. P. John, 2nd edn, pp. 277–299. New York: Guilford Press.

Plant, T. M. (1994). Puberty in primates. In *The Physiology of Reproduction*, vol. 2, ed. E. Knobil & J. D. Neill, 2nd edn, pp. 453–485. New York: Raven Press.

Plavcan, J. M. (2001). Sexual dimorphism in primate evolution. *Yearbook of Physical Anthropology*, **44**, 25–53.

Plavcan, J. & van Schaik, C. P. (1997). Intrasexual competition and body weight dimorphism in anthropoid primates. *American Journal of Physical Anthropology*, **103**, 37–68.

Pope, G. G., Brooks, A. S. & Delson, E. (2000). Asia, Eastern and Southern. In *Encyclopedia of Human Evolution and Prehistory*, 2nd edn, ed. E. Delson, I. Tattersall, J. A. van Couvering & A. S. Brooks, pp. 84–91. New York: Garland Press.

Pope, T. R. (1998). Effects of demographic change on group kin structure and gene dynamics of populations of red howling monkeys. *Journal of Mammalogy*, **79**, 692–712.

Popp, J. L. & DeVore, I. (1979). Aggressive competition and social dominance theory: synopsis. In *The Great Apes*, ed. D. A. Hamburg & E. R. McCown, pp. 317–338. Menlo Park, CA: Benjamin Cummings.

Povinelli, D. J., Parks, K. A. & Novak, M. A. (1991). Do rhesus monkeys (*Macaca mulatta*) attribute knowledge and ignorance to others? *Journal of Comparative Psychology*, **105**, 318–325.

(1992). Role reversal by rhesus monkeys, but no evidence of empathy. *Animal Behaviour*, **44**, 269–282.

Preuschoft, S. (1992). 'Laughter' and 'smile' in Barbary macaques (*Macaca sylvanus*). *Ethology*, **91**, 220–236.

(1995a). '*Laughter' and 'Smiling' in Macaques: an Evolutionary Perspective*. Utrecht: Tessel Offset B.V.

(1995b). 'Smiling' and 'laughter' in Tonkean macaques (*Macaca tonkeana*)? In '*Laughter' and 'Smiling' in Macaques: an Evolutionary Perspective*, pp. 137–170. Utrecht: Tessel Offset B.V.

(1999). Are primates behaviorists? Formal dominance, cognition, and free-floating rationales. *Journal of Comparative Psychology*, **113**, 91–95.

Preuschoft, S. & Beckmann, F. (1995). Smiling' and 'laughter' in lion-tailed macaques (*Macaca silenus*): a preliminary analysis of their silent-bared-teeth displays. In '*Laughter' and 'Smiling' in Macaques: an Evolutionary Perspective*, pp. 109–136. Utrecht: Tessel Offset B.V.

Preuschoft, S. & Paul, A. (2000). Dominance, egalitarianism, and stalemate: an experimental approach to male-male competition in Barbary macaques. In *Primate Males: Causes and Consequences of Variation in Group Composition*, ed. P. M. Kappeler, pp. 205–216. Cambridge: Cambridge University Press.

Preuschoft, S. & van Hooff, J. A. R. A. M. (1995a). Homologizing primate facial displays: a critical review of methods. *Folia Primatologica*, **65**, 121–137.

Preuschoft, S. & van Hooff, J. A. R. A. M. (1995b). Variation in primate affiliative displays: an exercise in behaviour phylogeny. In '*Laughter' and 'Smiling' in Macaques: an Evolutionary Perspective*, pp. 195–220. Utrecht: Tessel Offset B.V.

(1997). The social function of 'smile' and 'laughter': variations across primate species and societies. In *Nonverbal Communication: Where Nature Meets Culture*, ed. U. Segerstråle & P. Molnàr, pp. 171–189. Hillsdale, NJ: Erlbaum.

Preuschoft, S. & van Schaik, C. P. (2000). Dominance and communication. Conflict management in various social settings. In *Natural Conflict Resolution*, ed. F. Aureli & F. B. M. de Waal, pp. 77–105. Berkeley, CA: University of California Press.

Preuschoft, S., Gevers, E. & van Hooff, J. A. R. A. M. (1995). Functional differentiation in the affiliative facial displays of longtailed macaques (*Macaca fascicularis*). In '*Laughter' and 'Smiling' in Macaques: an Evolutionary Perspective*, pp. 59–88. Utrecht: Tessel Offset B.V.

Preuschoft, S., Paul, A. & Kuester, J. (1998). Dominance styles of female and male Barbary macaques (*Macaca sylvanus*). *Behaviour*, **135**, 731–755.

Price, T. & Langen, T. (1992). Evolution of correlated characters. *Trends in Ecology and Evolution*, **7**, 307–310.

Prud'homme, J. (1991). Group fission in a semifree-ranging population of Barbary macaques (*Macaca sylvanus*). *Primates*, **32**, 9–22.

Prud'homme, J. & Chapais, B. (1993a). Rank relations among sisters in semi-free ranging Barbary macaques (*Macaca sylvanus*). *International Journal of Primatology*, **14**, 405–420.

(1993b). Aggressive interventions and matrilineal dominance relations in semifree-ranging Barbary macaques (*Macaca sylvanus*). *Primates*, **34**, 271–283.

(1996). Development of intervention behavior in Japanese macaques: testing the targeting hypothesis. *International Journal of Primatology*, **17**, 429–243.

Pryce, C. R. (1996). Socialization, hormones, and the regulation of maternal behaviour in nonhuman simian primates. *Advances in the Study of Behavior*, **25**, 423–473.

Prychodko, W., Goodman, M., Poulik, E., Miki, T. & Tanaka, T. (1969). Geographic variations of transferrin allelic frequencies in continental and insular macaque populations. *Proceedings of the Second International Congress of Primatology*, vol. 2, ed. H. O. Hofer, pp. 103–108. Basel: Karger.

Purvis, A. (1995). A composite estimate of primate phylogeny. *Philosophical Transactions of the Royal Society of London B*, **348**, 405–421.

Pusey, A. E. (1987). Sex-biased dispersal and inbreeding avoidance in birds and mammals. *Trends in Ecology and Evolution*, **2**, 295–299.

Pusey, A. E. & Packer, C. (1987). Dispersal and philopatry. In *Primate Societies*, ed. B. B. Smuts, D. L. Cheney, R. M. Seyfarth, R. W. Wrangham & T. T. Struhsaker, pp. 250–266. Chicago: University of Chicago Press.

Qu, W. Y., Zhang, Y. Z., Manry, D. & Southwick, C. H. (1993). Rhesus monkeys (*Macaca mulatta*) in the Tailang mountains, Jiyuan Country, Henan, China. *International Journal of Primatology*, **14**, 607–621.

Quiatt, D. (1979). Aunts and mothers: adaptive implications of allomaternal behavior of nonhuman primates. *American Anthropologist*, **81**, 310–319.

Raff, R. A. (1996). *The Shape of Life*. Chicago: University of Chicago Press.

Rahaman, H. & Parthasarathy, M. D. (1969). Studies on the social behavior of bonnet monkeys. *Primates*, **10**, 149–162.

Rajecki, D. W., Lamb, M. E. & Obmascher, P. (1978). Toward a general theory of infantile attachment: a comparative review of aspects of the social bond. *Behavioral and Brain Sciences*, **1**, 417–464.

Raleigh, M. J. & McGuire, M. T. (1989). Female influence on male dominance acquisition in captive vervet monkeys, *Cercopithecus aethiops sabaeus*. *Animal Behaviour*, **38**, 59–67.

Ralls, K. (1977). Sexual dimorphism in mammals: avian models and unwarranted unanswered questions. *American Naturalist*, **111**, 917–938.

Ralls, K. & Ballou, J. (1982). Effects of inbreeding on infant mortality in captive primates. *International Journal of Primatology*, **3**, 491–505.

Ralls, K., Ballou, J. & Templeton, A. R. (1988). Estimates of lethal equivalents and the cost of inbreeding in mammals. *Conservation Biology*, **2**, 185–193.

Ralls, K., Harvey, P. H. & Lyles, A. M. (1986). Inbreeding in natural populations of birds and mammals. In *Conservation Biology: The Science of Scarcity and Diversity*, ed. M. E. Soulé, pp. 35–56. Sunderland, MA: Sinauer.

Ransom, T. W. (1981). *Beach Troop of the Gombe*. Lewisburg: Bucknell University Press.

Ransom, T. W. & Rowell, T. E. (1972). Early social development of feral baboons. In *Primate Socialization*, ed. F. E. Poirier, pp. 105–144. New York: Random House.

Rasmussen, K. L. R. & Suomi, S. J. (1989). Heart rate and endocrine responses to stress in adolescent male rhesus monkeys on Cayo Santiago. *Puerto Rico Health Science Journal*, **8**, 65–71.

Rasmussen, K. L. R., Fellowes, J. R., Byrne, E. & Suomi, S. J. (1988). Heart rate measures associated with early emigration in adolescent male rhesus macaques (*Macaca mulatta*). *American Journal of Primatology*, **14**, 439.

Ratnayeke, S. (1994). The behavior of postreproductive females in a wild population of toque macaques (*Macaca sinica*) in Sri Lanka. *International Journal of Primatology*, **15**, 445–469.

Rawlins, R. G. & Kessler, M. J. (1986). Demography of the free-ranging Cayo Santiago macaques (1976–1983). In *The Cayo Santiago Macaques*, ed. R. G. Rawlins & M. J. Kessler, pp. 47–72. Albany, NY: State University of New York Press.

Rawlins, R. G., Kessler, M. J. & Turnquist, J. E. (1984). Reproductive performance, population dynamics and anthropometrics of the free-ranging Cayo Santiago rhesus monkeys. *Journal of Medical Primatology*, **13**, 247–259.

Redican, W. K. (1975). Facial expressions in nonhuman primates. In *Primate Behavior: Developments in Field and Laboratory Research*, vol. 4, ed. L. A. Rosenblum, pp. 103–194. New York: Academic Press.

Reed, C., O'Brien, T. G. & Kinnaird, M. F. (1997). Male social behavior and dominance hierarchy in the Sulawesi crested black macaque (*Macaca nigra*). *International Journal of Primatology*, **18**, 247–260.

Reichler, S. (1996). *Untersuchungen zum Dominanzstil bei Geladas, Theropithecus gelada*. Diplomarbeit, Ruhr-Universität Bochum.

Reiman, E. M., Lane, R. D., Ahern, G. L., Schwartz, G. E. & Davidson, R. J. (2000). Positron Emission Tomography in the study of emotion, anxiety, and anxiety disorders. In *Cognitive Neuroscience of Emotion*, ed. R. D. Lane & L. Nadel, pp. 389–406. New York: Oxford University Press.

Rendall, D. & Taylor, L. L. (1991). Female sexual behavior in the absence of male–male competition in Japanese macaques. *Zoo Biology*, **10**, 319–328.

Resko, J. A., Goy, R. W., Robinson, J. A. & Norman, R. L. (1982). The pubescent rhesus monkey: some characteristics of the menstrual cycle. *Biology of Reproduction*, **27**, 354–361.

Reynolds, J. D. & Gross, M. R. (1990). Costs and benefits of female mate choice: is there a lek paradox? *American Naturalist*, **136**, 230–243.

Rhine, R. J., Cox, R. L. & Costello, M. B. (1989). A twenty-year study of long-term and temporary dominance relations among stumpailed macaques (*Macaca arctoides*). *Americal Journal of Primatology*, **19**, 69–82.

Richard, A. F. (1981). Changing assumptions in primate ecology. *American Anthropologist*, **83**, 517–533.

Richard, A. F., Goldstein, S. J. & Dewar, R. E. (1989). Weed macaques: the evolutionary implications of macaques feeding ecology. *International Journal of Primatology*, **10**, 569–594.

Richerson, P. J. & Boyd, R. (1989). The role of evolved predispositions in cultural evolution: or, human sociobiology meets Pascal's Wager. *Ethology and Sociobiology*, **10**, 195–219.

Robinson, J. A. & Goy, R. W. (1986). Steroid hormones and the ovarian cycle. In *Comparative Primate Biology*, vol. 3, ed. W. R. Dukelow & J. Erwin, pp. 63–91. New York: Alan R. Liss.

Robinson, J. A., Scheffler, G., Eisele, S. G. & Goy, R. W. (1975). Effects of age and season on sexual behavior and plasma testosterone and dihydrotestosterone concentrations of laboratory-housed male rhesus monkeys (*Macaca mulatta*). *Biology of Reproduction*, **13**, 203–210.

Robinson, J. G. (1988). Group size in wedge-capped capuchin monkeys *Cebus olivaceus* and the reproductive success of males and females. *Behavioral Ecology and Sociobiology*, **23**, 187–197.

Roeder, J. J., Fornasieri, I. & Gosset, D. (2002). Conflict and postconflict behaviour in two lemur species with different social organizations (*Eulemur fulvus* and *Eulemur macaco*): a study of captive groups. *Aggressive Behavior*, **28**, 62–74.

Rogers, J., Comuzzie, A. G., Martin, L., Mann, J. J. & Kaplan, J. R. (2001). Quantitative genetics of monoamine metabolites in pedigreed baboons, *Papio hamadryas*. *American Journal of Medical Genetics*, **105**, 582.

Rolls, E. T. (1990). A theory of emotion, and its application to understand the neural basis of emotion. *Cognition and Emotion*, **4**, 161–190.

   (1995). A theory of emotion and consciousness, and its application to understanding the neural basis of emotion. In *The Cognitive Neurosciences*, ed. M. S. Gazzaniga, pp. 1091–1106. Cambridge, MA: MIT Press.

   (1999). *The Brain and Emotion*. Oxford: Oxford University Press.

Rosati, L. (1996). *Function and Distribution of Reconciliation in Captive Japanese Macaques* (Macaca fuscata). Master Thesis, University La Sapienza, Rome.

Rosenblum, L. A. & Kaufman, I. C. (1967). Laboratory observations of early mother-infant relations in pigtail and bonnet macaques. In *Social Communication among Primates*, ed. S. A. Altmann, pp. 33–41. Chicago: University of Chicago Press.

Rosenblum, L. A. & Nadler, R. D. (1971). The ontogeny of sexual behavior in male bonnet macaques. In *Influence of Hormones on the Nervous System*, ed. D. H. Ford, pp. 388–400. Basel: Karger.

Rosenblum, L. A., Coe, C. L. & Bromley, L. J. (1975). Peer relations in monkeys: the influence of social structure, gender and familiarity. *Origins of Behavior*, **4**, 67–98.

Rosenblum, L. A., Kaufman, I. C. & Stynes, A. J. (1964). Individual distance in two species of macaque. *Animal Behaviour*, **12**, 338–342.

Rosenblum, L. A., Smith, E. L. P., Altemus, M. *et al.* (2002). Differing concentrations of corticotropin-releasing factor and oxytocin in the cerebrospinal fluid of bonnet and pigtail macaques. *Psychoneuroendocrinology*, **27**, 651–660.

Ross, C. (1988). The intrinsic rate of natural increase and reproductive effort in primates. *Journal of Zoology (London)*, **214**, 199–219.

Rowell, T. (1967). Female reproductive cycles and the behavior of baboons and rhesus macaques. In *Social Communication Among Primates*, ed. S. A. Altmann, pp. 15–32. Chicago: University of Chicago Press.

   (1974). The concept of social dominance. *Behavioral Biology*, **11**, 131–154.

   (1993). Reification of social systems. *Evolutionary Anthropology*, **2**, 135–137.

Rowell, T. E. & Hinde, R. A. (1962). Vocal communication by the rhesus monkey (*Macaca mulatta*). *Proceedings of the Zoological Society of London*, **138**, 279–294.

Rowell, T. E., Hinde, R. A. & Spencer-Booth, Y. (1964). "Aunt"-infant interaction in captive rhesus monkeys. *Animal Behaviour*, **12**, 219–226.

Russon, A. E. (2003). Developmental perspectives on great ape traditions. In *The Biology of Traditions: Models and Evidence*, ed. D. M. Fragaszy & S. Perry, pp. 329–364. Cambridge: Cambridge University Press.

Ruzzante, D. E. & Doyle, R. W. (1991). Rapid behavioral changes in medaka caused by selection for competitive and noncompetitive growth. *Evolution*, **45**, 1936–1946.

(1993). Evolution of social behaviour in a resource rich, structured environment: selection experiments with medaka. *Evolution*, **47**, 456–470.

Ryman, N., Reuterwall, C., Nygren, K. & Nygren, T. (1980). Genetic variation and differentiation in Scandinavian moose (*Alces alces*): are large mammals monomorphic? *Evolution*, **34**, 1037–1049.

Sade, D. S. (1967). Determinants of dominance in a group of free-ranging rhesus monkeys. In *Social Communication Among Primates*, ed. S. A. Altmann, pp. 99–111. Chicago: Chicago University Press.

(1968). Inhibition of son–mother mating among free-ranging rhesus monkeys. *Science and Psychoanalysis*, **12**, 18–38.

(1972a). A longitudinal study of social behavior in rhesus monkeys. In *The Functional and Evolutionary Biology of Primates*, ed. R. H. Tuttle, pp. 378–398. Chicago: Aldine-Atherton.

(1972b). Sociometrics of *Macaca mulatta*. I. Linkages and cliques in grooming matrices. *Folia Primatologica*, **18**, 196–223.

Sade, D. S., Cushing, K., Cushing, P. *et al.* (1976). Population dynamics in relation to social structure on Cayo Santiago. *Yearbook of Physical Anthropology*, **20**, 253–262.

Sadleir, R. M. S. (1969). *The Ecology of Reproduction in Wild and Domestic Mammals*. London: Methuen.

Sahlins, M. (1958). *Social Stratification in Polynesia*. Seattle: University of Washington Press.

Saito, C. (1996). Dominance and feeding success in female Japanese macaques, *Macaca fuscata*: effects of food patch size and inter-patch distance. *Animal Behaviour*, **51**, 967–980.

Saito, C., Sato, S., Suzuki, S. *et al.* (1998). Aggressive intergroup encounters in two populations of Japanese macaques (*Macaca fuscata*). *Primates*, **39**, 303–312.

Saj, T. L., Sicotte, P. & Paterson, J. D. (2001). The conflict between vervet monkeys and farmers at the forest edge in Entebbe, Uganda. *African Journal of Ecology*, **39**, 195–199.

Sapolsky, R. M. (2000). Physiological correlates of individual dominance style. In *Natural Conflict Resolution*, ed. F. Aureli & F. B. M. de Waal, pp. 114–116. Berkeley, CA: University of California Press.

Sapolsky, R. M. & Ray, J. C. (1989). Styles of dominance and their endocrine correlates among wild olive baboons (*Papio anubis*). *American Journal of Primatology*, **18**, 1–13.

Scheffrahn, W., Ménard, N., Vallet, D. & Gaci, B. (1993). Ecology, demography, and population genetics of Barbary macaques in Algeria. *Primates*, **34**, 381–394.

Scheurer, J. & Thierry, B. (1985). A further food-washing tradition in Japanese macaques (*Macaca fuscata*). *Primates*, **26**, 491–494.

Schino, G. & Troisi, A. (1992). Opiate receptor blockade in juvenile macaques: effect on affiliative interactions with their mothers and group companions. *Brain Research*, **576**, 125–130.

(2001). Relationship with the mother modulates the response of yearling Japanese macaques (*Macaca fuscata*) to the birth of a sibling. *Journal of Comparative Psychology*, **115**, 392–396.

Schino, G., Maestripieri, D., Scucchi, S. & Turillazzi, P. G. (1990). Social tension in familiar and unfamiliar pairs of long-tail macaques. *Behaviour*, **113**, 264–272.

Schino, G., Perretta, G., Taglioni, A. M., Monaco, V. & Troisi, A. (1996). Primate displacement activities as an ethopharmacological model of anxiety. *Anxiety*, **2**, 186–191.

Schino, G., Rosati, L. & Aureli, F. (1998). Intragroup variation in conciliatory tendencies in captive Japanese macaques. *Behaviour*, **135**, 897–912.

Schino, G., Scucchi, S., Maestripieri, D. & Turillazzi, P. G. (1988). Allogrooming as a tension-reduction mechanism: a behavioral approach. *American Journal of Primatology*, **16**, 43–50.

Schino, G., Speranza, L. & Troisi, A. (2001). Early maternal rejection and later social anxiety in juvenile and adult Japanese macaques. *Developmental Psychobiology*, **38**, 186–190.

Schino, G., Troisi, A., Perretta, G. & Monaco, V. (1991). Measuring anxiety in nonhuman primates: effects of lorazepam on macaque scratching. *Pharmacology, Biochemistry and Behavior*, **38**, 889–891.

Schlichting, C. D. & Pigliocci, M. (1998). *Phenotypic Evolution: a Reaction Norm Perspective*. Sunderland, MA: Sinauer.

Schmidt-Nielsen, K. (1997). *Animal Physiology*. Cambridge: Cambridge University Press.

Schneider, M. L. (1992). Prenatal stress exposure alters postnatal behavioral expression under conditions of novelty challenge in rhesus monkey infants. *Developmental Psychobiology*, **25**, 529–540.

Schneider, M. L., Moore, C. F., Suomi, S. J. & Champoux, M. (1991). Laboratory assessment of temperament and environmental enrichment in rhesus monkey infants (*Macaca mulatta*). *American Journal of Primatology*, **25**, 137–155.

Schulman, S. R. & Chapais, B. (1980). Reproductive value and rank relations among macaque sisters. *American Naturalist*, **115**, 580–593.

Schwartz, C. E., Snidman, N. & Kagan, J. (1999). Adolescent social anxiety as an outcome of inhibited temperament in childhood. *Journal of the American Academy of Child and Adolescent Psychiatry*, **38**, 1008–1015.

Schwartz, S. M. & Kemnitz, J. W. (1992). Age- and gender-related changes in body size, adiposity, and endocrine and metabolic parameters in free-ranging rhesus macaques. *American Journal of Physical Anthropology*, **89**, 109–121.

Schwartz, S. M., Wilson, M. E., Walker, M. E. & Collins, D. C. (1988). Dietary influences on growth and sexual maturation in premenarchial rhesus monkeys. *Hormones and Behavior*, **22**, 231–151.

Seidensticker, J. (1983). Predation by *Panthera* cats and measures of human influence. *International Journal of Primatology*, **4**, 323–326.

Selander, R. K., Yang, S. Y., Lewontin, R. C. & Johnson, W. E. (1970). Genetic variation in the horsehoe crab (*Limulus polyphemus*), a phylogenetic 'relic'. *Evolution*, **24**, 402–414.

Semple, S. (1998). The function of Barbary macaque copulation calls. *Proceedings of the Royal Society of London B*, **265**, 287–291.

Seth, P. K. & Seth, S. (1985). Ecology and feeding behaviour of the free ranging rhesus monkeys in India. *Indian Anthropology*, **15**, 51–62.

Seyfarth, R. M. (1977). A model of social grooming among adult female monkeys. *Journal of Theoretical Biology*, **65**, 671–698.

(1981). Do monkeys rank each other? *The Behavioral and Brain Sciences*, **4**, 447–448.

Sherman, P. W. (1998). The evolution of menopause. *Nature*, **393**, 759–761.

Shively, C. & Smith, D. G. (1985). Social status and reproductive success of male *Macaca fascicularis*. *American Journal of Primatology*, **9**, 129–135.

Shively, C., Clarke, S., King, N., Schapiro, S. & Mitchell, G. (1982). Patterns of sexual behavior in male macaques. *American Journal of Primatology*, **2**, 373–384.

Short, R., England, N., Bridson, W. E. & Bowden, D. M. (1989). Ovarian cyclicity, hormones, and behavior as markers of aging in female pigtailed macaques (*Macaca nemestrina*). *Journal of Gerontology: Biological Sciences*, **44**, B131–138.

Shotake, T. & Santiapilai, C. (1982). Blood protein polymorphisms in the troops of the toque macaque, *Macaca sinica*, in Sri Lanka. *Kyoto University Overseas Research Report of Studies on Asian Non-human Primates*, **2**, 79–95.

Shotake, T., Nozawa, K. & Santiapilai, C. (1991). Genetic variability within and between the troops of toque macaque, *Macaca sinica*, in Sri Lanka. *Primates*, **32**, 283–299.

Sicotte, P. (1993). Inter-group encounters and female transfer in mountain gorillas: influence of group composition on male behavior. *American Journal of Primatology*, **30**, 21–36.

Siddiqi, M. F. & Southwick, C. H. (1988). Food habits of rhesus monkeys (*Macaca mulatta*) in the north Indian plains. In *Ecology and Behavior of Food-enhanced Primate Groups*, ed. J. E. Fa & C. H. Southwick, pp. 113–123. New York: Alan R. Liss.

Silk, J. B. (1980). Kidnapping and female competition among captive bonnet macaques. *Primates*, **21**, 100–110.

(1982). Altruism among female *Macaca radiata*: explanations and analysis of patterns of grooming and coalition formation. *Behaviour*, **79**, 162–187.

(1983). Local resource competition and facultative adjustment of sex ratios in relation to competititve ability. *American Naturalist*, **121**, 56–64.

(1987). Social behavior in evolutionary perspective. In *Primate Societies*, ed. B. B. Smuts, D. L. Cheney, R. M. Seyfarth, R. W. Wrangham & T. T. Struhsaker, pp. 318–329. Chicago: University of Chicago Press.

(1988). Social mechanisms of population regulation in a captive group of bonnet macaques (*Macaca radiata*). *American Journal of Primatology*, **14**, 111–124.

(1990). Sources of variation in interbirth intervals among captive bonnet macaques (*Macaca radiata*). *American Journal of Physical Anthropology*, **82**, 213–230.

(1991). Mother–infant relationships in bonnet macaques: sources of variation in proximity. *International Journal of Primatology*, **12**, 21–38.

(1992). The patterning of intervention among male bonnet macaques: reciprocity, revenge, and loyalty. *Current Anthropology*, **33**, 318–325.

(1997). The function of peaceful post-conflict contacts among primates. *Primates*, **38**, 265–279.

(1999). Why are infants so attractive to others? The form and function of infant handling in bonnet macaques. *Animal Behaviour*, **57**, 1021–1032.

Silk, J. B., Samuels, A. & Rodman, P. (1981). Hierarchical organization of female *Macaca radiata* in captivity. *Primates*, **22**, 84–95.

Silk, J., Short, J., Roberts, J & Kusnitz, J. (1993). Gestation length in rhesus macaques. *International Journal of Primatology*, **14**, 95–104.

Sillen-Tullberg, B. & Moller, A. P. (1993). The relationship between concealed ovulation and mating systems in anthropoid primates: a phylogenetic analysis. *American Naturalist*, **141**, 1–25.

Simonds, P. (1973). Outcast males and social structure among bonnet macaques (*Macaca radiata*). *American Journal of Physical Anthropology*, **38**, 599–604.

(1965). The bonnet macaque in South India. In *Primate Behavior: Field Studies of Monkeys and Apes*, ed. I. DeVore, pp. 175–196. New York: Holt, Rinehart & Winston.

Simpson, G. G. (1983). *Fossils and the History of Life*. New York: Scientific American Library.

Simpson, M. J. A. & Simpson, A. E. (1982). Birth sex ratios and social rank in rhesus monkey mothers. *Nature*, **300**, 400–401.

Simpson, M. J. A., Simpson, A. E., Hooley, J. & Zunz, M. (1981). Infant-related influences on birth intervals in rhesus monkeys. *Nature*, **290**, 49–51.

Singh, M. & Vinathe, S. (1990). Inter-population differences in the time budgets of bonnet monkeys (*Macaca radiata*). *Primates*, **31**, 589–596.

Singh, M., Kumara, H. N., Kumar, M. A. & Sharma, A. K. (2001). Behavioral responses of lion-tailed macaques to rapidly changing habitat in a tropical rainforest fragment in Western Ghats, India. *Folia Primatologica*, **72**, 278–291.

Singh, M., Kumara, H. N., Kumar, M. A., Sharma, A. K. & DeFalco, K. (2000). Status and conservation of lion-tailed macaque and other arboreal mammals in tropical rainforests of Sringeri Forest Range, Western Ghats, Karnataka, India. *Primate Report*, **58**, 5–16.

Singh, M., Singh, M., Kumar, M. A., Kumara, H. N. Sharma, A. K. & Kaumans, W. (2002). Distribution, population structure, and conservation of lion-tailed macaques (*Macaca silenus*) in the Anaimalai Hills, Western Ghats, India. *American Journal of Primatology*, **57**, 91–102.

Singh, S. D. (1966). The effect of human environment on the social behavior of rhesus monkeys. *Primates*, **7**, 33–39.

(1969). Urban monkeys. *Scientific American*, **221**, 108–115.

Sinha, A. (1998). Knowledge acquired and decisions made: triadic interactions during allogrooming in wild bonnet macaques, *Macaca radiata*. *Philosophical Transactions of the Royal Society of London*, **353**, 619–631.

(2001). The monkey in the town's commons: a natural history of the Indian bonnet macaque. *NIAS Report R 2-01*. Bangalore: National Institute of Advanced Studies.

Sites, J. W. & Moritz, C. (1987). Chromosomal evolution and speciation revisited. *Systematic Zoology*, **36**, 153–174.

Slatkin, M. (1987). Gene flow and the geographic structure of natural populations. *Science*, **236**, 787–792.

Small, M. F. (1990a). Promiscuity in Barbary macaques (*Macaca sylvanus*). *American Journal of Primatology*, **20**, 267–282.

(1990b). Alloparental behaviour in Barbary macaques, *Macaca sylvanus*. *Animal Behaviour*, **39**, 297–306.

(1990c). Social climber: independent rise in rank by a female Barbary macaque (*Macaca sylvanus*). *Folia Primatologica*, **55**, 85–91.

Smith, D. G. (1981). The association between rank and reproductive success of male rhesus monkeys. *American Journal of Primatology*, **1**, 83–90.

(1986a). Inbreeding in the maternal and paternal lines of four captive groups of rhesus monkeys (*Macaca mulatta*). In *Current Perspective in Primate Biology*, ed. D. M. Taub & F. A. King, pp. 214–225. New York: van Nostrand Reinhold.

(1986b). Incidence and consequences of inbreeding in three captive groups of rhesus macaques (*Macaca mulatta*). In *Primates: the Road to Self-sustaining Populations*, ed. K. Benirschke, pp. 856–874. New York: Springer.

(1993). A 15-year study of the association between dominance rank and reproductive success of male rhesus macaques. *Primates*, **34**, 471–480.

(1994). Male dominance and reproductive success in a captive group of rhesus macaques (*Macaca mulatta*). *Behaviour*, **129**, 225–242.

(1995). Avoidance of close consanguineous inbreeding in captive groups of rhesus macaques. *American Journal of Primatology*, **35**, 31–40.

Smith, D. G. & Small, M. F. (1987). Mate choice by lineage in three captive groups of rhesus macaques (*Macaca mulatta*). *American Journal of Physical Anthropology*, **73**, 185–191.

Smith, D. G. & Smith, S. (1988). Parental rank and reproductive success of natal rhesus males. *Animal Behaviour*, **36**, 554–562.

Smith, O. A., Astley, C. A., Spelman, F. A. *et al.* (2000). Cardiovascular responses in anticipation of changes in posture and locomotion. *Brain Research Bulletin*, **53**, 69–76.

Smith, R. C. & Jungers, W. L. (1997). Body mass in comparative primatology. *Journal of Human Evolution*, **32**, 523–559.

Smucny, D. A., Price, C. S. & Byrne, E. A. (1997). Post-conflict affiliation and stress reduction in captive rhesus macaques. *Advances in Ethology*, **32**, 157.

Smuts, B. B. (1987). Sexual competition and mate choice. In *Primate Societies*, ed. B. B. Smuts, D. L. Cheney, R. M. Seyfarth, R. W. Wrangham & T. T. Struhsaker, pp. 385–399. Chicago: University of Chicago Press.

Smuts, B. B. & Smuts, R. W. (1993). Male aggression and sexual coercion of females in nonhuman primates and other mammals: evidence and theoretical implications. *Advances in the Study of Behavior*, **22**, 1–63.

Soltis, J. (1999). Measuring male–female relationships during the mating season in wild Japanese macaques (*Macaca fuscata yakui*). *Primates*, **40**, 453–468.

(2002). Do primate females gain non-procreative benefits by mating with multiple males? Theoretical and empirical considerations. *Evolutionary Anthropology*, **11**, 187–197.

Soltis, J. & McElreath, R. (2001). Can females gain paternal investment by mating with multiple males? A game theoretic approach. *American Naturalist*, **158**, 519–529.

Soltis, J., Mitsunaga, F., Shimizu, K. *et al.* (1997b). Sexual selection in Japanese macaques. II: Female mate choice and male–male competition. *Animal Behaviour*, **54**, 737–746.

Soltis, J., Mitusanaga, F., Shimizu, K., Yanagihara, Y. & Nozaki, M. (1997a). Sexual selection in Japanese macaques: I. female mate choice or male sexual coercion? *Animal Behaviour*, **54**, 725–736.

(1999). Female mating strategy in an enclosed group of Japanese macaques. *American Journal of Primatology*, **47**, 263–278.

Soltis, J., Thomsen, R., Matsubayashi, K. & Takenaka, O. (2000). Infanticide by resident males and female counter-strategies in wild Japanese macaques (*Macaca fuscata*). *Behavioral Ecology and Sociobiology*, **48**, 195–202.

Soltis, J., Thomsen, R. & Takenaka, O. (2001). The interaction of male and female reproductive strategies and paternity in wild Japanese macaques (*Macaca fuscata*). *Animal Behaviour*, **62**, 485–494.

Southwick, C. H. & Siddiqi, M. F. (1995). Primate in Asia: survival in a competitive world. *Primate Report*, **41**, 15–22.

Southwick, C. H., Beg, M. A. & Siddiqi, M. R. (1965). Rhesus monkeys in North India. In *Primate Behavior: Field Studies of Monkeys and Apes*, ed. I. DeVore, pp. 111–159. New York: Holt, Rinehart & Winston.

Southwick, C., Zhang, Y., Jiang, H., Liu, Z. & Qu, W. (1996). Population ecology of rhesus macaques in tropical and temperate habitats in China. In *Evolution and Ecology of Macaque Societies*, ed. J. E. Fa & D. J. Lindburg, pp. 95–105. Cambridge: Cambridge University Press.

Spencer-Booth, Y. (1968). The behaviour of group companions toward rhesus monkey infants. *Animal Behaviour*, **16**, 541–557.

Sprague, D. S. (1991a). Influence of mating on the troop choice on non-troop males among the Japanese macaques of Yakushima Island. In *Primatology Today*, ed. A. Ehara, T. Kimura, O. Takenaka & M. Iwamoto, pp. 207–210. Amsterdam: Elsevier.

(1991b). Mating by nontroop males among the Japanese macaques of Yakushima Island. *Folia Primatologica*, **57**, 156–158.

Sprague, D. S., Suzuki, S., Takahashi, H. & Sato, S. (1998). Male life history in natural populations of Japanese macaques: migration, dominance rank, and troop participation of males in two habitats. *Primates*, **39**, 351–363.

Sprague, D. S., Suzuki, S. & Tsukahara, T. (1996). Variation in social mechanisms by which males attained the alpha rank among Japanese macaques. In *Evolution and Ecology of Macaque Societies*, ed. J. E. Fa & D. G. Lindburg, pp. 444–458. New York: Cambridge University Press.

Stallman, R. R. & Froehlich, J. W. (2000). Primate sexual swellings as coevolved signal systems. *Primates*, **41**, 1–16.

Stanford, C. B. (1991). Social dynamics of intergroup encounters in the capped langur (*Presbytis pileata*). *American Journal of Primatology*, **25**, 35–47.

(1996). The hunting ecology of wild chimpanzees: implications for the evolutionary ecology of Pliocene hominids. *American Anthropologist*, **98**, 96–113.

Stanford, C. B., Wallis, J., Matama, H. & Goodall, J. (1994). Patterns of predation by chimpanzees on red colobus monkeys in the Gombe National Park, 1982–1991. *American Journal of Physical Anthropology*, **94**, 213–228.

Stearns, S. C. (1992). *The Evolution of Life Histories*, Oxford: Oxford University Press.

Steiner, R. A. (1987). Nutritional and metabolic factors in the regulation of reproductive hormone secretion in the primate. *Proceedings of the Nutrition Society*, **46**, 159–175.

Steklis, H. D. & Fox, R. (1988). Menstrual cycle phase and sexual behavior in semi-free ranging stumptail macaques (*Macaca arctoides*). *International Journal of Primatology*, **9**, 443–456.

Sterck, E. H. M. (1999). Variation in langur social organization in relation to the socioecological model, human habitat alteration, and phylogenetic constraints. *Primates*, **40**, 199–213.

Sterck, E. H. M., Watts, D. P. & van Schaik, C. P. (1997). The evolution of female social relationships in nonhuman primates. *Behavioral Ecology and Sociobiology*, **41**, 291–309.

Stern, B. R. & Smith, D. G. (1984). Sexual behaviour and paternity in three captive groups of rhesus monkeys (*Macaca mulatta*). *Animal Behaviour*, **32**, 23–32.

Stevenson-Hinde, J. & Zunz, M. (1978). Subjective assessment of individual rhesus monkeys. *Primates*, **19**, 473–482.

Stevenson-Hinde, J., Stillwell-Barnes, R. & Zunz, M. (1980a). Subjective assessment of rhesus monkeys over four successive years. *Primates*, **21**, 66–82.

(1980b). Individual differences in young rhesus monkeys: consistency and change. *Primates*, **21**, 498–509.

Stoner, K. E. (1996). Prevalence and intensity of intestinal parasites in mantled howling monkeys (*Alouatta palliata*) in northeastern Costa Rica: implications for conservation biology. *Conservation Biology*, **10**, 539–546.

Storz, J. F. (1999). Genetic consequences of mammalian social structure. *Journal of Mammalogy*, **80**, 553–569.

Strier, K. B. (1994). Myth of the typical primate. *Yearbook of Physical Anthropology*, **37**, 233–271.

(1997). Behavioral ecology and conservation biology of primates and other animals. *Advances in the Study of Behavior*, **26**, 101–158.

Struhsaker, T. T. (1969). Correlates of ecology and social organization among African cercopithecines. *Folia Primatologica*, **11**, 80–118.

(1973). A recensus of vervet monkeys in the Masai-Amboseli Game Reserve, Kenya. *Ecology*, **54**, 930–932.

Struhsaker, T. T. & Gartlan, J. S. (1970). Observations on the behavior and ecology of the patas monkey (*Erythrocebus patas*) in the Waza Reserve, Cameroon. *Journal of the Zoology (London)*, **161**, 49–63.

Struhsaker, T. T. & Leakey, M. (1990). Prey selectivity by crowned hawk-eagles on monkeys in the Kibale Forest, Uganda. *Behavioral Ecology and Sociobiology*, **26**, 435–443.

Struhsaker, T. T. & Leland, L. (1988). Group fission in red-tail monkeys (*Cercopithecus ascanius*) in the Kibale forest, Uganda. In *A Primate Radiation: Evolutionary*

Biology of the African Guenons, ed. A. Gautier-Hion, F. Bourlière & J. P. Gautier, pp. 363–388. Cambridge: Cambridge University Press.

Strum, S. B. & Latour, B. (1987). Redefining the social link: from baboons to humans. *Social Science Information*, **26**, 783–802.

Stuart, M. D. & Strier, K. B. (1995). Primates and parasites: a case for a multidisciplinary approach. *International Journal of Primatology*, **16**, 577–593.

Su, H. H. & Lee, L. L. (2001). Food habits of Formosan rock macaques (*Macaca cyclopis*) in Jentse, northeastern Taiwan, assessed by fecal analysis and behavioral observation. *International Journal of Primatology*, **22**, 359–377.

Suarez, B. & Ackerman, D. R. (1971). Social dominance and reproductive behavior in male rhesus monkeys. *American Journal of Physical Anthropology*, **35**, 219–222.

Sugardjito, J., Southwick, C. H., Supriatna, J. *et al.* (1989). Population survey of macaques in northern Sulawesi. *American Journal of Primatology*, **18**, 285–301.

Sugawara, K. (1988). Ethological study of the social behavior of hybrid baboons between *Papio anubis* and *P. hamadryas* in free-ranging groups. *Primates*, **29**, 429–448.

Sugiura, H., Saito, C., Sato, S. *et al.* (2000). Variation in intergroup encounters in two populations of Japanese macaques. *International Journal of Primatology*, **21**, 519–535.

Sugiyama, Y. (1960). On the division of a natural troop of Japanese monkeys at Takasakiyama. *Primates*, **2**, 109–148.

Sugiyama, Y. & Ohsawa, H. (1982a). Population dynamics of Japanese monkeys with special reference to the effect of artificial feeding. *Folia Primatologica*, **39**, 238–263.

(1982b). Population dynamics of Japanese macaques at Ryozenyama: III. Female desertion of the troop. *Primates*, **23**, 31–44.

Suomi, S. J. (1995). Influence of attachment theory on ethological studies of biobehavioral development in nonhuman primates. In *Attachment Theory: Social, Developmental, and Clinical Perspectives*, ed. S. Goldberg, R. Muir & J. Kerr, pp. 185–201. Hillsdale: Analytic Press.

(1999). Attachment in rhesus monkeys. In *Handbook of Attachment: Theory, Research and Clinical Applications*, ed. J. Cassidy & P. R. Shaver, pp. 181–197. New York: Guilford.

Supriatna, J., Manansang, J., Tumbelaka, L. *et al.*, eds. (2001). *Conservation Assessment and Management Plan for the Primates of Indonesia: Final Report.* Apple Valley, Minnesota: Conservation Breeding Specialist Group (SSC/IUCN).

Takahashi, H. (2002). Female reproductive parameters and fruit availability: factors determining onset of estrus in Japanese macaques. *American Journal of Primatology*, **51**, 141–153.

Takahata, Y. (1980). The reproductive biology of a free-ranging troop of Japanese monkeys. *Primates*, **21**, 303–329.

(1982). The sociosexual behavior of Japanese macaques. *Zeitschrift für Tierpsychologie*, **59**, 89–108.

Takahata, Y., Huffman, M. A., Suzuki, S., Koyama, N. & Yamagiwa, J. (1999). Why dominants do not consistently attain high mating and reproductive success: a review of longitudinal Japanese macaque studies. *Primates*, **40**, 143–158.

Takahata, Y., Koyama, N. & Suzuki, S. (1995). Do old aged females experience a long post-reproductive life span? The cases of Japanese macaques and chimpanzees. *Primates*, **36**, 169–180.

Takasaki, H. (1981). Troop size, habitat quality, and home range area in Japanese macaques. *Behavioral Ecology and Sociobiology*, **9**, 277–281.

Tanaka, I. (1995). Matrilineal distribution of louse egg-handling techniques during grooming in free-ranging Japanese macaques. *American Journal of Physical Anthropology*, **98**, 197–201.

(1998). Social diffusion of modified louse egg-handling techniques during grooming in free-ranging Japanese macaques. *Animal Behaviour*, **56**, 1229–1236.

Taub, D. M. (1980). Female choice and mating strategies among wild Barbary macaques (*Macaca sylvanus* L.). In *The Macaques: Studies in Ecology, Behavior, and Evolution*, ed. D. G. Lindburg, pp. 287–344. New York: van Nostrand Reinhold.

Teas, J., Richie, T., Taylor, H. & Southwick, C. (1980). Population patterns and behavioral ecology of rhesus monkeys (*Macaca mulatta*) in Nepal. In *The Macaques: Studies in Ecology and Evolution*, ed. D. G. Lindburg, pp. 247–62. New York: van Nostrand Reinhold.

te Boekhorst, I. J. A. & Hogeweg, P. (1994). Selfstructuring in artificial 'CHIMPS' offers new hypotheses for male grouping in chimpanzees. *Behaviour*, **130**, 229–252.

Teleki, G., Hunt, E. E. & Pfifferling, J. H. (1976). Demographic observations (1963–1973) on the chimpanzees of Gombe National Park, Tanzania. *Journal of Human Evolution*, **5**, 559–598.

Terasawa, E. & Fernandez, D. L. (2001). Neurobiological mechanisms of the onset of puberty in primates. *Endocrine Reviews*, **22**, 111–151.

Terborgh, J. (1983). *Five New World Primates: a Study in Comparative Ecology*. Princeton, NJ: Princeton University Press.

Terborgh, J. & Janson, C. H. (1986). The socioecology of primate groups. *Annual Review of Ecology and Systematics*, **17**, 111–136.

Thierry, B. (1984). Clasping behaviour in *Macaca tonkeana*. *Behaviour*, **89**, 1–28.

(1985a). Patterns of agonistic interactions in three species of macaque (*Macaca mulatta, M. fascicularis, M. tonkeana*). *Aggressive Behavior*, **11**, 223–233.

(1985b). Social development in three species of macaque (*Macaca mulatta, M. fascicularis, M. tonkeana*): a preliminary report on the first ten weeks of life. *Behavioural Processes*, **11**, 89–95.

(1985c). Le comportement d'étreinte dans un groupe de macaques de Java (*Macaca fascicularis*). *Biology of Behaviour*, **10**, 23–30.

(1986a). A comparative study of aggression and response to aggression in three species of macaque. In *Primate Ontogeny, Cognition, and Social Behaviour*, ed. J. G. Else & P. C. Lee, pp. 307–313. Cambridge: Cambridge University Press.

(1986b). Affiliative interference in mounts in a group of Tonkean macaques (*Macaca tonkeana*). *American Journal of Primatology*, **14**, 89–97.

(1990a). Feedback loop between kinship and dominance: the macaque model. *Journal of Theoretical Biology*, **145**, 511–521.

(1990b). L'état d'équilibre entre comportements agonistiques chez un groupe de macaques japonais (*Macaca fuscata*). *Comptes-rendus de l'Académie des Sciences de Paris*, **310 III**, 35–40.

(1994a). Emergence of social organizations in non-human primates. *Revue Internationale de Systémique*, **8**, 65–77.

(1994b). Social transmission, tradition and culture in primates: from the epiphenomenon to the phenomenon. *Techniques et Cultures*, **23–24**, 91–119.

(1997). Adaptation and self-organization in primate societies. *Diogenes*, **180**, 39–71.

(2000). Covariation of conflict management patterns across macaque species. In *Natural Conflict Resolution*, ed. F. Aureli & F. B. M. de Waal, pp. 106–128. Berkeley, CA: University of California Press.

Thierry, B., Anderson, J. R., Demaria, C., Desportes, C. & Petit, O. (1994). Tonkean macaque behaviour from the perspective of the evolution of Sulawesi macaques. In *Current Primatology*, vol. 2, ed. J. J. Roeder, B. Thierry, J. R. Anderson & N. Herrenschmidt, pp. 103–117. Strasbourg: Université Louis Pasteur.

Thierry, B., Aureli, F., de Waal, F. B. M. & Petit, O. (1997). Variation in reconciliation patterns and social organization across nine species of macaques. *Ethology*, **32** (Suppl.), S39.

Thierry, B., Demaria, C., Preuschoft, S. & Desportes, C. (1989). Structural convergence between silent bared-teeth display and relaxed open-mouth display in the Tonkean macaque (*Macaca tonkeana*). *Folia Primatologica*, **52**, 178–184.

Thierry, B., Gauthier, C. & Peignot, P. (1990). Social grooming in Tonkean macaques (*Macaca tonkeana*). *International Journal of Primatology*, **11**, 357–375.

Thierry, B., Heistermann, M., Aujard, R. & Hodges, J. K. (1996). Long-term data on basic reproductive parameters and evaluation of endocrine, morphological, and behavioral measures for monitoring reproductive status in a group of semifree-ranging Tonkean macaques (*Macaca tonkeana*). *American Journal of Primatology*, **39**, 47–62.

Thierry, B., Iwaniuk, A. N. & Pellis, S. M. (2000). The influence of phylogeny on the social behaviour of macaques (Primates: *Cercopithecidae*, genus *Macaca*). *Ethology*, **106**, 713–728.

Thierry, B. H., Milhaud, C. L. & Klein, M. J. (1984). Effect of d-amphetamine and diazepam on the greeting behavior of rhesus monkeys (*Macaca mulatta*). *Pharmacology, Biochemistry & Behavior*, **21**, 191–195.

Thomsen, R., Soltis, J. & Teltscher, C. (2003). Sperm competition and the function of male masturbation in non-human primates. In *Sexual Selection and Reproductive Competition in Primates: New Perspectives and Directions*, ed. C. B. Jones, pp. 437–443. Norman, OK: American Society of Primatologists.

Thomson, J. A., Hess, D. L, Dahl, K. D., Iliff-sizemore, S. A., Stouffer, R. L. & Wolf, D. P. (1992). The Sulawesi crested black macaque (*Macaca nigra*) menstrual cycle: changes in perineal tumescence and serum estradiol, progesterone, follicle-stimulating hormone, and luteinizing hormone. *Biology of Reproduction*, **46**, 879–884.

Tigges, J., Gordon, T. P., McClure, H. M., Hall, E. C. & Peters, A. (1988). Survival rate and life span of rhesus monkeys at the Yerkes Regional Primate Research Center. *American Journal of Primatology*, **15**, 263–273.

Timme, A. (1995). Sex differences in infant integration in a semifree-ranging group of Barbary macaques (*Macaca sylvanus*, L. 1758) at Salem, Germany. *American Journal of Primatology*, **37**, 221–231.

Tinbergen, N. (1952). "Derived" activities: their causation, biological significance, origin, and emancipation during evolution. *Quarterly Review of Biology*, **27**, 1–32.
(1963). On aims and methods of ethology. *Zeitschrift für Tierpsychologie*, **20**, 410–433.

Tokuda, K. (1961). A study on the sexual behavior on the Japanese monkey troop. *Primates*, **3**, 1–40.

Tokuda, K. & Jensen, G. (1968). The leader's role in controlling aggressive behavior in a monkey group. *Primates*, **9**, 319–322.

Tokuda, K., Simons, R. C. & Jensen, G. (1968). Sexual behavior in a captive group of pigtailed monkeys (*Macaca nemestrina*). *Primates*, **9**, 283–294.

Tomasello, M. & Call, J. (1997). *Primate Cognition*. Oxford: Oxford University Press.

Tomasello, M., Call, J. & Hare, B. H. (1998). Five primate species follow the visual gaze of conspecifics. *Animal Behaviour*, **55**, 1063–1069.

Trefilov, A., Berard, J., Krawczak, M. & Schmidtke, J. (2000). Natal dispersal in rhesus macaques is related to serotonin transporter gene promoter variation. *Behavior Genetics*, **30**, 295–301.

Trivers, R. L. (1972). Parental investment and sexual selection. In *Sexual Selection and the Descent of Man, 1871–1971*, ed. B. G. Campbell, pp. 136–179. Chicago: Aldine de Gruyter.

Trivers, R. L. & Hare, H. (1976). Haplodiploidy and the evolution of the social insects. *Science*, **191**, 249–263.

Trivers, R. L. & Willard, D. E. (1973). Natural selection of parental ability to vary the sex ratio of offspring. *Science*, **179**, 90–92.

Troiden, R. R. (1989). The formation of homosexual identities. *Journal of Homosexuality*, **17**, 43–73.

Troisi, A. (2002). Displacement activities as a behavioral measure of stress in nonhuman primates and human subjects. *Stress*, **5**, 47–54.

Troisi, A. & Carosi, M. (1998). Female orgasm rate increases with male dominance in Japanese macaques. *Animal Behaviour*, **56**, 1261–1256.

Troisi, A. & Schino, G. (1987). Environmental and social influences on autogrooming behaviour in a captive group of Java monkeys. *Behaviour*, **100**, 292–303.

Troisi, A., Schino, G., d'Antoni, M., Pandolfi, N., Aureli, F. & d'Amato, F. R. (1991). Scratching as a behavioral index of anxiety in macaque mothers. *Behavioral and Neural Biology*, **56**, 307–313.

Trollope, J. & Blurton Jones, N. G. (1975). Aspects of reproduction and reproductive behaviour in *Macaca arctoides*. *Primates*, **16**, 191–205.

Tutin, C. E. G. (1980). Reproductive behaviour of wild chimpanzees in the Gombe National Park. *Journal of Reproduction and Fertility*, **28** (Suppl.), 43–57.

van Honk, C. & Hogeweg, P. (1981). The ontogeny of the social structure in a captive *Bombus terrestris* colony. *Behavioral Ecology and Sociobiology*, **9**, 111–119.

van Hooff, J. A. R. A. M. (1967). The facial displays of catarrhine monkeys and apes. In *Primate Ethology*, ed. D. Morris, pp. 7–68. London: Weidenfeld & Nicolson.
(1972). A comparative approach to the phylogeny of laughter and smiling. In *Nonverbal Communication*, ed. R. A. Hinde, pp. 209–241. London: Cambridge University Press.

van Hooff, J. A. R. A. M. & Aureli, F. (1994). Social homeostasis and the regulation of emotion. In *Emotions: Essays on Emotion Theory*, ed. S. H. M van Goozen, N. E. van de Poll & J. A. Sergeant, pp. 197–217. Hillsdale, NJ: Lawrence Erlbaum.

van Hooff, J. A. R. A. M. & Preuschoft, S. (2003). 'Laughter' and 'smiling': the intertwining of nature and culture. In *Animal Social Complexity*, ed. F. B. M. de Waal & P. L. Tyack, pp. 260–287. Cambridge, MA: Harvard University Press.

van Noordwijk, M. A. & van Schaik, C. P. (1985). Male migration and rank acquisition in wild long-tailed macaques (*Macaca fascicularis*). *Animal Behaviour*, **33**, 849–861.

(1999). The effects of dominance rank and group size on female lifetime reproductive success in wild long-tailed macaques, *Macaca fascicularis*. *Primates*, **40**, 105–130.

van Rhijn, J. G. & Vodegel, R. (1980). Being honest about one's intentions: an evolutionary stable strategy for animal conflicts. *Journal of Theoretical Biology*, **85**, 623–641.

van Schaik, C. P. (1983). Why are diurnal primates living in groups? *Behaviour*, **87**, 120–144.

(1989). The ecology of social relationships amongst female primates. In *Comparative Socio-ecology: The Behavioral Ecology of Humans and Other Animals*, ed. V. Standen & R. A. Foley, pp. 195–218. Oxford: Blackwell.

(1996). Social evolution in primates: the role of ecological factors and male behaviour. *Proceedings of the British Academy*, **88**, 9–31.

van Schaik, C. P. & Hrdy, S. B. (1991). Intensity of resource competition shapes the relationship between maternal rank and sex ratios at birth in cercopithecine primates. *American Naturalist*, **138**, 1555–1562.

van Schaik, C. P. & Paul, A. (1996). Male care in primates: does it ever reflect paternity? *Evolutionary Anthropology*, **5**, 152–156.

van Schaik, C. P. & van Hooff, J. A. R. A. M. (1983). On the ultimate causes of primate social systems. *Behaviour*, **85**, 91–117.

(1996). Toward an understanding of the orangutan's social system. In *Great Ape Societies*, ed. M. C. McGrew, L. F. Marchant & T. Nishida, pp. 3–15. Cambridge: Cambridge University Press.

(1985). Evolutionary effect of the absence of felids on the social organization of the macaques on the island of Simeulue (*Macaca fascicularis fusca*, Miller 1903). *Folia Primatologica*, **44**, 138–147.

(1986). The hidden costs of sociality: intra-group variation in feeding strategies in Sumatran long-tailed macaques (*Macaca fascicularis*). *Behaviour*, **99**, 296–315.

(1988). Scramble and contest in feeding competition among female long-tailed macaques (*Macaca fascicularis*). *Behaviour*, **105**, 77–98.

van Schaik, C. P., Assink, P. R. & Salapsky, N. (1992). Territorial behavior in Southeast Asian langurs: resource defense or mate defense? *American Journal of Primatology*, **26**, 233–242.

van Schaik, C. P., Hodges, J. K. & Nunn, C. L. (2000). Paternity confusion and the ovarian cycles of female primates. In *Infanticide by Males and its Consequences*, ed. C. P. van Schaik & C. H. Janson, pp. 361–387. Cambridge: Cambridge University Press.

van Schaik, C. P., Preuschoft, S. & Watts, D. P. (2004). Great ape social structures. In *The Evolution of Thought: Evolutionary Origins of Great Ape Intelligence*, ed. A. E. Russon & D. R. Begun. Cambridge: Cambridge University Press. (In press.)

van Schaik, C. P., van Noordwijk, M. A. & Nunn, C. L. (1999). Sex and social evolution in primates. In *Comparative Primate Socioecology*, ed. P. C. Lee, pp. 204–240. Cambridge: Cambridge University Press.

Vasey, P. L. (1995). Homosexual behavior in primates: a review of evidence and theory. *International Journal of Primatology*, **16**, 173–204.

(1996). Interventions and alliance formation between female Japanese macaques, *Macaca fuscata*, during homosexual consortships. *Animal Behaviour*, **52**, 539–551.

(1998). Female choice and inter-sexual competition for female sexual partners in Japanese macaques. *Behaviour*, **135**, 579–597.

(2002a). Sexual partner preference in female Japanese macaques. *Archives of Sexual Behavior*, **31**, 45–56.

(2002b). Same-sex sexual partner preference in hormonally and neurologically unmanipulated animals. *Annual Review of Sex Research*, **13**, 141–179.

(2004). Pre- and post-conflict interactions among female Japanese macaques during homosexual consortships. *International Journal of Comparative Psychology*, **17**. (In press.)

Vasey, P. L. & Gauthier, C. (2000). Skewed sex ratios & female homosexual activity in Japanese macaques: an experimental analysis. *Primates*, **41**, 17–25.

Vasey, P. L., Chapais, B. & Gauthier, C. (1998). Mounting interactions between female Japanese macaques: testing the influence of dominance and aggression. *Ethology*, **104**, 387–398.

Veenema, H. C., Das, M. & Aureli, F. (1994). Methodological corrections for the study of reconciliation. *Behavioural Processes*, **31**, 29–38.

Vehrencamp, S. L. (1983). A model for the evolution of despotic versus egalitarian societies. *Animal Behaviour*, **31**, 667–682.

Vermeulen, A. (1990). Androgens and male senescence. In *Testosterone: Action, Deficiency, Substitution*, ed. S. Nieschlag, pp. 261–276. Berlin: Springer.

Vessey, S. H. & Meikle, D. B. (1987). Factors affecting social behavior and reproductive success of male rhesus monkeys. *International Journal of Primatology*, **8**, 281–292.

Visalberghi, E. & Fragaszy, D. M. (1990). Food-washing behaviour in tufted capuchin monkeys, *Cebus apella*, and crabeating macaques, *Macaca fascicularis*. *Animal Behaviour*, **40**, 829–836.

von Holst, D. (1998). The concept of stress and its relevance for animal behavior. *Advances in the Study of Behavior*, **27**, 1–131.

von Segesser, F., Ménard, N., Gaci, B. & Martin, R. D. (1999). Genetic differentiation within and between isolated Algerian subpopulations of Barbary macaques (*Macaca sylvanus*): evidence from microsatellites. *Molecular Ecology*, **8**, 433–442.

von Uexküll, J. (1956). *Streifzüge durch die Umwelten von Tieren und Menschen: ein Bilderbuch unsichtbarer Welten*. Hamburg: Rowohlt.

Wada, K. & Tokida, E. (1981). Habitat utilization by wintering Japanese monkeys (*Macaca fuscata fuscata*) in the Shiga heights. *Primates*, **22**, 330–348.

Waddington, C. H. (1957). *The Strategy of the Genes*. London: Allen & Unwin.

Walker, M. L. (1995). Menopause in female rhesus monkeys. *American Journal of Primatology*, **35**, 59–71.

Wallen, K. (1990). Desire and ability: hormones and the regulation of female sexual behavior. *Neuroscience and Biobehavioral Reviews*, **14**, 233–241.

(1995). The evolution of sexual desire. In *Sexual Nature and Sexual Culture*, ed. P. R. Abramson & S. D. Pinkerton, pp. 57–79. Chicago: University of Chicago Press.

(2001). Sex and context: hormones and primate sexual motivation. *Hormones and Behavior*, **40**, 339–357.

Wallen, K. & Parsons, W. A. (1997). Sexual behavior in same-sexed nonhuman primates: is it relevant to understanding homosexuality? *Annual Review of Sex Research*, **7**, 195–223.

Wallen, K., Winston, L. A., Gaventa, S., Davis-DaSilva, M. & Collins, D. C. (1984). Periovulatory changes in female sexual behavior and patterns of ovarian steroid secretion in group-living rhesus monkeys. *Hormones and Behavior*, **18**, 431–450.

Walters, J. R. (1980). Interventions and the development of dominance relationships in female baboons. *Folia Primatologica*, **34**, 61–89.

Waser, P. M., Austad, S. N. & Keane, B. (1986). When should animals tolerate inbreeding? *American Naturalist*, **128**, 529–537.

Watanabe, K. (1989). Fish: a new addition to the diet of Japanese macaques on Koshima Island. *Folia Primatologica*, **52**, 124–131.

Watanabe, K. & Muroyama, Y. (2004). Recent expansion of distribution and management problems of Japanese macaques. In *Commensalism and Conflict: the Primate – Human Interface*, ed. J. D. Paterson. Winnipeg: Hignell Printing, American Society of Primatologists. (In press.)

Watanabe, K., Mori, A. & Kawai, M. (1992). Characteristic features of the reproduction of Koshima monkeys (*Macaca fuscata fuscata*): a summary of thirty-four years of observation. *Primates*, **33**, 1–32.

Watts, D. P., Colmenares, F. & Arnold, K. (2000). Redirection, consolation, and male policing: how targets of aggression interact with bystanders. In *Natural Conflict Resolution*, ed. F. Aureli & F. B. M. de Waal, pp. 281–301. Berkeley, CA: University of California Press.

Weber, M. (1947). *The Theory of Social and Economic Organizations*. Oxford: Oxford University Press.

Wehrenberg, W. B., Dyrenfurth, I. & Ferin, M. (1980). Endocrine characteristics of the menstrual cycle in the Assamese monkey (*Macaca assamensis*). *Biology of Reproduction*, **23**, 522–525.

Weiss, K. (2002). Is the medium the message? Biological traits and their regulation. *Evolutionary Anthropology*, **11**, 88–93.

West-Eberhard, M. J. (1989). Phenotypic plasticity and the origins of diversity. *Annual Review of Ecology and Systematics*, **20**, 249–278.

Westergaard, G. C., Suomi, S. J., Higley, J. D. & Mehlman, P. T. (1999). CSF 5-HIAA and aggression in female macaque monkeys: species and interindividual differences. *Psychopharmacology*, **146**, 440–446.

Western, D. & Ssemakula, J. (1982). Life history patterns in birds and mammals and their evolutionary interpretation. *Oecologia*, **54**, 281–290.

Wheatley, B. P. (1980). Feeding and ranging of east Bornean *Macaca fascicularis*. In *The Macaques: Studies in Ecology, Behavior and Evolution*, ed. D. G. Lindburg, pp. 215–246. New York: van Nostrand Reinhold.

Wheatley, B. P., Hariyaputra, D. K. & Gonder M. K. (1996). A comparison of wild and food-enhanced long-tailed macaques (*Macaca fascicularis*). In *Evolution and Ecology of Macaque Societies*, ed. J. E. Fa & D. G. Lindburg, pp. 182–206. Cambridge: Cambridge University Press.

Wheeler, R. L. (1986). Development of sex differences in neonatal pigtail macaques. In *Current Perspectives in Primate Social Dynamics*, ed. D. M. Taub & F. A. King, pp. 111–119. New York: van Nostrand Reinhold.

White, M. J. D. (1978). *Modes of Speciation*. San Francisco: Freeman.

Whiten, A. (1996). When does smart behaviour-reading become mind-reading? In *Theories of Theories of Mind*, ed. P. Carruthers & P. K. Smith, pp. 277–292. Cambridge: Cambridge University Press.

Whiten, A. & Ham, R. (1992). On the nature and evolution of imitation in animal kingdom: reappraisal of a century of research. *Advances in the Study of Behavior*, **21**, 239–283.

Whitten, P. L. (1987). Infants and adult males. In *Primate Societies*, ed. B. B. Smuts, D. L. Cheney, R. M. Seyfarth, R. W. Wrangham & T. T. Struhsaker, pp. 343–357. Chicago: University of Chicago Press.

(1988). Biology and primatology. *American Journal of Primatology*, **14**, 305–308.

Wich, S. A., Fredriksson, G. & Sterck, E. H. M. (2002). Measuring fruit patch size for three sympatric Indonesian primate species. *Primates*, **43**, 19–27.

Wickler, W. (1967). Socio-sexual signals and their intra-specific imitation among primates. In *Primate Ethology*, ed. D. Morris, pp. 69–79. Chicago: Aldine.

Widdig, A. (2000). Coalition formation among male Barbary macaques (*Macaca fascicularis*). *American Journal of Primatology*, **50**, 37–51.

Widdig, A., Nürnberg, P., Krawczak, M., Streich, W. J. & Bercovitch, F. B. (2001). Paternal relatedness and age proximity regulate social relationships among adult females rhesus macaques. *Proceedings of the National Academy of Sciences of the USA*, **98**, 13769–13773.

Wildt, D., Bush, M., Goodrowe, K. L. *et al.* (1987). Reproductive and genetic consequences of founding isolated lion populations. *Nature*, **329**, 328–331.

Williams, G. C. (1966). *Adaptation and Natural Selection*. Princeton, NJ: Princeton University Press.

Wilson, A. C., Bush, G. L., Case, S. M. & King, M. C. (1975). Social structuring of mammalian populations and rate of chromosomal evolution. *Proceedings of the National Academy of Sciences of the USA*, **72**, 5061–5065.

Wilson, D. S. & Sober, E. (1994). Reintroducing group selection to the human behavioral sciences. *Behavioral and Brain Sciences*, **17**, 585–654.

Wilson, E. O. (1975). *Sociobiology: The New Synthesis*. Harvard, MA: Harvard University Press.

Wilson, E. O. & Bossert, W. H. (1971). *A Primer of Population Biology*. Stamford, CA: Sinauer Associates.

Wilson, M. E. (1981). Social dominance and female reproductive success in rhesus monkeys (*Macaca mulatta*). *Animal Behaviour*, **29**, 472–482.

Wilson, M. E., Gordon, T. P. & Bernstein, I. S. (1978). Timing of births and reproductive success in rhesus monkey social groups. *Journal of Medical Primatology*, **7**, 202–212.

Wilson, M. E., Gordon, T. P. & Collins, D. C. (1982). Serum 17-β-estradiol and progesterone associated with mating behavior during early pregnancy in female rhesus monkeys. *Hormones and Behavior*, **6**, 94–106.

(1986). Ontogeny of luteinizing hormone secretion and first ovulation in seasonal breeding rhesus monkeys. *Endocrinology*, **118**, 293–301.

Wimsatt, W. C. & Schank, J. C. (1988). Two constraints on the evolution of complex adaptations and the means for their avoidance. In *Evolutionary Progress*, ed. M. H. Nitecki, pp. 231–275. Chicago: University of Chicago Press.

Wingfield, J. C., Jacobs, J. D., Tramontin, A. D. *et al.* (2000). Toward an ecological basis of hormone-behavior interactions in reproduction of birds. In *Reproduction in Context*, ed. K. Wallen & J. E. Schneider, pp. 85–128. Cambridge, MA: MIT Press.

Winter, D. G., John, O. P., Stewart, A. J. & Klohnen, E. C. (1998). Traits and motives: toward an integration of two traditions in personality research. *Psychological Review*, **105**, 230–250.

Witt, R., Schmidt, C. & Schmitt, J. (1981). Social rank and Darwinian fitness in a multimale group of Barbary macaques (*Macaca sylvana* Linnaeus, 1758). *Folia Primatologica*, **36**, 201–211.

Wolf, J. B., Brodie, E. D., Cheverud, J. M., Moore, A. J. & Wade, M. J. (1998). Evolutionary consequences of indirect genetic effects. *Trends in Ecology and Evolution*, **13**, 64–69.

Wolfe, L. D. (1979). Behavioral patterns of estrous females of the Arashiyama West troop of Japanese macaques (*Macaca fuscata*). *Primates*, **20**, 525–534.

(1984). Japanese macaque female sexual behavior: a comparison of Arashiyama East and West. In *Female Primates: Studies by Women Primatologists*, ed. M. F. Small, pp. 141–157. New York: Alan R. Liss.

Wong, C. L. & Ni, I. H. (2000). Population dynamics of the feral macaques in the Kowloon Hills of Hong Kong. *American Journal of Primatology*, **50**, 53–66.

Worlein, J. M., Eaton, G. G., Johnson, D. F. & Glick, B. B. (1988). Mating season effects on mother–infant conflict in Japanese macaques, *Macaca fuscata*. *Animal Behaviour*, **36**, 1472–1481.

Worthman, C. M. (1993). Biocultural interactions in human development. In *Juvenile Primates*, ed. M. E. Pereira & L. A. Fairbanks, pp. 339–358. Oxford: Oxford University Press.

Wrangham, R. W. (1980). An ecological model of female-bonded primate groups. *Behaviour*, **75**, 262–300.

Wrangham, R. W. & Bergmann Riss, E. (1990). Rates of predation on mammals by Gombe chimpanzees, 1972–1975. *Primates*, **31**, 157–170.

Wrangham, R. W. & Rubenstein, D. I. (1986). Social evolution in birds and mammals. In *Ecology and Social Evolution: Birds and Mammals*, ed. D. I. Rubenstein & R. W. Wrangham, pp. 452–470. Princeton, NJ: Princeton University Press.

Wright, S. (1951). The genetical structure of populations. *Annals of Eugenics*, **15**, 323–354.

(1978). *Evolution and the Genetics of Populations*, vol. 4. Chicago: University of Chicago Press.

Wu, H. Y. & Lin, Y. S. (1992). Life history variables of wild troops of Formosan macaques (*Macaca cyclopis*) in Kenting, Taiwan. *Primates*, **33**, 85–97.

Yamagiwa, J. (1985). Socio-sexual factors of troop fission in wild Japanese monkeys (*Macaca fuscata yakui*) on Yakushima Island, Japan. *Primates*, **26**, 105–120.

Yamagiwa, J. & Hill, D. A. (1998). Intraspecific variation in the social organization of Japanese macaques: past and present scope of field studies in natural habitats. *Primates*, **39**, 257–273.

Yeager, C. P. (1996). Feeding ecology of the long-tailed macaque (*Macaca fascicularis*) in Kalimantan Tengah, Indonesia. *International Journal of Primatology*, **17**, 51–62.

Yerkes, R. M. (1939). Social dominance and sexual status in chimpanzees. *Quarterly Review of Biology*, **14**, 115–136.

(1940). Social behavior of chimpanzees: dominance between mates in relation to sexual status. *Journal of Comparative Psychology*, **30**, 147–186.

Zeh, J. A. & Zeh, D. W. (1996). The evolution of polyandry. I. Intragenomic incompatibility. *Proceedings of the Royal Society of London B*, **263**, 1711–1717.

(1997). The evolution of polyandry. II. Post-copulatory defenses against genetic incompatibility. *Proceedings of the Royal Society of London B*, **264**, 69–75.

Zhang, Y. Z., Chen, L. W., Qu, W. Y. & Coggins, C. (2002). The primates of China: biogeography and conservation status. *Asian Primates*, **8**, 20–22.

Zhao, Q. (1993). Sexual behavior of Tibetan macaques at Mt. Emei, China. *Primates*, **34**, 431–444.

(1996). Etho-ecology of Tibetan macaques at Mount Emei, China. In *Evolution and Ecology of Macaque Societies*, ed. J. E. Fa & D. G. Lindburg, pp. 263–289. Cambridge: Cambridge University Press.

(1997). Intergroup interactions in Tibetan macaques at Mt. Emei, China. *American Journal of Physical Anthropology*, **104**, 459–470.

Zhao, Q. & Deng, Z. (1988a). *Macaca thibetana* at Mt. Emei, China: II. Birth seasonality. *American Journal of Primatology*, **16**, 261–168.

(1988b). *Macaca thibetana* at Mt. Emei, China: III. Group composition. *American Journal of Primatology*, **16**, 269–273.

Ziegler, T. E. & Bercovitch, F. B., eds. (1990). *Socioendocrinology of Primate Reproduction*. New York: Wiley-Liss.

Zuckerman, M. (1991). *Psychobiology of Personality*. Cambridge: Cambridge University Press.

Zumpe, D. & Michael, R. P. (1987). Relation between the dominance rank of female rhesus monkeys and their access to males. *American Journal of Primatology*, **13**, 155–169.

Zupanc, G. K. H. & Lamprecht J. (2000). Towards a cellular understanding of motivation: structural reorganization and biochemical switching as key mechanisms of behavioral plasticity. *Ethology*, **106**, 467–477.

Index

414